CHEMICAL SCIENCE AND ENGINEERING TECHNOLOGY

Perspectives on Interdisciplinary Research

CHEMICAL SCIENCE AND ENGINEERING TECHNOLOGY

Perspectives on Interdisciplinary Research

Edited by
Devrim Balköse, PhD
Ana Cristina Faria Ribeiro, PhD
A. K. Haghi, PhD
Suresh C. Ameta, PhD
Tanmoy Chakraborty, PhD

APPLE ACADEMIC PRESS

Apple Academic Press Inc.
3333 Mistwell Crescent
Oakville, ON L6L 0A2
Canada

Apple Academic Press Inc.
1265 Goldenrod Circle NE
Palm Bay, Florida 32905
USA

ISBN 13: 978-1-77463-394-6 (pbk)
ISBN 13: 978-1-77188-705-2 (hbk)

Library and Archives Canada Cataloguing in Publication

Chemical science and engineering technology : perspectives on interdisciplinary research / edited by Devrim Balköse, PhD, Ana Cristina Faria Ribeiro, PhD, A.K. Haghi, PhD, Suresh C. Ameta, PhD, Tanmoy Chakraborty, PhD.

Includes bibliographical references and index.

Issued in print and electronic formats.

ISBN 978-1-77188-705-2 (hardcover).--ISBN 978-1-351-04832-3 (PDF)

1. Chemical engineering. I. Balköse, Devrim, editor II. Ribeiro, Ana Cristina Faria, editor III. Haghi, A. K., editor IV. Ameta, Suresh C., editor V. Chakraborty, Tanmoy, editor

| TP146.C54 2018 | 660 | C2018-905482-4 | C2018-905483-2 |

Library of Congress Cataloging-in-Publication Data

Names: Balköse, Devrim, editor. | Ribeiro, Ana Cristina Faria, editor. | Haghi, A. K., editor. | Ameta, Suresh C., editor. | Chakraborty, Tanmoy, editor.

Title: Chemical science and engineering technology : perspectives on interdisciplinary research / Devrim Balköse, PhD, Ana Cristina Faria Ribeiro, PhD, A. K. Haghi, PhD, Suresh C. Ameta, PhD, Tanmoy Chakraborty, PhD.

Description: First edition. | Waretown, NJ : Apple Academic Press, 2019. | Includes bibliographical references and index.

Identifiers: LCCN 2018043181 (print) | LCCN 2018044034 (ebook) | ISBN 9781351048323 (ebook) | ISBN 9781771887052 (hardcover : alk. paper)

Subjects: LCSH: Chemical engineering.

Classification: LCC TP146 (ebook) | LCC TP146 .C47 2019 (print) | DDC 660--dc23

LC record available at https://lccn.loc.gov/2018043181

ABOUT THE EDITORS

Devrim Balköse, PhD

Devrim Balköse, PhD, is currently a retired faculty member in the Chemical Engineering Department at Izmir Institute of Technology, Izmir, Turkey. She graduated from the Middle East Technical University in Ankara, Turkey, with a degree in chemical engineering. She received her MS and PhD degrees from Ege University, Izmir, Turkey, in 1974 and 1977, respectively. She became Associate Professor in macromolecular chemistry in 1983 and Professor in process and reactor engineering in 1990. She worked as a research assistant, assistant professor, associate professor, and professor between 1970 and 2000 at Ege University. She was the Head of Chemical Engineering Department at Izmir Institute of Technology, Izmir, Turkey, between 2000 and 2009. She is now a retired faculty member in the same department. Her research interests are in polymer reaction engineering, polymer foams and films, adsorbent development, and moisture sorption. Her research projects are on nanosized zinc borate production, ZnO polymer composites, zinc borate lubricants, antistatic additives, and metal soaps.

Ana Cristina Faria Ribeiro, PhD

Ana C. F. Ribeiro, PhD, is a researcher in the Department of Chemistry at the University of Coimbra, Portugal. Her area of scientific activity is physical chemistry and electrochemistry. Her main areas of research interest are transport properties of ionic and nonionic components in aqueous solutions. She has experience as a scientific adviser and teacher of different practical courses. Dr. Ribeiro has supervised master degree theses as well as some PhD theses and has been a theses jury member. She has been referee for various journals as well as an expert evaluator of several research programs funded by the Romanian government through the National Council for Scientific Research. She has been a member of the organizing committee of scientific conferences, and she is an editorial member of several journals. She has received several grants, consulted for a number of major corporations and is a frequent speaker for national and international audiences. She is a member of the Research Chemistry Centre, Coimbra, Portugal.

A. K. Haghi, PhD

A. K. Haghi, PhD, is the author and editor of 165 books, as well as 1,000 published papers in various journals and conference proceedings. Dr. Haghi has received several grants, consulted for a number of major corporations, and is a frequent speaker to national and international audiences. Since 1983, he served as professor at several universities. He is currently Editor-in-Chief of the *International Journal of Chemoinformatics and Chemical Engineering* and *Polymers Research Journal* and has served on the editorial boards of many international journals. He is also a member of the Canadian Research and Development Center of Sciences and Cultures (CRDCSC), Montreal, Quebec, Canada. He holds a BSc in urban and environmental engineering from the University of North Carolina (USA), an MSc in mechanical engineering from North Carolina A&T State University (USA), a DEA in applied mechanics, acoustics and materials from the Université de Technologie de Compiègne (France), and a PhD in engineering sciences from Université de Franche-Comté (France).

Suresh C. Ameta, PhD

Suresh C. Ameta, PhD, is currently Dean of the Faculty of Science at PAHER University, Udaipur, India. He has served as Professor and Head of the Department of Chemistry, and as Head of the Department of Polymer Science, at North Gujarat University, Patan and M. L. Sukhadia University, Udaipur. He also served as Dean of Postgraduate Studies. Prof. Ameta has held the position of President of the Indian Chemical Society, Kolkata and is now a life-long Vice President. He received a number of prestigious awards during his career such as national prizes twice for writing chemistry books in Hindi. He also received the Prof. M. N. Desai Award (2004), the Prof. W. U. Malik Award (2008), the National Teacher Award (2011), the Prof. G. V. Bakore Award (2007) and a Life-time Achievement Award from the Indian Chemical Society (2011) and the Indian Council of Chemist (2015). He has successfully guided 81 PhD students. Having more than 350 research publications to his credit in journals of national and international repute, he is also the author of many undergraduate- and postgraduate-level books. He has published three books with Apple Academic Press: *Chemical Applications of Symmetry and Group Theory*, *Microwave-Assisted Organic Synthesis*, and *Green Chemistry: Fundamentals and Applications* and two with Taylor and Francis: *Solar Energy Conversion and Storage* and *Photocatalysis*. He has also written chapters in books published by several other international publishers. Prof. Ameta has delivered lectures and chaired sessions at national conferences and is a reviewer of number of international journals. In addition, he has completed five major research projects from different funding agencies, such as DST, UGC, CSIR, and Ministry of Energy, Govt. of India.

Tanmoy Chakraborty, PhD

Tanmoy Chakraborty, PhD, is Associate Professor in the Department of Chemistry at Manipal University, Jaipur, India. He has been working in the challenging field of computational and theoretical chemistry for the last six years. He completed his PhD with the University of Kalyani, West Bengal, India, in the field of application of QSAR/QSPR methodology in bioactive molecules. He has published many international research papers in peer-reviewed international journals with high-impact factors. Dr. Chakraborty is serving as an editorial board member of the *International Journal of Chemoinformatics and Chemical Engineering*. He is also reviewer of the *World Journal of Condensed Matter Physics* (WJCMP). Dr. Chakraborty received the prestigious Paromeswar Mallik Smawarak Padak from Hooghly Mohsin College, Chinsurah (University of Burdwan) in 2002.

CONTENTS

CONTRIBUTORS

Sacide Alsoy Altınkaya
Department of Chemical Engineering, Izmir Institute of Technology, Gülbahçe, 35437 Urla-Izmir, Turkey

Rakshit Ameta
Department of Chemistry, PAHER University, Udaipur 313003, Rajasthan, India

Suresh C. Ameta
Department of Chemistry, PAHER University, Udaipur 313003, Rajasthan, India

Ashima Bagaria
Department of Physics, Manipal University Jaipur, Jaipur, Rajasthan, India

Devrim Balköse
Department of Chemical Engineering, Izmir Institute of Technology, Gülbahçe, 35437 Urla-Izmir, Turkey

Surbhi Benjamin
Department of Chemistry, PAHER University, Udaipur 313003, Rajasthan, India

Didem Berkün
Department of Chemical Engineering, Izmir Institute of Technology, Gülbahçe, 35437 Urla-Izmir, Turkey

Jayesh Bhatt
Department of Chemistry, PAHER University, Udaipur 313003, Rajasthan, India

Gloria Castellano
Departamento de Ciencias Experimentales y Matemáticas, Facultad de Veterinaria y Ciencias Experimentales, Universidad Católica de Valencia San Vicente Mártir, Guillem de Castro-94, E-46001 València, Spain

Tanmoy Chakraborty
Department of Chemistry, Manipal University Jaipur, Dehmi Kalan, Jaipur, Rajasthan, India

Mehmet Gönen
Department of Chemical Engineering, Engineering Faculty, Süleyman Demirel University, E-13 Blok, Batı Yerleşkesi, Isparta 32260, Turkey

M. A. Khadar
Centre for Nanoscience and Nanotechnology, University of Kerala, Trivandrum 695037, India

Poonam Khullar
Department of Chemistry, B.B.K. D.A.V. College for Women, Amritsar 143005, Punjab, India

Obey Koshy
International and Inter University Centre for Nanoscience and Nanotechnology, Mahatma Gandhi University, Kottayam, 686560, India
Centre for Nanoscience and Nanotechnology, University of Kerala, Trivandrum 695037, India

Divya Mandial
Department of Chemistry, B.B.K. D.A.V. College for Women, Amritsar 143005, Punjab, India

Sukanchan Palit
Department of Chemical Engineering, University of Petroleum and Energy Studies, Energy Acres, Post-Office: Bidholi via Prem Nagar, Dehradun 248007, Uttarakhand, India
43, Judges Bagan, Post-Office: Haridevpur, Kolkata 700082, West Bengal, India

Jagdish Parihar
Department of Physics, Manipal University Jaipur, Jaipur, India

Yasir Beeran Pottathara
International and Inter University Centre for Nanoscience and Nanotechnology, Mahatma Gandhi University, Kottayam, 686560, India
Centre for Nanoscience and Nanotechnology, University of Kerala, Trivandrum 695037, India

L. Arul Pragasan
Department of Environmental Sciences, Bharathiar University, Coimbatore 641046, India

Prabhat Ranjan
Department of Mechatronics Engineering, Manipal University Jaipur, Jaipur, Rajasthan, India

Ana C. F. Ribeiro
Department of Chemistry, University of Coimbra, 3004-535 Coimbra, Portugal

R. Sarath
Department of Environmental Sciences, Bharathiar University, Coimbatore 641046, India

Mukesh Saran
Department of Physics, Manipal University Jaipur, Jaipur, India

Cecília I. A. V. Santos
Department of Chemistry, University of Coimbra, 3004-535 Coimbra, Portugal
MRC, CP-165/62, Université Libre de Bruxelles, 50, Av. F.D. Roosevelt, B-1050 Brussels, Belgium

Shalini
Department of Chemistry, Alankar P.G. Girls College, Jaipur, Rajasthan, India

Valentina Shevtsova
MRC, CP-165/62, Université Libre de Bruxelles, 50, Av. F.D. Roosevelt, B-1050 Brussels, Belgium

Dipti Soni
Department of Chemistry, PAHER University, Udaipur 313003, Rajasthan, India

Lavanya Tandon
Department of Chemistry, B.B.K. D.A.V. College for Women, Amritsar 143001, Punjab, India

Funda Tıhmınlıoğlu
Department of Chemical Engineering, Izmir Institute of Technology, Gülbahçe, 35437 Urla-Izmir, Turkey

Francisco Torrens
Institut Universitari de Ciència Molecular, Universitat de València, Edifici d'Instituts de Paterna, P. O. Box 22085, E-46071 València, Spain

Monika Trivedi
Department of Chemistry, PAHER University, Udaipur 313003, Rajasthan, India

Ali Yalçın
Department of Chemical Engineering, Engineering Faculty, Süleyman Demirel University, E-13 Blok, Batı Yerleşkesi, Isparta 32260, Turkey

ABBREVIATIONS

7DG	7-deazaguanine
9DG	9-deazaguanine
ACS	American Chemical Society
ADMET	absorption, distribution, metabolism, excretion, and toxicity
AFM	atomic force microscope
AI	avian influenza
AMOSA	annealing-based multiobjective optimization
APG	adenyl 3′–5′ guanosine
BBB	blood–brain barrier
BBI	Bowman–Birk inhibitor
BSA	bovine serum albumin
CFUs	colony forming units
CONPs	cerium oxide nanoparticles
CTAB	cetyl trimethyl ammonium bromide
DDP	2,5-diamino-4,6-dihydroxypyrimidine
DL	detection limit
DMSS	dual microporous silica sphere
DPV	differential pulse voltammetry
DSC	differential scanning calorimetry
EC	evolutionary computation
EDS	energy-dispersive X-ray spectroscopy
EDX	energy-dispersive X-ray
EPR	enhanced permeability and retention effect
FCC	fluid catalytic cracking
FDA	Food and Drug Administration
FFAs	free fatty acids
FFT	fast Fourier transform
FMP	formycin-5′-monophosphate
FRET	transfer of fluorescence resonance energy
FTIR	Fourier-transform infrared spectroscopy
FWHM	full width at half maximum
GA	genetic algorithm

GC	gas chromatography
GI-XRD	grazing incidence X-ray diffraction
GOD	glucose oxidase
HBPH	hyperbranched polyhydroxyl
HMSs	hollow mesoporous silica spheres
HPC	hydroxypropyl cellulose
HSA	human serum albumin
HTS	high throughput screening
IBB	isobutylbenzene
LBTs	lanthanoid binding tags
MBE	molecular beam epitaxy
MD	molecular dynamics
MHT	magnetic hyperthermia
MOEAs	multiobjective evolutionary algorithms
MOG	8-methyl-9-oxoguanine
MOO	multiobjective optimization
MRD	mean relative deviation
MRI	magnetic resonance imaging
MS	mass spectrometric
NaCMC	sodium salt of carboxy methyl cellulose
NBE	near band edge emission
NEM	northeast monsoon
NEO	neopterin
NIR	near-infrared
ODI	optical digital interferometry
PAP	pokeweed antiviral protein
PDB	Protein Data Bank
PL	photoluminescence
PLD	pulsed laser deposition
PLGA	polylactic-co-glycolic acid
PRO	propranolol
PS-b-PAA	polystyrene-b-poly acryl acid
PSS	polysodium-4-styrenesulfonate
PTA	pteroic acid
PV	peroxide value
QSAR	quantitative structure–activity relationship
RFCC	resid fluidized catalytic cracking
RH	relative humidity

RIPs	ribosome inactivating proteins
RMS	root mean square
ROSs	reactive oxygen species
SAED	selected area electron diffraction
SAR	structure–activity relationship
SC	sodium caseinate
SE	standard error
SEM	scanning electron microscopy
SERS	surface enhanced Rayleigh scattering
SGF	simulated gastric fluid
SIF	simulated intestinal fluid
SIMs	single ion magnets
SPI	soy protein isolate
SPION	supraparamagnetic iron oxide nanoparticles
SPR	surface plasmon resonance
SS	summer season
SWM	southwest monsoon
TEM	transmission electron microscopy
TEOS	tetraorthosilicate
TGA	thermal gravimetric analysis
THN	1,2,3,4-tetrahydronaphthalene
TPs	tea polyphenols
TQ	thymoquinone
TSP	travelling salesman problem
UV	ultraviolet
WS	winter season
WVP	water vapor permeability
XRD	X-ray diffraction

PREFACE

One of the major areas of emphasis in academia in recent years has been "interdisciplinary research," a trend that promises new insights and innovations rooted in cross-disciplinary collaboration. This book is designed for stepping beyond traditional disciplinary boundaries and applying knowledge and insights from multiple fields. This book provides a practical guide for researchers who are seeking to develop interdisciplinary research in chemical science and engineering technology. It provides comprehensive interdisciplinary studies with an approach that is conceptual and practical. This comprehensive volume presents academic, industry, and other complementary viewpoints on interdisciplinary research. This is an important reference publication that provides new research and updates on a variety of interdisciplinary perspectives in chemical science and engineering technology. The book uses case studies and supporting technologies, and it also explains the conceptual thinking behind current uses and potential uses not yet implemented. International experts with countless years of experience lend this volume credibility.

The book is divided into four major sections.

The first section presents "Frontiers in Computational Chemistry."

In the first chapter, the author guides enthusiastic readers on the pathways of success in evolutionary computation, particularly genetic algorithms, simulated annealing and multiobjective optimization, and its vast applications in chemical engineering and petroleum engineering science. "Theoretical Computation of Periodic Descriptors Invoking Periodic Properties" is discussed in Chapter 2 in detail.

The second section comprises four chapters on "Analytical Nanoscience and Nanotechnology."

"Antimicrobial Activity of Nanosized Photocatalytic Materials" is discussed in Chapter 3.

Chapter 4 covers a wide range of "Nanodevices and Organization of Single Ion Magnets and Spin Qubits." The aim of Chapter 5 is to initiate a debate by suggesting a number of questions. The discussion deals with these questions from the perspective of historical research in science. The viewpoint is from the history and sociology of science, making a comparison between *historicism* and *presentism*.

Synthesis, characterization, and bioapplicability of protein conjugated nanomaterials are reviewed in Chapter 6. This chapter aims to showcase recent found applications of biopolymers with a marked focus on BSA, soy protein, and Cyt C.

In third section, the "Key Issues in Industrial Chemistry and Chemical Engineering" are discussed in detail in eight chapters.

In Chapter 7, extraction techniques for oil separation from *Nigella sativa* seeds were summarized and discussed. The use of thymoquinone and its derivatives for the treatment of diseases were investigated based on the recent studies.

Simultaneously, evaporated Al-doped Zn films for optoelectronic applications are discussed in Chapter 8. In this chapter, simultaneously evaporated and oxidized thin films of aluminum and zinc containing different atomic weight percentages of Al (2, 4, and 6%) were fabricated, and their structure, morphology, and optical and electrical properties were investigated. Apart from the synthesis, characterization, and material quality of the Al-doped Zn films, the present study highlights the response of transmittance, photoluminescent, and electrical properties to varying Al concentrations. By addressing such concentration-dependent physical property variations within existing crystal structures, we can aim to improve the various technological applications of the same.

The objective of Chapter 9 is to produce and characterize cellulose-based edible films that are water vapor permeable. For this purpose, sodium NaCMC and HPC were used as the cellulose-based film forming materials. Films were characterized by using scanning electron microscopy (SEM), differential scanning calorimetry (DSC), X-ray diffraction analysis (XRD), and thermal gravimetric analysis (TGA) techniques. In addition, the water vapor sorption and permeability characteristics and mechanical properties of the films were determined. Increasing shelf life of carrots by coating with NaCMC was also investigated.

In Chapter 10, "Recognition of Adenine-like Rings by the Abrin—a Binding Site: A Flexible Docking Approach" was developed in detail.

Biocarbon storage potential of tea plantation of Nilgiris, India, in relation to leaf chlorophyll and soil parameters is reviewed in Chapter 11.

"Synthesis, Characterization, and Applications of Silica Spheres" are presented in Chapter 12.

The author in Chapter 13 pointedly focuses on the deep scientific success, the vast technological vision, and the much-needed scientific profundity in the research pursuit in industrial fixed bed reactors, fluidized bed catalytic reactors, and the vast world of energy sustainability and petroleum engineering science.

Some important aspects of diffusion and thermodiffusion in hydrocarbon mixtures are discussed in Chapter 14 in detail.

In the last section of this book, new insights in the "Role of Chemoinformatics in Modern Drug Discovery and Development" are presented.

PART I
Frontiers in Computational Chemistry

CHAPTER 1

EVOLUTIONARY COMPUTATION: A CRITICAL OVERVIEW AND A VISION FOR THE FUTURE

SUKANCHAN PALIT*

Department of Chemical Engineering, University of Petroleum and Energy Studies, Energy Acres, Post-Office: Bidholi via Prem Nagar, Dehradun 248007, Uttarakhand, India

E-mail: sukanchan68@gmail.com; sukanchan92@gmail.com

ABSTRACT

The vast world of technology and engineering science are witnessing immense scientific challenges and deep scientific farsightedness. Science and technology are today in the process of scientific rejuvenation and scientific revamping. Chemical process engineering, petroleum engineering science, and other diverse areas of science and engineering are the scientific challenges of today's research pursuit. Evolutionary computation (EC) and applied mathematical tools are the necessities of scientific endeavor today. EC encompasses genetic algorithm (GA), multiobjective optimization (MOO), and multiobjective simulated annealing. The science of MOO and GA are witnessing immense scientific challenges and barriers. In this chapter, the author deeply comprehends the vast scientific genesis, the scientific profundity, and the scientific fortitude behind EC. EC is the next generation science and engineering endeavor. In this chapter, the author rigorously focuses on the scientific intricacies, the scientific hindrances, and the vast needs of EC and evolutionary programming. Human ingenuity and scientific farsightedness are the forerunners toward a newer visionary era in the EC application to human society and science and technology in general. The vast vision and the vast scientific provenance behind EC are leading human civilization and human scientific research pursuit toward a newer realm. Today, petroleum engineering systems and chemical engineering systems are at deep stake due to environmental disasters, unsound environmental engineering tools, and depletion of fossil-fuel resources. This chapter pointedly focuses on the immense scientific success and vast scientific efficiency of EC in designing chemical and petroleum engineering systems. Scientific vision and scientific subtleties are the veritable hallmarks of this well-researched chapter.

1.1   INTRODUCTION

The world of challenges in evolutionary computation (EC) and evolutionary programming is today witnessing drastic and dramatic challenges. Human civilization and human scientific research pursuit today stands in the midst of scientific comprehension and deep technological advancements. Technology and engineering science of computation or EC are the true focuses and true targets of scientific emancipation in present day human civilization. In a similar vein, EC is highly challenged and needs to be envisioned as scientific endeavor crosses vast frontiers. In this chapter, the author pointedly focuses on the vast scientific vision, the scientific fortitude, and the scientific girth behind application of EC to human society and human scientific endeavor. Technology and engineering science are today in the path of immense scientific regeneration and vast scientific rejuvenation. This chapter also elucidates on the recent advances in the field of EC, genetic algorithm (GA), multiobjective optimization (MOO), and multiobjective simulated annealing. These are the areas of intense scientific research pursuit today. Applied mathematics and applied science are today replete with vision, girth, and determination. In the similar vein, chemical process engineering and petroleum engineering science need to be revamped and re-envisioned with the passage of scientific history and time. Human vision and human scientific endeavor today are in the path of deep scientific revelation and vast scientific endurance. This chapter willfully pronounces and presents the scientific history and the genesis of engineering science behind EC applications. The aisles of this chapter involve deep introspection into the field of application of EC in chemical process engineering and petroleum engineering science. The author remarkably guides the enthusiastic reader the pathways of success in EC particularly GA, simulated annealing, and MOO and its vast applications in chemical engineering, and petroleum engineering science.

1.2   THE VISION OF THIS STUDY

Mankind's immense scientific prowess, the vast technological profundity, and the futuristic vision of applied mathematics and applied science are the forerunners toward a true realization and true emancipation of EC today. The vision of this study surpasses scientific frontiers and scientific imagination as science and technology trudge forward in a new century. EC was propounded in the 1960s yet far-reaching and crossing scientific boundaries. The purpose and the mission of this scientific endeavor are to present before the scientific domain the deep scientific vision, the scientific genesis, and the critical analysis behind EC and evolutionary programming. Human scientific generations are today in a state of immense revival. The subtleties of engineering science of GA and optimization are rigorously presented in minute details as science moves toward a newer century.

1.3   LITERATURE REVIEW

EC and computational techniques are today in a state of immense scientific regeneration and deep scientific forbearance. Human civilization and human scientific endeavor today stands in the midst of scientific introspection and vast challenges. In this section, the author deeply comprehends the vast scientific potential, the scientific genesis, and the scientific success behind the application of EC in design of chemical engineering and petroleum engineering systems. Rangaiah[1] elucidated with deep and cogent insight techniques and applications of MOO in chemical engineering. This is a watershed text in the field of process systems engineering. Scientific history, scientific progeny, and deep scientific articulation are the utmost imperatives of research pursuit in chemical process engineering and petroleum engineering science today. MOO today is in the threshold of a new beginning and a newer scientific rejuvenation. The author treads a visionary as well as weary path in process optimization, basics and methods of MOO, some examples in the scientific endeavor in chemical process optimization, vast MOO examples in biotechnology and food industry, petroleum refining and petrochemicals, pharmaceuticals and other products, applications in polymerization, and a whole gamut of visionary applications. Human scientific research pursuit and deep scientific understanding are the hallmarks of progress in scientific and academic rigor today. This chapter veritably presents before the vast scientific domain multiobjective evolutionary algorithms (MOEAs) and the review of the state of the art and some of their vast applications in chemical engineering. The author in this entire chapter gleans and unravels the success of multiobjective GA and simulated annealing, surrogate assisted EA, interactive MOO in chemical process design, and diverse areas of GA paradigm. Technological advancements, the deep scientific forays, and the success of human scientific research pursuit are the veritable cornerstones of scientific and academic rigor today. The vision and the challenge surpass scientific imagination and scientific fortitude. Rangaiah[1] successfully unravels the technological stance, the deep scientific revelation, and the scientific hindrances in MOO in chemical process engineering. Technological innovations, scientific sagacity, and scientific determination are all leading a long and visionary way in the true emancipation of computational science today. The vast scientific barriers, the scientific hindrances, and the innovative challenges in research pursuit are today veritably opening new avenues in future thoughts and future scientific candor. Human scientific research pursuit needs to be re-envisioned and revamped as science marches ahead. In this chapter, the author pointedly focuses on the scientific intricacies, the vast scientific divination, and the genesis of EC with the sole aim and mission for the furtherance of science and technology.[1–4,17]

1.4 WHAT DO YOU MEAN BY EC?

EC, MOO, and multiobjective simulated annealing are today in the midst of immense scientific comprehension and deep vision. Human scientific endeavor and human scientific pragmatism in research trends in the field of EC are changing the immense scientific landscape. Technological sagacity and vast scientific discernment are needs of today's research pursuit. In computer science, EC is a family of algorithms for global optimization inspired by biological evolution and the subdomain of artificial intelligence and soft computing studying these algorithms. In technical parlance, they are a family of population-based trial and error problem solvers with a metaheuristic or stochastic optimization character.[1–4] Technology of EC today is highly advanced and needs to be re-envisioned with the passage of scientific history and time. In EC, an initial set of candidate solutions is generated and iteratively updated. Each new generation is produced by stochastically removing less-desired solutions and introducing small random changes. In biological terminology, a population of solutions is subjected to natural selection (or artificial selection) and mutation. As a result, the population will gradually evolve to increase in fitness, in this case the chosen fitness function of the algorithm.

EC techniques are changing the scientific landscape of computer science and applied mathematics. The vast scientific vision, the scientific forbearance, and the scientific discerning are the cornerstones of this well-researched chapter. EC techniques can produce highly optimized solutions in a wide range of problem settings, making them popular in computer science. EC is also sometimes used in evolutionary biology as an in silico experimental procedure to study common aspects of general evolutionary processes.[1,2,17]

The use of evolutionary principles for problem solving originated in the 1950s. Evolutionary programming was introduced by Lawrence J. Fogel in the United States, while John Henry Holland called his method a GA. In Germany, Ingo Rechenberg and Hans-Paul Schwefel introduced evolution strategies. These areas developed separately for about 15 years. Thus evolutionary principles ushered in a new era of evolutionary computing. Since the 1990s, nature-inspired algorithms have become immensely popular and fiercely competitive in the world of evolutionary programming, computer science, and applied mathematics.[1–3,17]

EC techniques today include

- Ant colony optimization
- Artificial bee colony optimization
- Bees algorithm
- Differential algorithm
- Evolutionary algorithms
- Evolutionary programming
- Evolution strategy
- Gene expression programming
- GA
- Genetic programming
- Particle swarm optimization
- Swarm intelligence[17]

1.5 SCIENTIFIC DOCTRINE, SCIENTIFIC VISION, AND THE SUCCESS OF EC

EC and advanced computer science today veritably stand in the midst of deep scientific questions and vast scientific introspection. This technology is not new yet vastly unraveled. In this chapter, the author rigorously points out toward the scientific farsightedness, the scientific fortitude, and the futuristic vision in the pursuit of engineering science and technology. Today, chemical process engineering, petroleum engineering science, environmental engineering, and other diverse avenues of research pursuit are veritably opening up new generation of scientific thoughts and scientific innovations. In every branch of challenge of science, EC and optimization are today ushering in a new era in science and engineering. Human scientific endeavor and the challenges of science need to be veritably addressed and envisioned as science and engineering moves forward. Science of computation today stands in the midst of deep scientific introspection and deep insight. Technology and engineering science today are huge colossus with a vast vision of its own. Since the 1950s, EC has undergone immense challenges and deep fortitude. Human scientific rejuvenation and regeneration are today in the path of success and revelation. These technologies of EC, GA, and the interfaces of MOO are today challenging the avenues of scientific endeavor.[1,2,17]

Vision of science and the challenges behind it are changing the scientific landscape and the scientific mindset of chemical engineers and petroleum engineers today. Applied mathematics and computational techniques are in a state of immense scientific introspection and scientific overhauling. In this chapter, the author repeatedly and rigorously points out toward the immense scientific potential, the scientific vision, and the scientific advancements behind EC.

1.6 MULTIOBJECTIVE OPTIMIZATION AND GENETIC ALGORITHM

Scientific endeavor in the field of MOO and GA today stands in the midst of deep scientific regeneration. In the 1990s, optimization and GA witnessed drastic challenges and visionary changes. Engineering science and technology are today huge colossus with a vast vision of its own. In the similar manner, GA and MOO are in the path of newer innovations and newer scientific rejuvenation.

MOO is an area of multicriteria decision-making, which is concerned with mathematical optimization problems involving more than one objective function to be optimized simultaneously. MOO has been applied in many fields of science, including engineering, economics, and logistics where optimal decisions need to be taken in the presence of trade-offs between two or more conflicting objectives.[1,2,17] Human scientific endeavor and human scientific vision in the field of MOO are today crossing vast scientific frontiers. Today is the era of challenge of MOO application in the design of chemical engineering and petroleum engineering systems. In this chapter, the author rigorously points out toward the diverse applications of MOO in engineering and science.[1,2,17]

GA falls under the vast domain of EC. The vision of science, the mankind's technological prowess and the challenges of engineering science will all lead a long and visionary way in the true emancipation of newer mathematical techniques. Applied mathematics and applied computer science are today ushering in a new era in the field of engineering science, particularly chemical engineering and petroleum engineering. Human civilization and human scientific endeavor are highly challenged today with the passage of scientific history and time. Scientific cognizance, scientific vision, and deep scientific understanding are today in a state of immense regeneration as regards application of GA. In computer science and operations research, a GA is a metaheuristic inspired by the process of natural selection that belongs to the larger class of EAs.[17] GAs are commonly used to generate high-quality solutions to optimize and search problems by relying on bio-inspired operators such as mutation, crossover, and selection. In a GA, a population of candidate solutions (called individuals, creatures, and phenotypes) to an optimization problem is evolved toward better solutions.[1,2,17] Science and technology of GA and its application in optimization problems are highly advanced today. In every avenues of scientific research pursuit and scientific vision of GA, the futuristic vision and the challenges need to be restructured and reorganized. In GA, the evolution starts from a population of randomly generated individuals and is an iterative process, with the population in each iteration called generation. In each generation, the fitness of every individual in the population is evaluated; the fitness is usually the value of the objective function in the optimization problem being solved. A typical GA requires (1) a genetic representation of the solution domain and (2) a fitness function to evaluate the solution domain.[1,2,17]

There are limitations of the use of GA compared to alternative optimization algorithms:

1. Repeated fitness function evaluation for complex problems is often the most prohibitive and limiting segment of artificial EAs.
2. GAs do not scale well with complexity.
3. The "better solution" is only in comparison to other solutions.
4. In many problems, GA may have a tendency to converge to a local optima rather than the global optimum of the problem.[1,2,17]

Applied mathematics and optimization techniques are challenging the vast scientific landscape. In this chapter, the author pointedly focuses on the deep scientific vision behind optimization science. Optimization science and GA have application areas in diverse domains such as chemical engineering and petroleum engineering. Besides these branches, GA and MOO have visionary applications in environmental engineering science. Human scientific endeavor today stands in the midst of deep scientific introspection and vast scientific vision. Process design engineering and chemical process engineering in the similar manner lies in the visionary gulf of deep comprehension and vision. The application of optimization and GA in chemical engineering needs to be restructured and re-envisaged as human scientific endeavor gears forward.

1.7   SCIENTIFIC DOCTRINE BEHIND GA

GA and applications of GA in optimization are witnessing today drastic challenges. Science and engineering are huge colossus with a definite vision and a definite forbearance of its own. Scientific research pursuit needs to be re-envisioned and reorganized as regards application of GA today. GA is of utmost necessity in the path toward successful scientific research pursuit in computer science and mathematics today. EC encompasses GA, MOO, and multiobjective simulated annealing. The robustness of the computational method of GA is veritably changing the scientific landscape today. Mitchell[2] discussed in detail in a seminal work the vast scientific doctrine of GA. Technology and engineering science of GA are today undergoing vast and drastic challenges. Science and technology arise from the basic human desire to discern and control the human society.[2,17] Over the course of human history, humans have slowly and steadily built up a grand structure of knowledge insemination that enables human to predict, to varying extents, the weather, the motion of the planets, solar and lunar eclipses, the courses of diseases, the rise and fall of economic growth, the stages of language development in children, and a vast landscape of other natural, social, and cultural phenomena.[1,2,17] Science and technology are today immense edifices with a vast and ever-growing vision of its own. The advent and use of computers has arguably been the most revolutionary development in the history of science and technology and the march of academic rigor. The ongoing revolution is veritably increasing the human ability to predict and control nature in definite ways that were barely conceived of even a half century ago. The mission and goals of creating artificial intelligence and artificial life can be traced back to the very beginnings of the computer age. The success of human civilization, the scientific success, and the futuristic vision of human life has urged scientists to delve deep into computer science especially EC. Mitchell[2] rewrote human history and brought before the scientific domain the human intellect and the machine-learning process of EC and GA. The earliest proponents of computer science and information technology—Alan Turing, John von Neumann, Norbert Weiner and others—were motivated in large portion by visions of imbuing computer programs with intelligence, with the life-like ability to replicate and with adaptive capability to learn and veritably control their environments. Human scientific endeavor was at its zenith when this chapter was presented to the scientific domain. It is of no surprise that from the earliest days computers were applied not only to calculate missile trajectories and deciphering military codes but also to modeling the brain, mimicking human learning, and simulating biological evolution. Since the 1980s, biologically motivated computing activities have emerged as a major scientific endeavor. The first activity has grown in the field of neural networks, the second into machine learning, the third is the fundamental scientific activities in "EC."[1,2,17]

GAs were discovered by John Holland in the 1960s and were developed by Holland and his students and colleagues at the University of Michigan in the 1960s and the 1970s. In contrast with evolution strategies and evolutionary programming, Holland's original vision was not to design algorithms to solve specific problems but rather to formally study the phenomenon of adaptation of human life. Holland's 1975 book "Adaptation in Natural and Artificial Systems" presented the GA as an abstraction of biological evolution and with immense lucidity gave a theoretical framework for adaptation under the GA.[1,2,17] Holland's introduction of a population-based algorithm with crossover, inversion, and mutation was a major innovation and a visionary direction of computer science. Scientific genesis and scientific cognizance reached its helm as GA was presented with deep lucidity to the scientific domain. In the last several years, there has been a revolution and rejuvenation among researchers studying various EC methods and the scientific boundaries between GAs, evolution strategies, evolutionary programming, and other evolutionary approaches.[1,2,17] Human scientific endeavor are today ushering in a new era in the field of evolutionary programming as well. The vision and the challenge of science needs to be revamped today as regards application of GA in optimization and applied mathematics. GA and EC are today changing the face of human scientific research pursuit. Science has millions of answers toward EC and computer science as a whole today. Machine learning and artificial intelligence are changing the scientific landscape of human civilization. In this chapter, the author rigorously points out toward the vast scientific potential, the scientific expanse, and the deep scientific vision behind human intelligence, artificial intelligence, and GA.

1.8   VISIONARY SCIENTIFIC ENDEAVOR IN GA

EC, GA, and MOO are witnessing immense scientific challenges with the passage of scientific history and the visionary time frame. Scientific vision and scientific excellence are today in a state of immense regeneration as regards application and scientific emancipation. Multiobjective simulated annealing is another wide branch of today's scientific endeavor. Research trends and research profundity in EC are witnessing immense challenges and hindrances.

Mitchell[2] redefined and revisited the vast and visionary domain of GA. EA and GA are the cornerstones of this well-researched chapter. Scientific endeavor in GA are replete with vision and profundity. Today, evolution is an inspiration for solving computational problems. The author deeply delineated the scientific genesis and the scientific progeny of GA and EC. To EC researchers, the mechanisms of evolution seem well suited for most pressing computational problems in many fields. Many computational problems require searching through a huge number of possibilities for solutions, and the procedures are tedious. One example is the problem of computational protein engineering, in which an algorithm is sought that will search among the vast number of possible amino acid sequences for a protein with specified properties. This is the vast appeal of EC. The author deeply discussed the genesis and the vision behind GA and its wide applications. Many computational problems require a computer program to continue to perform well in a changing scientific environment. This is the intricacy of computer science and EC. This is typified and identified by problems in robot control in which a robot has to perform a task in a variable environment.

Coley[3] discussed with deep details an introduction to GAs for scientists and engineers. GAs are numerical optimization algorithms inspired by both natural selection and natural genetics.[3] Scientific vision, scientific restructuring, and deep scientific cognizance are the hallmarks of scientific endeavor in computational techniques today. This method of GA is a general one, capable of being applied to an extremely wide range of problems. Unlike some approaches, their promise has rarely been over-sold, and they are being used to help solve practical problems on a daily basis.[3] The algorithms are simple to understand and the required computer code is easy to write. A typical algorithm might consist of the following: (1) a number, or population, of guesses of the solution to the problem; (2) a way of calculating how good or bad the individual solutions within the population are; (3) a method of fixed fragments of the better solutions to form new, on average better solutions; (4) a mutation operator to avoid permanent loss of diversity.[3] A deep investigation into the intricacies of GA are dealt within this chapter. The importance of GA over other traditional methods is that it simply has proved themselves capable of solving many large complex problems where other methods have experienced many difficulties. The challenge and the vision of GA are effectively dealt within this chapter.[3] Many practical problems of GA are dealt within areas of (1) image processing; (2) prediction of three dimensional protein structures; (3) VLSI electronic chip layouts; (4) laser technology; (5) medicine; (6) spacecraft trajectories; (7) analysis of time series; (8) solid-state physics; (9) aeronautics; (10) robotics; (11) liquid crystals; (12) water networks; (13) facial recognition; (14) job shop scheduling; (15) control.[3] Rather than starting from a single point (or guess) within the search space, GAs are initialized with the population of guesses. These are veritably random and will be spread throughout the search space. A typical algorithm then uses three operators, selection, crossover, and mutation, to direct the population toward convergence at the global optimum.[3] There are many imperatives which need to be addressed before applying a GA to a particular problem.[3] They include (1) the method of encoding the unknown parameters; (2) how to exchange information contained between the strings; (3) the population size; (4) how to apply the concept of mutation to the representation; (5) the termination criterion.[3] Technological challenges and the vast and wide scientific validation are the cornerstones of this fundamental approach in GA applications. This is a watershed text in the field of applied computing and optimization.[3] In this chapter, GAs have been introduced as general search algorithms based on metaphors with natural selection and natural genetics. Technological and scientific validations are the utmost need of the hour as GA advances with immense scientific vision.

Evolutionary inspired computing can be traced back to the earliest days of computer science and GAs were invented in the 1960s by John Holland. The science of EC needs to be redefined and readdressed with the passage of scientific forays, scientific history, and time.[3]

1.9   VISIONARY SCIENTIFIC ENDEAVOR IN EC

Technological advancements today are in a state of immense scientific rejuvenation. The vision of GA and EC are changing the entire scientific landscape and surpassing scientific frontiers. GA and bio-inspired evolution techniques are changing the mathematical modeling scenario in chemical engineering, petroleum engineering, and diverse avenues of engineering science. Human scientific endeavor and the progress of scientific and academic rigor today stand in the midst of deep scientific introspection and vision.

Back et al.[4] discussed with deep and cogent insight basic algorithms and operators in EC. This technology is highly advanced and far reaching. The basic idea of EC, which came onto the scene in the 1950s, has been to make use of the powerful process of natural evolution as a problem-solving paradigm, either by simulating it in a laboratory or by simulating in a computer.[4] The challenge and the vision of scientific and academic rigor in the field of EC are delineated in lucid details in this well-researched chapter. The authors in this entire chapter deeply comprehend GAs, evolution strategies, and evolutionary programming in a true vision toward the furtherance of science and engineering of EC.[4] In the 1960s, visionary researchers developed these mainstream methods of EC, namely, J. H. Holland (1962) at Ann Arbor, Michigan, USA, H. J. Bremermann (1962) at Berkeley, California, USA, and A. S. Fraser (1957) at Canberra, Australia, for GAs, L. J. Fogel (1962) at San Diego, California, for evolutionary programming, and I. Rechenberg (1965) and H. P. Schwefel (1965) at Berlin, Germany, for evolutionary strategies.[4] The capabilities of EC were redefined with the progress of scientific vision, scientific rigor, and the visionary time frame.[4] Similar in some aspects to the earlier positive research advances toward imitating nature's powerful problem-solving tools, such as artificial neural network techniques and fuzzy systems, EC and related areas of scientific computation also had to go a long way of ignorance and recognition before receiving international acclaim and worldwide scientific validation.[4] The great success that these methods had, in extremely complex optimization problems from various avenues of research pursuit, has facilitated the undeniable breakthrough of EC as a robust and excellent problem-solving methodology.[4] Human scientific research pursuits in computer science and applied mathematics slowly and steadily evolved in a new era of scientific regeneration.[4] The progress in the theory of EC methods since 1990 vastly confirms the strength of the algorithms as well as their limitations. Validation and vision of engineering science and technology are the challenging features of EC and evolutionary programming today.[4]

Fogel[5] discussed with deep and cogent insight an introduction to EC. This vast field is relatively young and was invented as recently as 1991. It represents a vast effort to bring together researchers who have been following different approaches in stimulating various aspects of evolution.[5] The vast vision of computer science and applied mathematics needs to be revamped and redefined as engineering science moves forward. Today, human society is highly technology driven. The vision and the vast challenge of applied mathematics and computer science will all lead a long and visionary way in the true realization of engineering science today. EC today is in the path of immense scientific overhauling. This chapter repeatedly urges the scientific success, the scientific splendor, and the vast scientific vision behind evolutionary programming and genetic programming as well.[5]

Schwefel[6] delineated in a detailed research paper the advantages and disadvantages of EC over other approaches. Since, according to the no-free-lunch (NFL) theorem, there cannot exist any algorithm for solving all problems that is generally superior to any competitor, the question of whether EAs are inferior/superior to any alternative approach is senseless.[6] What can be envisioned is that EAs behave better than other methods with respect to solving a specific class of problems.[6] The NFL theorem can be analyzed in the case of EAs versus many classification optimization methods insofar as the latter is vastly more efficient in solving linear, quadratic, strongly convex, unimodal, separable, and many other special problems.[6] Technological profundity and scientific validation of EC are in the path toward newer scientific innovation and newer scientific rejuvenation.[6] According to the authors, the best one can think about EAs, is therefore, that they present a methodological framework that is easy to understand and handle and is moving toward more sophistication, specialization, or hybridization.[6]

Jong et al.[7] discussed with vast foresight a comprehensive history of EC. This is a historical summary of EC rather arbitrarily at a stage as recent as the mid-1950s. Technological challenges, mankind's immense scientific prowess, and the futuristic vision of applied mathematics will all go a long and visionary way in the true emancipation of EC and evolutionary programming.[7]

Beasley[8] briefly discussed possible applications of EC. Applications of EC encompass a wide continuum of scientific research area. They have been split into five major areas which are as follows: (1) planning, (2) design, (3) simulation and identification, (4) control, and (5) classification.[8] Validation of science and technological motivation are the hallmarks of this well-researched endeavor. Applications in planning involve routing, combinatorial optimization problems, or the travelling salesman problem (TSP). A salesman must visit number of cities and then return home. Optimizing the trade-off between speed and accuracy of the solution has been one of the visions of the problem.[8] A veritable generalization of the TSP occurs when there is more than one salesman. The vehicle routing problem is a similar scientific research pursuit in EC problem. Closely related to this is the transportation problem, in which a single commodity must be distributed to a number of customers from a number of depots. The challenge and the vision of science need to be re-envisioned and re-envisaged with the passage of scientific history and time. Scheduling involves devising a plan to carry out a number of activities over a period of time, where the activities require resources which are limited and there are various constraints. The challenge and the scientific sagacity of EC and evolutionary programming are immense and far reaching as engineering science and technology enter into a newer era and newer innovations. Job shop scheduling is a complex EC problem. The scenario is a manufacturing plant, with machines of different types. Technology and applied mathematics need to be revamped and re-envisioned as EC witnesses major scientific thrusts and major scientific restructuring. Applications in design involve the design of filters which has received considerable attention in scientific field.[8] EAs have been used to design electronic or digital systems which implement a desired frequency response. This chapter opens a new avenue of scientific research pursuit in the field of EAs. Both finite impulse response and infinite impulse response filter structures have been employed with EAs. Applications in simulation and identification involve taking a design or model for a system, and determining how the system will behave.[8] In some cases, this is effectively done because we are unsure about the behavior (the most glaring example is the design of an aircraft). In other cases, the behavior is known. But we wish to test the accuracy of the model. Validation of science and engineering are changing the face of EC today. EC has been applied to difficult problems in chemistry and biology also. The vision of EC is today opening up new challenges and newer scientific intellect in decades to come. Applications in control involve two distinct approaches: off-line and on-line. The off-line approach uses an EA to design a controller, which is then used to control the system. Some scientists have sought to use the adaptive qualities of EAs in order to build on-line controllers for dynamic and unsteady systems. There are vast advantages of an evolutionary controller. It can cope with systems whose characteristics change over time, whether the change is gradual or sudden. Scientific vision, the vast scientific challenges, and the deep scientific forbearance are opening up a new chapter in the field of EC and EA applications. A significant amount of EC research forays has encompassed the theory and practice of classifier systems. Classifier systems are at the heart of many other types of systems. Human scientific endeavor and the progress of scientific rigor in EC are today ushering in new vision and scientific fortitude. Many control systems rely on being able to classify characteristics of their environment before an appropriate control decision needs to be made. Robotics application in EC is one such vast area.[8] EC today has vast application domain. They are more refined than the conventional computation tools.[8] The success of this scientific research pursuit today ushers in a new scientific imagination and newer vision.[8]

Deb[9] delineated with lucid details representations. Every search and optimization algorithms deals with solutions, each of which represents an involvement of the underlying problem.[9] Science and technology of EC are highly advanced today. In most engineering problems, a solution is a real-valued vector specifying dimensions to the key parameters of the problem.[9] As the structure of a solution varies from problem to problem, a solution of a particular problem can be represented in a number of ways.[9] This is the concept of representations.[9] Usually, a search method is most efficient in dealing with a particular representation and is not so effective in dealing with other representations. Technology and engineering science of EC are witnessing immense challenges as human civilization and human scientific endeavor move forward toward a newer realm and a newer era.[9] Thus, the choice of an efficient representation scheme depends not only on the underlying problem but also on the chosen search method.[9] The efficiency and the vast complexity of a search algorithm largely depend on how the solutions have been represented and how suitable the representation is in the context of the underlying search operators.[9] In a classical search and optimization method, all decision variables are usually represented as vectors of real numbers and the algorithm works on one vector of solution to create a new vector of solution.[9] Science of EC is today ushering in a new era in the field of scientific validation and technological vision and motivation. The author rigorously points out toward the scientific success, the scientific potential, and the vast scientific forbearance in tackling the concepts of EC.[9]

Human civilization and human scientific endeavor are today in the path of immense rejuvenation and deep introspection. The deep challenges of technological validation needs to be readdressed and restructured as applied mathematics and computational science ushers in a new era in the field of validation and vision. In this chapter, the author pointedly focuses on the immense scientific ingenuity, the comprehension of engineering science, and the scientific success of EC with the sole aim of furtherance of science and technology.

1.10   THE VAST SCIENTIFIC VISION BEHIND GA AND ITS APPLICATIONS

GA has diverse applications in various avenues of scientific research pursuit. The scientific vision behind GA and its applications are far reaching and opening up new avenues of scientific forbearance, scientific fortitude, and deep scientific grit. This technology is not new yet unexplored. The intricacies of GA are replete with scientific vision and might. GA today is in a state of immense scientific regeneration. The challenges and the vast vision of science and engineering of optimization science have no bounds. In this chapter, the author repeatedly pronounces the scientific success, the vast scientific prowess, and the technological vision behind GA and its diverse applications. The use of GAs for problem-solving is not new. The pioneering work of J. H. Holland in the 1970s proved to be a significant scientific contribution.[3,17] Today, in the domain of scientific research pursuit, the vast vision and the vast scientific forbearance are of immense importance. The challenge which Dr. Holland put forward to the scientific community in the area of GA needs to be redefined and revamped. Coley[3] discussed with deep and cogent insight the success of GA applications in diverse areas of science and engineering. The advancements of science and engineering in applied mathematics and computer science are today reshaping the vast scientific landscape.[3] Although the roots of evolutionary inspired computing can be traced back to the earliest days of computer science and applied mathematics, GAs themselves were invented in the 1960s by John Holland.[3,17] Deep scientific cognizance, scientific sagacity, and the technological motivation are the hallmarks toward a newer generation of scientific forays in GA today. The mathematical concept of schema theory is of immense importance as GA applications tread forward. The schema theory was invented by Dr. John Holland and was popularized by Dr. Goldberg.[3]

In this chapter, the author pointedly focuses on the vast scientific regeneration and the difficulties in the application of GA in diverse areas of engineering such as petroleum engineering and chemical engineering science.

1.11   THE SCIENTIFIC DOCTRINE BEHIND MULTIOBJECTIVE OPTIMIZATION AND MULTIOBJECTIVE SIMULATED ANNEALING

Multiobjective simulated annealing is a new area of deep scientific endeavor. This is a major area of EC today. Research endeavor needs to be envisioned and redefined as science and engineering of GA and MOO surges ahead in the vast scientific landscape. MOO and multiobjective simulated annealing are the visionary avenues of introspection and validation in today's world of engineering science. Process systems engineering and chemical process engineering are today the vast applications of MOO. Petroleum engineering and environmental engineering science are the crucial areas of scientific endeavor in MOO. Depletion of fossil fuel resources and strains in environmental biodiversity are the vexing issues facing human civilization today. The challenge and the vision of science today go beyond scientific imagination and deep introspection. This chapter is a vast visionary pursuit toward the goal of scientific emancipation and technological realization of computational science.

Rangaiah[10] gave a detailed introduction to process optimization. Optimization refers to finding the values of decision variables, which correspond to and provide the maximum or minimum of one or more desired objectives. Science and technology of MOO are ushering in a new era in the field of chemical process engineering and petroleum engineering science. The author in this chapter deeply traverses into the intricacies of process optimization with immense emphasis on chemical engineering and petroleum engineering systems.[10] Optimization is essential for reducing material and energy requirements as well as the harmful environmental impact of chemical processes. This area of applied science and applied mathematics is highly advanced and crossing visionary scientific boundaries. Its endeavor leads to better design and operation of chemical processes as well as to sustainable processes. Energy and environmental sustainability are today the hallmarks of human civilization and human scientific endeavor. In every branch of scientific research pursuit, sustainability needs to be linked with diverse areas of science and engineering. Many applications of optimization involve several objectives, some of which are veritably conflicting. This chapter addresses these problems.[10] MOO is required to solve the resulting problems in these applications. The first chapter of this book provides an introduction to MOO with a realistic application, namely, the alkylation process optimization for two objectives.[10] The second chapter reviews nearly 100 chemical engineering applications of MOO since the year 2000 to mid-2007. This chapter enumerated on the selected MOO techniques; they include (1) review of MOEAs in the context of chemical engineering, (2) multiobjective GA and simulated annealing as well as their jumping gene adaptations, (3) surrogate assisted MOEA, (4) interactive MOO in chemical process design, and (5) two methods for ranking the Pareto solutions.[10] This entire chapter gleans on the vast importance of MOO in the furtherance of chemical process design, chemical process engineering, and petroleum engineering.[10] Technological vision and vast scientific adjudication and truth today stand in the midst of immense regeneration. Mathematical tools and its applications are in the forefront of scientific understanding today. The authors in this chapter successfully pinpoints on the immense success and immense scientific judgment in the application of MOO in design of chemical engineering systems. Optimization refers to finding the values of decision (or free) variables which correspond to and provide the maximum or minimum of one or more desired objectives. Optimization has immense applications in engineering, science, business, and economics, except that, in these applications, quantitative models and methods are employed unlike qualitative assessment of choices in daily life. The main focus of optimization of chemical processes so far has been optimization for one objective at a time (i.e., single-objective optimization).[10] These objectives include capital cost/investment, operating cost, profit, payback period, selectivity, quality and/or recovery of the product, conversion, energy required, efficiency, process safety and/or complexity, operation time, robustness, etc.[10] Science and engineering of optimization and GA are ushering in a new era and opening new windows of innovation in the field of scientific vision and scientific forbearance. Optimization science today is witnessing immense challenges as regards efficiency of the mathematical techniques.[10]

Masuduzzaman et al.[11] deeply discussed and delineated with lucidity multiobjective applications in chemical engineering. Technology and engineering science of MOO and chemical engineering science are crossing vast and versatile scientific boundaries. It has attracted immense interest from researchers throughout the world. Technological vision, scientific profundity, and the immense scientific prowess of human civilization are all gearing forward toward newer innovations and opening up new chapters of scientific instinct in decades to come.[11] This challenge and this vision of multiobjective applications are delineated in this chapter.[11] In this chapter, nearly 100 multiobjective applications in chemical engineering reported in journals from 2000 until mid-2007 are deeply reviewed. They are categorized into five groups: (1) process design and operation, (2) biotechnology and food industry, (3) petroleum refining and petrochemicals, (4) pharmaceuticals and other products, and (5) polymerization.[11] Chemical engineering and unit operations in chemical engineering are witnessing drastic challenges today. MOO is needed in every sphere of science and engineering applications today. GA stands between deep introspection and vision.[11] In the similar manner, MOO needs to be re-envisioned and restructured as human civilization faces environmental and energy crisis of immense proportions.[11] Mankind's immense scientific prowess, the futuristic vision of applied mathematics, and computing the world of challenges and barriers will all lead a long and visionary way in the true realization of energy sustainability today. Optimization refers to obtaining the values of decision variables, which correspond to the maximum and minimum of one or more objective functions. The foundations of optimization science are laid in this well-researched chapter.[11] Major part of research pursuit in optimization and its applications in chemical engineering considers only one objective function, probably due to the computational resources include methods.[11] However, most real-world chemical engineering problems involve one or more objectives which are conflicting in nature. The procedure of finding solutions of such problem is known as MOO.[11] Over the last few decades, this field has grown significantly, and many chemical engineering applications of MOO have been reported in various research forays. Optimization science and GA today stands in the midst of scientific vision and scientific restructuring. The immense scientific prowess of human civilization, the vast scientific profundity, and the futuristic vision are all the torchbearers toward a new generation of scientific research in MOO in decades to come.[11] There are many reviews of MOO applications in chemical engineering and petroleum engineering science.[11]

Jaimes et al.[12] reviewed the entire domain of MOEAs in a state-of-the-art treatise. The authors in this chapter pointedly focus on the immense applications of MOEAs in chemical engineering. In this chapter, the authors present a well-researched overview of evolutionary MOO with particular emphasis on algorithms in current use. Several applications of these algorithms in chemical engineering are deeply discussed and analyzed. The authors provided a deep scientific understanding and additional information about public domain resources available for those interested in research endeavor in this area.[12] Human civilization and human scientific research pursuit today stands in the gulf of deep scientific rejuvenation and scientific revival.[12] Applied mathematics and computer science are the challenging areas of scientific sagacity and scientific vision today.[12] The solution of problems having two or more (normally conflicting) objectives has become very common in the last few years in a wide variety of research forays. Such problems are called "multiobjective" and can be solved using either mathematical programming techniques or metaheuristics.[12] In either of the cases, the concept of Pareto optimality is normally adopted. When using this concept, the authors aimed to obtain the best possible trade-offs among all the objectives.[12] The first MOEAs were introduced in the 1980s, but they became popular only in the mid-1990s.[12] Nowadays, the use of MOEAs in all disciplines has become widespread, and chemical engineering and petroleum engineering, is by no means, an exception. Technological vision, scientific objectives, and the needs of optimization science are today leading a long and visionary way in the true realization of MOEAs and computational science. A historical perspective in EAs is widely analyzed and described in this chapter.[12]

Ray et al.[13] analyzed and described surrogate assisted EA for MOO.[13] EAs are population-based approaches that start with an initial population of candidate solutions and evolve them over a number of generations to finally arrive at a set of desired solutions.[13] Such population-based algorithms are particularly attractive for MOO problems as they can result in a set of nondominated solutions in a single run. A radial basis function network is used as a surrogate model.[13] The algorithm performs actual evaluations of objectives and constraints for all the members of the initial population and periodically evaluates all the members of the population after every generation. Five multiobjective test problems are presented in this study, and a comparison with Nondominated Sorting GA–II are deeply gleaned and analyzed in this well-researched chapter. EAs have been successfully applied to a range of multiobjective problems.[13] Human scientific endurance, the immense scientific grit, and sagacity are slowly ushering in a new era in the field of EC.[13] The scientific success of research in MOO and GA are today replete with immense vision and scientific profundity.[13] EAs are particularly suitable for multiobjective problems as they result in a set of nondominated solutions in a single run. Furthermore, EAs do not rely on functional and slope continuity and thus can be readily applied to optimization problems with mixed variables. Science and technology of EC are witnessing immense challenges and hurdles today.[13]

Bandyopadhyay et al.[14] deeply discussed with vast foresight the domain of a simulated annealing-based multiobjective optimization (AMOSA).[14] This chapter describes a simulated AMOSA algorithm that incorporates the concept of archive in order to provide a set of trade-off solutions for the problem under consideration.[14] To veritably determine the acceptance probability of a new solution vis-à-vis the current solution, an elaborate procedure is followed that takes into account the domination status of the new solution with the current solution, as well as those in the archive.[14]

1.12   A BRIEF IDEA OF CHEMICAL ENGINEERING AND PETROLEUM ENGINEERING SYSTEMS

Chemical engineering and petroleum engineering systems are today surpassing vast visionary scientific frontiers. Environmental engineering science and petroleum engineering today stand in the midst of deep crisis. Depletion of fossil fuel resources and frequent environmental catastrophes are driving human civilization toward a massive crisis. Scientific vision, deep scientific forbearance, and vast technological farsightedness are the necessities of human civilization today. Chemical engineering and petroleum engineering research questions today are crossing vast and versatile scientific boundaries. The unit operations in chemical engineering which are the heart of chemical engineering techniques today stands in the midst of deep scientific introspection and vision. The science of petroleum engineering also stands in the midst of deep crisis. In such a critical juncture, application of mathematical tools in process design and design of chemical and petroleum engineering systems assumes immense importance. There are vast technical issues which need to be solved such as design of a particular refining unit in petroleum refinery. Here, MOO and GA are applied. Palit[15] deeply and lucidly comprehended the application of evolutionary MOO in Fluidized Catalytic Cracking unit and chemical engineering systems in a well-researched review. Petroleum refining industry is moving in the direction of visionary challenges today. Chemical engineering and its revolutionary techniques in petroleum engineering science today stand in the midst of deep introspection and vast insight. Here in this chapter, the author took a visionary example of design of Fluidized Catalytic Cracking Unit in a petroleum refinery. MOO involves optimizing a number of objectives simultaneously. The problem becomes highly challenging when the objectives are of conflict to each other, that is, the optimal solution of an objective function is different from each other. The purpose and the aim of this study are to unravel the intricacies of the hidden world of evolutionary MOO and its vast and varied future. The Fluidized Catalytic Cracking unit is a pillar of a petroleum refinery. There are immense problems today in the domain of Resid Fluidized Catalytic Cracking operation.[15] The main problems associated with the processing of FCC units are the higher boiling ranges of the resid feed, the thermolysis of the larger molecules leading to the coke formation, and the larger molecular size of resid components. During FCC operation, the partially vaporized feed comes in contact with the hot regenerated catalyst. At these conditions, the feed is not only vaporized but also undergo thermal cracking (thermolysis). These scientific issues can be overcome with the implementation of mathematical tools such as MOO and GA. Modeling and simulation of Fluidized Catalytic Cracking Units is a vast area of scientific endeavor. Modeling of catalytic processes combines vast scientific knowledge and visionary scientific forays. Varshney et al.[16] with deep and cogent insight dealt with the modeling of the riser reactor in a resid

Fluidized-bed catalytic cracking unit using a multigrain model for an active matrix-zeolite catalyst.[16] The riser reactor in a resid fluidized-bed catalytic cracking (resid-FCC) unit is simulated in detail using a multigrain model of the catalyst in which the amorphous matrix as well as the embedded Y-zeolite crystals are porous and reactive.[16] The model is used to study the effect of important process variables and catalyst properties such as the catalyst-to-oil ratio, feedstock properties, matrix activity, and surface area on the yield of the different lumps.[16] Design of a petroleum refining unit such as Fluidized Cracking Unit is of immense importance in today's scientific research pursuit. The challenge and the vast vision of application of GA or EC in petroleum refining are reaching new importance.

In this entire chapter, the author stresses upon the success of computer science, the computational techniques, and the mathematical tools used in the application of design of chemical engineering and petroleum engineering systems. The targets of energy and environmental sustainability also need to be emphasized and re-envisioned with the progress of academic rigor.

Scientific vision, the scientific discernment, and the scientific profundity behind chemical engineering and petroleum engineering systems are veritably challenging the vast landscape of science and technology. Mathematical modeling of diffusion and reaction in petrochemical and petroleum refining systems is a very strong tool for the world of design and research endeavor. Today unit operations of chemical engineering and chemical process design are highly advanced. In this world of energy sustainability, petroleum refining assumes immense importance. This challenge and vision are deliberated in this well-researched chapter.

1.13   ENERGY SUSTAINABILITY AND THE FUTURE OF CHEMICAL AND PETROLEUM ENGINEERING SCIENCE

Energy and environmental sustainability today are replete with scientific vision and deep scientific fortitude. Human civilization and human scientific endeavor are in the path of newer innovation and newer regeneration. Energy and environmental crisis are challenging the scientific landscape today. The frequent environmental disasters and the loss of ecological biodiversity are the causes of immense concern today. These factors have tremendously urged the scientific community to target newer innovations and newer challenges. Global energy scenario is in a state of immense upheavals. Petroleum refining and the vast domain of renewable energy stand tall in the midst of the global energy crisis. Design of chemical engineering and petroleum engineering systems are today linked with advanced mathematical tools such as EC, MOO, and multiobjective simulated annealing. The vision of Dr. Gro Harlem Brundtland, former Prime Minister of Norway, on the concept of sustainability needs to be redefined and restructured. The application of evolutionary MOO in designing Fluidized Catalytic Cracking unit and chemical engineering systems are the versatile avenues of research pursuit today.

1.14   FUTURE FRONTIERS IN EC AND FUTURE TRENDS IN RESEARCH

Human scientific research pursuit today stands in the midst of deep scientific vision and girth. Technology and engineering science today need to be restructured and revamped as science and technology treads a weary and visionary path toward a perfect human destiny. The challenge and the vision of EC are immense and pathbreaking. Scientific revelation, scientific provenance, and scientific determination are today veritably challenged as chemical process engineering, environmental engineering, petroleum engineering science, and other diverse areas of engineering are faced with immense challenges and hindrances. Energy and environmental sustainability are in the state of immense disaster and at the same time scientific revival. Technological revamping, scientific validation, and deep scientific farsightedness are the cornerstones of scientific research today. The challenge and the vision need to be highly revived as computational techniques in engineering passes through a reorganized phase of human scientific determination. Environmental engineering and petroleum engineering today stand in the midst of deep disaster with the ever-growing concerns of environmental disasters and depletion of fossil fuel resources. Applied mathematics and computational science are regenerating itself with immense vision as these two branches of scientific research pursuit need to be redefined and re-envisioned. Future trends in research should be in the area of nontraditional EC. Science and technology are today huge colossus with a definite and purposeful vision of its own. Space science, nuclear technology, and the science of renewable energy are challenging the veritable fabric of scientific endeavor. Scientific ingenuity, scientific articulation, and the immense prowess of human scientific endeavor will lead a long and visionary way in the true scientific and engineering emancipation. In this chapter, the author redefines the course of scientific history in EC with the sole vision of furtherance of science and engineering.

1.15   FUTURE RECOMMENDATIONS AND THE FUTURE AVENUES OF RESEARCH

Future recommendations and future vision of research in EC are changing the face of scientific research pursuit and the deep scientific determination. Today, science and engineering need to be redefined with respect to applications and futuristic vision. Human genesis and scientific progeny are changing the face of research pursuit in computational techniques such as EC. Validation of science, technological profundity scientific genesis is the need of the hour as computational techniques and advanced computer science usher in a new era in science. Future recommendations of this study are immensely groundbreaking as science and engineering of EC enter into a new eon of scientific regeneration. Design of chemical and petroleum engineering science are the utmost need of the hour for deep scientific emancipation. Mankind's immense scientific prowess, the needs for scientific girth and determination, and the futuristic vision of technology will all lead a long and visionary way in the true emancipation of scientific truth today. Avenues in research in energy sustainability needs to be redefined and readjudicated as engineering science and technology tread a weary path toward today's scientific destiny. Future recommendations in research in EC are vast, versatile, and far reaching. Technological vision and scientific profundity are the utmost need of the hour as EC moves from one visionary paradigm toward another. Engineering science, chemical process engineering, petroleum engineering science, and other diverse branches of scientific endeavor are the imperatives of vision today. The world and the human civilization today stand in the midst of deep comprehension and vast scientific understanding. Science and technology are huge colossus with a definite and sound vision of its own. Nuclear science, space technology, and the avenues of renewable energy are veritably changing the scientific fabric of human scientific endeavor. In such a situation, applied mathematical tools and EC will lead a long and visionary way in the true emancipation and true realization of applied engineering science.

1.16   SUMMARY, CONCLUSION, AND SCIENTIFIC PERSPECTIVES

The challenges and vision of scientific endeavor in science and engineering are vast and varied. Scientific perspectives and scientific profundity today are veritably challenged and need to be redefined. In this chapter, the author rigorously points toward the scientific success, the technological vision, and the scientific genesis behind EC applications in human scientific endeavor. Design of chemical engineering and petroleum engineering systems are the scientific imperatives today as engineering science and its scientific rigor move forward. The entire crux of this chapter goes beyond scientific imagination and scientific progeny as the author deeply delves into the unknown world of computation science and more specifically EC. Human civilization and human scientific research pursuit in today's difficult global situation needs to be re-envisioned as science and technology treads forward. In today's scientific world, technological validation in scientific endeavor is an utmost need with the emancipation of scientific rigor. Human scientific challenges in applied mathematics and computation are diverse, vast, and far reaching. The question of energy and environmental sustainability needs to be addressed with vigor and might as science and engineering enters a newer eon. This chapter gives a wider glimpse to the scientific success and deep scientific profundity in EC and evolutionary programming research and presents a deeper scientific understanding in the domain of MOO, GA, and multiobjective simulated annealing. Technology of EC needs to be vehemently addressed and envisioned as human civilization witnesses drastic challenges as regards challenges in petroleum engineering science and chemical process engineering. In this chapter, the author rigorously states the solutions, the innovations, and the intricacies in deep scientific discernment in EC and computing as a whole. The success and efficiency of mathematical tools in designing chemical and petroleum engineering systems are today veritably ushering in a new knowledge dimension in optimization and EC. Thus, this scientific challenge and the scientific truth open new avenues of research pursuit in computation and scientific progress in years to come.

KEYWORDS

- genetic
- algorithm
- evolution
- optimization
- engineering
- computer

REFERENCES

1. Rangaiah, G. P. Multi-Objective Optimization—Techniques and Applications in Chemical Engineering. In *Advances in Process Systems Engineering*; World Scientific Publishing Co. Pte. Ltd.: Singapore, 2009; Vol. 1.
2. Mitchell, M. An Introduction to Genetic Algorithms. In *A Bradford Book*; The MIT Press: USA, 1996.
3. Coley, D. A. *An Introduction to Genetic Algorithms for Scientists and Engineers*; World Scientific Publishing Co. Pte. Ltd.: Singapore, 1999.
4. Back, T.; Fogel, D. B.; Michalewicz, Z. *Evolutionary Computation-1-Basic Algorithms and Operators*; Institute of Physics Publishing: U.K., U.S.A., 2000.
5. Fogel, D. B. Introduction to Evolutionary Computation. In *Evolutionary Computation-1—Basic Algorithms and Operators*; Back, T., Fogel, D. B., Michalewicz, Z., Eds.; Institute of Physics Publishing: U.K., U.S.A., 2000, pp 1–3.
6. Schwefel, H.-P. Advantages (and Disadvantages) of Evolutionary Computation Over Other Approaches. In *Evolutionary Computation-1—Basic Algorithms and Operators*; Back, T., Fogel, D. B., Michalewicz, Z., Eds., Institute of Physics Publishing: U.K., U.S.A., 2000; pp 20–22.
7. De Jong, K.; Fogel, D. B.; Schwefel, H.-P. A History of Evolutionary Computation. In *Evolutionary Computation-1—Basic Algorithms and Operators*; Back, T., Fogel, D. B., Michalewicz, Z., Eds.; Institute of Physics Publishing: U.K., U.S.A., 2000; pp 40–51.
8. Beasley, D. Possible Applications of Evolutionary Computation. In *Evolutionary Computation-1—Basic Algorithms and Operators*, Back, T., Fogel, D. B., Michalewicz, Z., Eds.; Institute of Physics Publishing: U.K., U.S.A., 2000; pp 4–18.
9. Deb, K. Introduction to Representations. In *Evolutionary Computation-1—Basic Algorithms and Operators*; Back, T., Fogel, D. B., Michalewicz, Z., Eds. Institute of Physics Publishing: U.K., U.S.A., 2000; pp 127–131.
10. Rangaiah, G. P. Introduction. In *Multi-Objective Optimization—Techniques and Applications in Chemical Engineering, Advances in Process Systems Engineering*; Rangaiah, G. P., Ed.; World Scientific Publishing Co. Pte. Ltd.: Singapore, 2009; Vol. 1, pp 1–23.
11. Masuduzzaman; Rangaiah, G. P. Multi-Objective Optimization Applications in Chemical Engineering. In *Multi-Objective Optimization—Techniques and Applications in Chemical Engineering, Advances in Process Systems Engineering*; Rangaiah, G. P., Ed.; World Scientific Publishing Co. Pte. Ltd.: Singapore. 2009; Vol. 1, pp 27–52.
12. Jaimes, A. L.; Coello Coello, C. A. Multi-Objective Evolutionary Algorithms: A Review of the State-of-the-Art and Some of Their Applications in Chemical Engineering. In *Multi-Objective Optimization—Techniques and Applications in Chemical Engineering, Advances in Process Systems Engineering*; Rangaiah, G. P., Ed.; World Scientific Publishing Co. Pte. Ltd.: Singapore. 2009; Vol. 1, pp 61–86.
13. Ray, T.; Isaacs, A.; Smith, W. Surrogate Assisted Evolutionary Algorithm for Multi-Objective Optimization. In *Multi-Objective Optimization—Techniques and Applications in Chemical Engineering, Advances in Process Systems Engineering*; Rangaiah, G. P., Ed.; World Scientific Publishing Co. Pte. Ltd.: Singapore. 2009; Vol. 1, pp 131–150.
14. Bandyopadhyay, S.; Saha, S.; Maulik, U.; Deb, K. A Simulated Annealing-Based Multiobjective Optimization Algorithm. *IEEE Trans. Evol. Comput.* **2009**, *12* (3), 269–283.

15. Palit, S. Application of Evolutionary Multi-Objective Optimization in Designing Fluidised Catalytic Cracking Unit and Chemical Engineering Systems—A Scientific Perspective and a Critical Overview. *Int. J. Comput. Intell. Res.* **2016**, *12* (1), 17–34.

16. Varshney, P.; Kunzru, D.; Gupta, S. K. Modelling of the Riser Reactor in a Resid Fluidised-Bed Catalytic Cracking Unit Using a Multigrain Model for an Active Matrix-Zeolite Catalyst. *Indian Chem. Eng.* **2014**, *57* (2), 115–135.

17. www.wikipedia.com

CHAPTER 2

THEORETICAL COMPUTATION OF PERIODIC DESCRIPTORS INVOKING PERIODIC PROPERTIES

SHALINI[1], PRABHAT RANJAN[2], and TANMOY CHAKRABORTY[3*]

[1]*Department of Chemistry, Alankar P.G. Girls College, Jaipur, Rajasthan, India*

[2]*Department of Mechatronics Engineering, Manipal University Jaipur, Jaipur, Rajasthan, India*

[3]*Department of Chemistry, Manipal University Jaipur, Dehmi Kalan, Jaipur, Rajasthan, India*

Corresponding author. E-mail: tanmoychem@gmail.com; tanmoy.chakraborty@jaipur.manipal.edu

ABSTRACT

Hardness is one of the important periodic descriptors used by scientists to correlate the physicochemical properties of atoms, molecules, and condensed phases of matter. Fundamentally, chemical hardness resists the deformation of the electron cloud of the atoms, ions, or molecules under perturbation. The concept of electronegativity and hardness is conceived from electrostatic attraction of nucleus on an electron of the periphery of atom. Electronegativity has already been defined in terms of energy and force concept as well. Though a number of researchers have defined the atomic hardness in terms of energy concept, the force model for hardness is still unexplored. Since the commonality of atomic hardness and atomic electronegativity is a well-established fact, we have defined the hardness as force model in this venture. As the atomic hardness is a force exerted by the screened nucleus on an electron at the periphery of an atom, the following ansatz is proposed to compute global hardness of periodic table, $\eta = a\,(Z_{eff}\,e^2/r^2) + b$; which is used for computing hardness of 103 elements of the periodic table, where *a* and *b* represent regression coefficients. We have used absolute radii (*r*) computed by Chakraborty et al. in the above equation. The new scale satisfies all the sine qua non of a scale of atomic hardness. A new electronic structure principle, namely, the principle of hardness equalization principle, is also proposed in the force model. Our data successfully establish hardness equalization principle.

2.1   INTRODUCTION

Electronegativity and hardness are two of the most important descriptors used by scientists to compute the various physical and chemical parameters in chemistry and physics.[1-11] The term electronegativity was defined by Pauling in the year 1960 as "the power of an atom in a molecule to attract electrons toward itself."[2] The electronegativity and chemical hardness are two different theoretical descriptors having manifold applications in science and technology.[12] These two quantities are basically hypothetical concept and cannot be computed in laboratory.[13-16] Some of the noted researchers in this arena have already explored their applications in the real field, especially in computing the physical and chemical properties like in evaluating chemical bonding.[12-22] These two quantities are periodic in nature and no validation has been put forward alleviating their periodicity. The periodic trend of electronegativity and hardness are related to the size of the atom.

A number of researchers have defined the term electronegativity in various ways; conceptually, it is an electrostatic energy with which an atom holds the electrons of outermost orbit.[13-16,23-25] Zhan et al.[26] have studied ionization potential, electron affinity, electronegativity, hardness, and first electron excitation energy, which are in well agreement with the experimental data. Electronegativity is an important parameter in the Q-e system for better understanding of the reactivity of a monomer, which comprises a double bond in free-radical copolymerizations.[27-29] Murphy et al.[30] have investigated Pauling's bond energy–bond polarity equations, which reveal the additivity of energy. Louzguine et al.[31] have studied the impact of electronegativity of rare earth metals on the supercooled liquid area in the Al–(Gd, Dy, or Er)–Ni–Co metallic glasses using X-ray diffractometry and differential scanning calorimetry technique. Electronegativity can also be used to categorize the hardness of crystal materials.[32] Cong et al.[33] have studied structure of double bond for prediction of charge distribution of polypeptides using density functional theory. Noorizadeh et al.[34] have investigated a new scale of electronegativity on the basis of electrophilicity index. The computed electronegativity is in good agreement with the electronegativity scales like Pauling and Allred–Rochow. Recently, Jadhao et al.[35] have studied effect of electronegativity on geometry, spectrophotometric, and thermochemical properties of fluorine and chlorine substituted isoxazoles by density functional theory methodology. Ruthenberg et al.[36] have investigated that electronegativity is a unique chemical concept, which correlates the ability of chemical species to attract electrons during their contact with other species with measurable quantities.

Atomic hardness is an important periodic descriptor in molecular system which helps in predicting and correlating physical and chemical properties of atoms. The minimum energy of molecular system is related to maximum value of hardness and molecular valency.[37] Kaya et al.[38] have investigated a new technique to compute molecular hardness using density functional theory and Datta's global hardness equalization principle. Labbè et al.[39] have studied a theoretical model based on the theory of maximum hardness and the Hammond postulate; it has been observed that chemical potential and hardness play an important role in activation and relaxation process. Ayers et al.[40] have explored the necessity of molecular properties on the chemical hardness. It has been already observed that size, polarizability, charge, and electronegativity are related with the molecular hardness. Berkowitz et al.[41] have explored the local hardness in molecular system and also establish an overview of the classical electrostatic potential. Harbola et al.[42] have studied local and global hardness and softness using density functional theory. In this chapter, it has been validated that hardness is an average of orbital contributions. Pearson et al.[43] have investigated the applications of electronegativity and hardness in organic chemistry, which is steady with the frontier orbital theory.

Conceptually, the electronegativity and molecular hardness express the electron holding power of atoms and molecules. Their fields of applications are different but their origin and objectification are same. They differ in labels but they represent the same fundamental quantity. Both hardness and electronegativity are periodic properties and are related to atomic size and controlled by the physical process of screening and the nature of variation. These can be measured in terms of energy as well as force concept. In this venture, we have proposed a new scale of atomic hardness of 103 elements of the periodic table for the first time in terms of force concept. A new scale of atomic hardness is also proposed invoking effective nuclear charge and absolute radii. The scale has been designed considering the fact that atomic hardness is a force exerted by the screened nucleus on an electron at the periphery of an atom.

2.2 METHOD OF COMPUTATION

It has been already established that the fundamental descriptors of atomic hardness and electronegativity have same source and origin. They rely upon the electron attracting powers of valence electrons.[1,12–16] Putz et al.[19] have also pointed out that electronegativity and hardness are directly proportional to each other, that is, $\chi \propto \eta$. Later on, Ghosh et al.[16] have established that electronegativity and atomic hardness are numerically and dimensionally same for atoms and molecules, that is, $\chi = \eta$. In this venture, we have established a new scale of atomic hardness of 103 elements of the periodic table in terms of force concept. Allred et al.[44] first suggested a scale of electronegativity based on electrostatic force using the following ansatz:

$$\chi \propto Z_{eff} \frac{e^2}{r^2} \tag{2.1}$$

where χ is electronegativity, r is the distance between an electron and nucleus, and Z_{eff} is charge at the electron due to the nucleus and its surrounding electrons.

The term Z_{eff} is more precise, and it involves shielding that prevents electron from nuclear charge Z.

Considering molecular hardness is equal to the electronegativity, we can propose a new ansatz as follows:

$$\eta \propto Z_{eff} \frac{e^2}{r^2} \tag{2.2}$$

It is assumed that hardness is directly proportional to effective nuclear charge $(\eta \propto Z_{eff})$ and inversely proportional to atomic radii $(\eta \propto 1/r^2)$.

Invoking this concept, we can define atomic hardness as follows:

$$\eta = a\left(Z_{eff} \frac{e^2}{r^2} \right) + b \tag{2.3}$$

where a and b represent regression coefficients. The value of a and b are presented in Table 2.1. As absolute atomic radius is not a finite and observable descriptor, to find out the right size of ions and atoms, we have considered the absolute radii of periodic elements calculated by Chakraborty et al.[45]

Invoking our computed hardness data, we have finally established hardness equalization principle, assuming molecular hardness as geometric mean of hardness of atomic components.[46–48]

TABLE 2.1 Calculated Atomic Hardness in mdyne.

Atomic hardness (mdyne)
Atomic number
Symbol of element

1	2	3	4	5	6	7	8	9	10	11	12	13	14	15	16	17	18
1 H 5.96																	2 He 46.31
3 Li 0.687	4 Be 3.50											5 B 3.42	6 C 8.43	7 N 18.17	8 O 17.80	9 F 36.43	10 Ne 68.44
11 Na 1.02	12 Mg 3.09	3	4	5	6	7	8	9	10	11	12	13 Al 2.18	14 Si 5.05	15 P 10.18	16 S 11.15	17 Cl 20.70	18 Ar 36.04
19 K 0.708	20 Ca 1.89	21 Sc 2.29	22 Ti 2.63	23 V 2.68	24 Cr 2.82	25 Mn 3.61	26 Fe 4.25	27 Co 4.42	28 Ni 4.27	29 Cu 4.53	30 Zn 7.24	31 Ga 3.09	32 Ge 6.31	33 As 11.31	34 Se 12.24	35 Br 20.54	36 Kr 32.80
37 Rb 6.53	38 Sr 1.6	39 Y 2.17	40 Zr 2.65	41 Nb 2.80	42 Mo 3.13	43 Te 3.44	44 Ru 3.68	45 Rh 3.92	46 Pd 5.19	47 Ag 4.34	48 Cd 6.57	49 In 2.86	50 Sn 5.38	51 Sb 8.52	52 Te 10.26	53 I 15.56	54 Xe 23.57
55 Cs 0.562	56 Ba 1.33	57–71	72 Hf 9.96	73 Ta 15.06	74 W 16.15	75 Re 16.40	76 Os 21.24	77 Ir 24.49	78 Pt 24.77	79 Au 26.95	80 Hg 36.68	81 Tl 3.2	82 Pb 5.49	83 Bi 5.87	84 Po 8.85	85 At 12.99	86 Rn 17.93
87 Fr 0.644	88 Ra 1.37	89–103															

57 La 1.88	58 Ce 2.11	59 Pr 2.39	60 Nd 2.77	61 Pm 3.17	62 Sm 3.61	63 Eu 4.00	64 Gd 5.15	65 Tb 5.01	66 Dy 5.53	67 Ho 6.1.6	68 Er 6.68	69 Tm 7.29	70 Yb 7.89	71 Lu 6.15
89 AC 1.37	90 Th 2.05	91 Pa 2.57	92 U 3.27	93 Np 3.79	94 Pu 4.22	95 Am 4.50	96 Cm 4.64	97 Bk 5.73	98 Cf 6.30	99 Es 7.00	100 Fm 7.65	101 Md 8.32	102 No 8.99	103 Lr 4.18

2.3 RESULTS AND DISCUSSION

In this study, the atomic hardness of 103 elements has been computed in force concept and presented in Table 2.2. It is well known that the molecule possessing the highest hardness value will be the least prone to response against external perturbation.[43] Chemical inertness of noble gas elements is reflected in Figure 2.1. It is also observed that hardness of Cs is significantly small compared to other elements. The half-filled stable electronic configuration of nitrogen over oxygen is also pronounced in our computed data. Nitrogen (18.171) possesses higher hardness value than oxygen (17.806). Computed hardness data exhibit easy deformability of transition elements. Periodicity is nicely reproduced through our result. In absence of any quantitative benchmark, we have made a comparative analysis between our computed atomic hardness and hardness computed in energy model (eV) by Ghosh et al.[15] in Figure 2.1.

TABLE 2.2 Value of Regression Coefficients (*a* and *b*) of Elements.

Period	a	b
First	10.57	2.035
Second	0.033	0.138
Third	0.037	0.102
Fourth	0.029	0.112
Fifth	0.038	0.105
Sixth	−20.83	1.586
Seventh	34.78	−2.55
Lanthanoids	0.018	0.095
Actinides	0.022	0.098

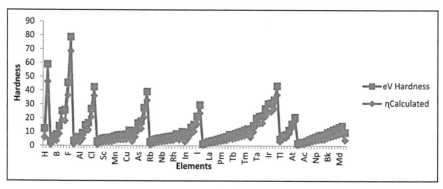

FIGURE 2.1 **(See color insert.)** Comparative analysis between our computed data (mdyne) with atomic hardness of Ghosh et al.

To apply our computed data in the real field, we have tried to establish hardness equalization principle invoking our result.

2.3.1 HARDNESS EQUALIZATION PRINCIPLE

Electronegativity equalization principle is a well-established chemical fact. In 1951, Sanderson introduced this fundamental concept to understand the formation of chemical bond.[46,47] After that, a lot of results are reported about this fundamental phenomenon.[49,50] Although hardness is an important periodic descriptor and has origin of construct similar to the electronegativity, a few reports are available on hardness equalization principle.[51–53] Hardness equalization principle emphasizes that magnitude of the hardness kernel is a variable quantity during the chemical reaction leading to the bond formation. In this process, magnitude of hardness values of atomic fragments will equalize to some intermediate values. After the formation of the molecule, atomic hardness is equalized,[54] that is,

$$\eta_{AB} = \eta_A'' = \eta_B'' \tag{2.4}$$

where η_A'' and η_B'' are the hardness of the atoms A and B, respectively, in the molecule AB.

Invoking geometric mean of atomic hardness to compute equalized molecular hardness is well established.[55]

$$\eta_{GM} = \sqrt{\eta_A \cdot \eta_B}, \quad \text{where } \lambda \equiv \eta \tag{2.5}$$

We have calculated equalized molecular hardness of some chemical species based on eq 2.5. A comparative analysis between our computed data with experimental counterparts[56] is presented in Figure 2.2.

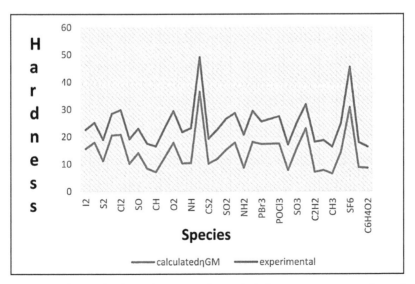

FIGURE 2.2 (See color insert.) Comparative analysis between our computed equalized hardness and experimental counterparts.

2.4   CONCLUSION

Establishment of any new concept depends on its capability to explain observations found in the experiment. Atomic hardness is neither experimentally observable quantity nor can be evaluated quantum mechanically. In absence of any experimental benchmark, success of establishing new scale of absolute hardness relies upon periodic behavior of hardness, comparative analysis with already existing scale and real field application. In this venture, relying upon a general relationship between global hardness of atoms and the absolute radii, we have suggested an ansatz to compute global hardness of atoms in terms of force model. Atomic hardness of 103 elements of periodic table has been computed invoking our proposed ansatz. Computed data follow all sine qua non of periodic behavior. Half and full filled shell structures are established. Successful comparative study with existing scales of hardness also validates our model. Hardness equalization principle has been established invoking our computed hardness data. Our computed equalized molecular hardness runs hand in hand with experimental counterparts.

KEYWORDS

- effective nuclear charge
- electronegativity
- hardness
- chemical periodicity
- regression coefficient

REFERENCES

1. Islam, N.; Ghosh, D. C. *Int. J. Mol. Sci.* **2012**, *13*, 2160.
2. Pauling, L. *The Nature of the Chemical Bond*, 3rd ed.; Cornell University: Ithaca, New York, 1960.
3. Iczkowski, R. P.; Margrave, J. L. *J. Am. Chem. Soc.* **1961**, *83*, 3547.
4. Parr, R. G.; Donnelly, R. A.; Levy, M.; Palke, W. E. *J. Chem. Phys.* **1978**, *68*, 3801.
5. Sen, K. D.; Jørgensen, C. K., Eds. *Electronegativity, Structure and Bonding*; Springer: Heidelberg, 1987; Vol. 66.
6. Pearson, R. G. *J. Am. Chem. Soc.* **1963**, *85*, 3533.

7. Parr, R. G.; Yang, W. *Density Functional Theory of Atoms and Molecules*; Oxford University Press: New York, 1989.
8. Sen, K. D.; Mingos, D. M. P., Eds. *Chemical Hardness, Structure and Bonding*; Springer-Verlag: Berlin, 1992; Vol. 80.
9. Chattaraj, P. K. *J. Indian Chem. Soc.* **1992**, *69*, 173.
10. Pearson, R. G. *Proc. Natl. Acad. Sci. U.S.A.* **1986**, *83*, 8440.
11. Nath, S.; Chattaraj, P. K. *Pramana—J. Phys.* **1995**, *45*, 65.
12. Islam, N.; Ghosh, D. C. *J. Quantum Inf. Sci.* **2011**, *1*, 135.
13. Ghosh, D. C. *J. Theor. Comput. Chem.* **2005**, *4*, 21.
14. Ghosh, D. C.; Chakraborty, T. *J. Mol. Struct. THEOCHEM* **2009**, *906*, 87.
15. Ghosh, D. C.; Islam, N. *Int. J. Quantum Chem.* **2010**, *110*, 1206.
16. Ghosh, D. C.; Islam, N. *Int. J. Quantum Chem.* **2011**, *111*, 40.
17. Putz, M. V.; Russo, N.; Sicilia, E. *Theor. Chem. Acc.* **2005**, *114*, 38.
18. Putz, M. V. *Int. J. Quantum Chem.* **2006**, *106*, 361.
19. Putz, M. V. *Absolute and Chemical Electronegativity and Hardness*; Nova Science: New York, 2008.
20. Putz, M. V.; Russo, N.; Sicilia, E. *J. Phys. Chem. A* **2003**, *107*, 5461.
21. Tarko, L.; Putz, M. V. *J. Math. Chem.* **2010**, *47*, 487.
22. Putz, M. V. *J. Mol. Struct. THEOCHEM* **2009**, *900*, 64.
23. Allred, L.; Rochow, E. G. *J. Inorg. Nucl. Chem.* **1958**, *5*, 264.
24. Mulliken, R. S. *J. Chem. Phys.* **1934**, *2*, 782.
25. Gordy, W. *Phys. Rev.* **1946**, *69*, 604.
26. Zhan, C. G.; Nichols, J. A.; Dixon, D. A. *J. Phys. Chem. A* **2003**, *107*, 4184.
27. Alfrey, T. Jr.; Price, C. C. *J. Polym. Sci.* **1947**, *2*, 101.
28. Rogers, S. C.; Mackrodt, W. C.; Davis, T. P. *Polymer* **1994**, *35*, 1258.
29. Jenkins, A. D. *J. Polym. Sci., Part A: Polym. Chem.* **1999**, *37*, 113.
30. Murphy, L. R.; Meek, T. L.; Allred, A. L.; Allen, L. C. *J. Phys. Chem. A* **2000**, *104*, 5867.
31. Louzguine, D. V.; Inoue, A. *Appl. Phys. Lett.* **2001**, *79*, 3410.
32. Li, K.; Wang, X.; Zhang, F.; Xue, D. *Phys. Rev. Lett.* **2008**, *100*, 235504.
33. Cong, Y.; Yang, Z. Z. *Chem. Phys. Lett.* **2000**, *316*, 324.
34. Noorizadeh, S.; Shakerzadeh, E. *J. Phys. Chem. A* **2008**, *112*, 3486.
35. Jadhao, N. U.; Naik, A. B.; Senn, H. M. *Cogent Chem.* **2017**, *3*, 1296342.
36. Ruthenberg, K.; González, J. C. M. *Found. Chem.* **2017**, *19*, 61.
37. Ghanty, T. K.; Ghosh, S. K. *J. Phys. Chem. A* **2000**, *104*, 2975.
38. Kaya, S.; Kaya, C. *Comput. Theor. Chem.* **2015**, *1060*, 66.
39. Labbè, A. T. *J. Phys. Chem. A* **1999**, *103*, 4398.
40. Ayers, P. W. *Faraday Discuss.* **2007**, *135*, 161.
41. Berkowitz, M.; Ghosh, S. K.; Parr, R. G. *J. Am. Chem. Soc.* **1985**, *107*, 6811.
42. Harbola, M. K.; Chattaraj, P. K.; Parr, R. G. *Isr. J. Chem.* **1991**, *31*, 395.
43. Pearson, R. G. *Inorg. Chem.* **1988**, *27*, 734.
44. Allred, A. L.; Rochow, E. G. *J. Inorg. Nucl. Chem.* **1958**, *5*, 264.
45. Chakraborty, T.; Gazi, K.; Ghosh, D. C. *Mol. Phys.* **2010**, *108*, 2081.
46. Sanderson, R. T. *Science* **1951**, *114*, 670.
47. Sanderson, R. T. *Science* **1951**, *121*, 207.
48. Ghosh, D. C.; Chakraborty, T. *Theor. Chem. Acc.* **2009**, *124*, 295.
49. Mortier, W. J.; Genechten, K. V.; Gasteiger, J. *J. Am. Chem. Soc.* **1985**, *107*, 829.
50. Mortier, W. J.; Ghosh, S. K.; Shankar, S. *J. Am. Chem. Soc.* **1986**, *108*, 4315.

51. Dutta, D. *J. Phys. Chem.* **1986**, *90*, 4211.
52. Berkowitz, M.; Ghosh, S. K.; Parr, R. G. *J. Am. Chem. Soc.* **1985**, *107*, 6811.
53. Ghosh, S. K.; Berkowitz, M.; Parr, R. G. *Proc. Natl. Acad. Sci. U.S.A.* **1984**, *81*, 8028.
54. Ghosh, D. C.; Islam, N. *Int. J. Quantum Chem.* **2010**, *111*, 1961.
55. Chattaraj, P. K.; Giri, S.; Duley, S. *J. Phys. Chem. Lett.* **2010**, *1*, 1064.
56. Chattaraj, P. K.; Nandi, P. K.; Sannigrahi, A. B. *Proc. Indian. Acad. Sci. (Chem. Sci.)* **1991**, *103*, 583.

PART II
Analytical Nanoscience
and Nanotechnology

CHAPTER 3

ANTIMICROBIAL ACTIVITY OF NANOSIZED PHOTOCATALYTIC MATERIALS

RAKSHIT AMETA*, MONIKA TRIVEDI, JAYESH BHATT, DIPTI SONI, SURBHI BENJAMIN, and SURESH C. AMETA

Department of Chemistry, PAHER University, Udaipur 313003, Rajasthan, India

Corresponding author. E-mail: rakshit_ameta@yahoo.in

ABSTRACT

Photocatalysis is emerging as a promising area with varied applications like wastewater treatment, self-cleaning, energy conversion and storage, antifogging, artificial photosynthesis, etc. Recently, a new frontier has been added to it and that photokilling of bacteria, and also antitumor activities. Search is still on for newer and effective photocatalyst, which can have still better antimicrobial activity. Since nanomaterials have high surface-to-volume ratio, and therefore, photocatalytic materials with nanosize may prove to be more active than their micro- and macrocounterparts. Here, photocatalytic activity of nanosized photocatalytic materials has been reviewed.

3.1 NANOMATERIALS

Nanomaterials is the term used for those materials, which have at least one external dimension in the size ranging from approximately 1 to 100 nm.[6] Nanosized metals, oxides, and semiconductors are important because of their mechanical, electrical, magnetic, optical, and chemical properties. Nanoparticles have their own significance as they are a bridge between bulk materials and atomic or molecular structures. Therefore, size-dependent properties play a significant role in nanomaterials. Nanomaterials are characterized by their surface-to-volume ratio, which provides the driving force for diffusion, particularly at high temperatures.

Nowadays, nanomaterials are also used as photocatalysts apart from its other applications. These can be applied to varied fields, and one of them is treatment of microbial infections. It has been reported that the antimicrobial activity of a photocatalyst is enhanced, when taken in nanometer range. Herein, many nanosized photocatalysts and their role as antimicrobial agent have been discussed. The nanosized photocatalytic antimicrobial agents are likely to inactivate microorganisms more successfully due to their smaller size and larger surface area than their micro- or macrosized counterparts.

As our environment is getting infected at a rapid pace due to the increasing number of microbes in the atmosphere, and therefore, there is a pressing demand to develop certain agents, which could combat against some dreadful diseases caused by these microbes. Nanomaterial may retard the growth of microbes causing least harm to the host. Postsurgical high mortality rate is mainly due to infections arising from postoperative complications. This could be reduced by coating surgical instruments or implants with antibacterial or antifungal substances, particularly photocatalytic materials of nanosize. Sometimes, unhygienic hospital environment also worsens the patients' health. This can be controlled by using antibacterial and antifungal agents in paints, sterilization of medical instruments, use of infection proof tiles, etc.

3.2   PHOTOCATALYSIS

Photocatalysis is the process of acceleration of a reaction in the presence of light and a catalyst (semiconductor in particular). The substrate which absorbs light and acts as a catalyst for that chemical reaction is called a photocatalyst. All the photocatalysts are basically semiconductors, but the reverse may or may not be true. When a photocatalyst is illuminated by light of appropriate wavelength, electrons in valence band absorbs this energy and jumps to the conduction band leaving behind a hole in the valence band (Fig. 3.1). Thus, the photocatalytic activity of a semiconductor depends on the ability of the catalyst to create electron–hole (e^-–h^+) pairs. This e^-–h^+ pair is used for redox reactions.

Photocatalysis finds application in wastewater treatment, decolorization, conservation and storage of energy, antifouling, antifogging, air purification, self-cleaning, sterilization, etc.

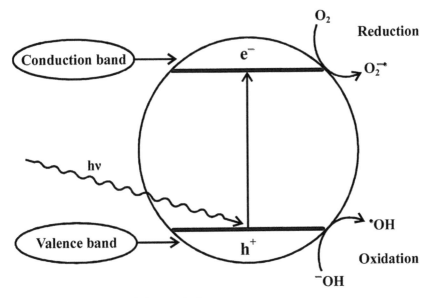

FIGURE 3.1 Mechanism of photocatalysis.

3.3 ANTIMICROBIAL ACTIVITY

The whole world is suffering from the harmful and fatal effects of various types of microbial diseases today. Most of the microorganisms such as bacteria, fungi, algae, and viruses are often overlooked as they cannot be seen with naked eyes, but they play a very damaging role for human beings, animals, avians, and plants. These effects of microbes are usually associated with an illness, which may be sometimes quite severe also. Therefore, there is an urgent need to keep away all the germs and microbes either out of our environment or at least in control by using certain antimicrobial agents.

An antimicrobial agent is used for killing or inhibiting the growth of microorganisms. The most important requirement is that it causes little or no damage to the host. The Greek words "anti" meaning against, "mikros" meaning little, and "bios" meaning life; taking all these terms together, the term antimicrobial is originated. These antimicrobial substances can have natural, semisynthetic, or synthetic origin. One of the most important God-gifted antimicrobial agents is the Sunlight. It can be used by a variety of semiconductors (known as photocatalyst also), which can prove to be a very efficient tool to wipe out the microbes from our environment. Nanosized photocatalysts are particularly effective as antimicrobial agents because of their high surface-to-volume ratio and improved surface reactivity. This enables them to inactivate microorganisms more effectively than their micro- or macroscale counterparts.

Antimicrobial medicines can be classified depending on the microorganism. If they act against bacterial infection, these are called antibacterials. These are termed as antifungals, if they act against fungi. In general, the agents that kill microbes are known as microbicidal, while those agents that only inhibit microbial growth are called biostatic. Antimicrobial chemotherapy is the use of antimicrobial medicines to treat infection, whereas antimicrobial prophylaxis is the use of antimicrobial medicines to prevent infection.

The antimicrobial agents can also be classified as disinfectants, antiseptics, and antibiotics. Disinfectants are used to kill microbes on nonliving surfaces to stop the spread of illness. Antiseptics are applied to living tissue and during surgery to lower infection rate, whereas antibiotics kill microorganisms within the body.

3.3.1   ANTIBACTERIAL AGENTS

Antibacterials act against bacterial infections. These antibacterial agents can be further divided into bactericidal and bacteriostatic agents. The bactericidal agents destroy the bacteria by killing them, whereas the bacteriostatic agents only work to retard the bacterial growth.

Antibacterials are the most commonly used pharmaceutical drugs. With a widespread use of antibacterials, an antibiotic-resistance has been developed in pathogens, and as a result, they pose a serious threat to global public health. Therefore, there is an urgent need to develop new antibacterial agents, which may be effective against these stubborn pathogenic bacteria.

3.3.2   ANTIFUNGAL AGENTS

Antifungal agents kill or slow down the growth of fungi. Fungi and humans are both eukaryotes, thus showing similarity at the molecular level and making the synthesis of antifungal drugs a difficult task. Other than medicines, antifungal agents also find applications in paints for use in high humidity areas like bathrooms or kitchens.

3.3.3 ANTIVIRAL AGENTS

Antivirals are used for treating infections caused by different viruses. They can be differentiated from viricides, which only deactivate viruses outside the body. These antivirals are relatively harmless to the host and therefore find applications in medical field.

3.4 ANTIMICROBIAL APPLICATIONS

Li et al. reviewed the merits, limitations, and applicability of various nano-materials having strong antimicrobial properties through different mechanisms like photocatalytic production of reactive oxygen species (ROSs) that are responsible to damage cells and viruses.[28] Some nanomaterials have also been used as antimicrobial agents in commercial products including home purification systems. They have highlighted applications of nanomaterials for water disinfection and biofouling control. Rasmussen et al. reviewed the biomedical applications of metal oxide nanomaterial like ZnO.[39] The proposed mechanism of cytotoxic activity as well as current researches has also been discussed so as to improve their targeting and cytotoxicity even against cancer cells. Lipovsky et al. also reviewed visible light induced anti-bacterial properties in metal oxide nanoparticles (NPs) instead of harmful ultraviolet (UV) radiation.[29]

$$\text{Photocatalyst} \xrightarrow{hv} \text{photocatalyst} \left(e^-_{CB} + h^+_{VB} \right)$$

where e^-_{CB} = conduction band electron and h^+_{VB} = valence band hole

$$H_2O + h^+_{VB} \rightarrow {}^\bullet OH + H^+$$

$$H^+ + e^-_{CB} \rightarrow H^\bullet$$

$$O_2 + e^-_{CB} \rightarrow O_2^{\bullet -} \quad H^+ \rightarrow HO_2^\bullet$$

For a cell or virous in contact with the semiconductor surface, these also be direct electron or hole transfer to the organism or one of its component. If semiconductor particles are small size, they may penetrate into the cell, and this process could be also in the interior of the cell. Light is essential for a photocatalytic reaction to proceed. Hydroxyl radicals are highly reactive, and therefore, these are short lived. Superoxide ions are relatively more long lived, but due to negative charge, they cannot penetrate the cell membrane. Upon their production on the semiconductor surface, both hydroxyl radicals and superoxide anion radicals will interact immediately with the outer surface of an organism, unless semiconductor particle penetrate into the cell.[5]

H_2O_2 is less detriment as compared to $^{\cdot}OH$ and superoxide radical anion. The important part of killing of bacteria is that H_2O_2 can enter the cell, and it is activated by ferrous ion via the Fenton reaction:

$$Fe^{2+} + H_2O_2 \rightarrow {}^{\cdot}OH + OH^- + Fe^{3+}$$

Photokilling of microbial cell-irradiated semiconductor plays an important role in because the ROSs such as OH^-, $O_2^{\cdot-}$, and H_2O_2 generated on irradiated photocatalyst surface; these species have been proposed to attach with polyunsaturated phospholipids in the cell membrane of microorganism. Iron levels on the cell surface, in the periplasmic space or inside the cell (either as iron clusters or in iron storage proteins such as Ferritin) are significant and serve as sources of ferrous ion. Therefore when the semiconductor is being illuminated to produce H_2O_2, the Fenton reaction may take place in vivo and produce more damaging hydroxyl radicals.

When light is turned off, any residual H_2O_2 would continue to interact with the iron species and generate $^{\cdot}OH$ radicals through Fenton reaction, but when both $^{\cdot}H_2O_2$ and $O_2^{\cdot-}$ are present, the iron catalyzed Haber–Weiss reaction can provide a second pathway to form additional $^{\cdot}OH$ radicals.

$$Fe^{3+} + O_2^{\cdot-} \rightarrow Fe^{2+} + O_2$$
$$Fe^{2+} + H_2O_2 \rightarrow {}^{\cdot}OH + OH^- + Fe^{3+}$$

Lipid peroxidation reaction takes place thus causes a breakdown of the cell membrane structure and its associated functions; this is the mechanism responsible for cell death. Because all life forms have cell membrane, and therefore, such proposed mechanism is applicable to all cell types.[31,40]

Apart from the rupture of cell wall, there is another possibility of death of microbial cell, this is destructive effect of oxidative catalyst on degradation of RNA and DNA molecules, mainly due to $^{\cdot}OH$ radicals (Fig. 3.2).

Nithya et al.[35] synthesized a chitosan–silver (CS–Ag) nanocomposite by green pathways without using any external chemical (reducing agent). This nanocomposite was characterized by UV–visible spectroscopy, X-ray diffraction (XRD), Fourier-transform infrared (FT-IR), thermogravimetric analysis (TGA), differential scanning calorimetry (DSC), field emission-scanning electron microscopy (FE-SEM), energy-dispersive X-ray (EDX), atomic force microscopy (AFM), high-resolution transmission electron microscopy (HR-TEM), selected area electron diffraction (SAED), X-ray photoelectron spectroscopy (XPS), and zeta potential analysis, etc. techniques. It was revealed that the particle size of the synthesized CS–Ag nanocomposite was about 20 nm, and it was found to be thermally more stable than pure chitosan. As-prepared nanocomposite assisted in photocatalytic bleaching of dye like methyl orange in presence of visible light. The nanocomposite also showed an excellent antimicrobial efficiency against both Gram-positive and Gram-negative bacteria. The highest antibacterial activity was exhibited against Gram-positive *Staphylococcus aureus* by CS–Ag nanocomposite.

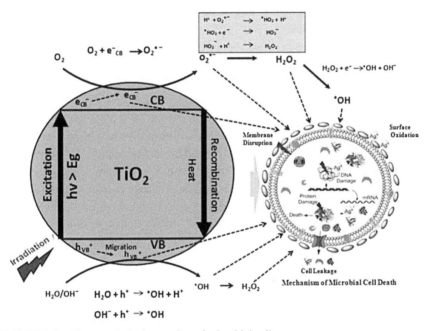

FIGURE 3.2 Photocatalytic destruction of microbial cell.

Pant et al.[37] synthesized ZnO nanoflowers doped with TiO_2 and Ag NPs by facile one-pot hydrothermal process. Antibacterial and photocatalytic properties of $Ag/TiO_2/ZnO$ nanoflowers were also studied. Sangari et al.[41] prepared carbon nanocone and disc-fluorine-co-doped TiO_2 nanocomposites via the solid state method. The efficiency of these nanocomposites has been tested by observing the antimicrobial activity and photocatalytic activity. It was observed that the carbon cone and disc-fluorine-TiO_2 nanocomposites behaved as an efficient antibacterial agent.

Joost et al. synthesized TiO_2 nanoparticles using sol–gel method. As-prepared TiO_2 NPs were used to prepare nanostructured films.[24] Rapid inactivation of *Escherichia coli* cells on these films was observed in presence of UV-A illumination. Scanning electron microscopic studies on the bacterial cells exhibited swelling of the cells, distortion of cellular membrane, and leakage of cytoplasm occur just after 10 min of exposure to photo activated TiO_2. No viable cells of *E. coli* were found after 20 min of exposure to UV-A activated TiO_2. The reason for cell death was suggested due to peroxidation and decomposition of fatty acids of membrane.

Gondal et al. synthesized nano-NiO photocatalyst using sol–gel method and employed it for the process of disinfecting water polluted with *E. coli* microorganism.[17] The as-prepared nano-NiO material was used as a photocatalyst for bacterial decay. The rate constant of antimicrobial activity of nano-NiO was found to be higher than the bacterial decay rate constant for TiO_2 as a photocatalyst under same conditions. The effect of concentration of nano-NiO and the irradiating laser pulse energy was observed on the depletion rate of bacterial count.

Tobaldi et al.[51] prepared Ag-modified titania NPs by a green aqueous sol–gel method (Fig. 3.3). The product was thermally treated and its photocatalytic and antibacterial properties were determined using UV visible light exposure. Greater antibacterial activity toward *E. coli* (Gram-negative) was observed as compared to methicillin-resistant *S. aureus* (Gram-positive) on using UV-light source. It was found that UV light changed the oxidation state of Ag from Ag(I) to Ag(0). These NPs had a detrimental effect on the antibacterial activity. However, when artificial white light was used, no deterioration in activity was found, and the material is suitable for use in health care, depleting the number of Gram-negative type bacteria such as *E. coli*.

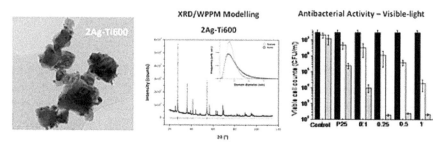

FIGURE 3.3   Antimicrobial activity of Ag-modified titania (Reprinted from Tobaldi, D. M.; Piccirillo, C.; Pullar, R. C.; Gualtieri, A. F.; Seabra, M. P.; Castro, P. M. L.; Labrincha, J. A. *J. Phys. Chem. C* **2014**, *118* (9), 4751–4766. With permission).

Armelao et al.[3] synthesized TiO_2 and Au/TiO_2 nanosystems by a sol–gel route and by an innovative hybrid radio frequency (RF)-sputtering/sol–gel approach, respectively. The photocatalytic activity of the as-prepared nanosystems in the degradation of the azo-dye plasmocorinth B and their antibacterial performance in the elimination of *Bacillus subtilis* was studied in comparison with films derived from standard Degussa P25. The dispersion of Au nanoparticles on the TiO_2 matrix had a positive effect on the dye photodegradation, whereas it was observed that the antimicrobial activity of Au/TiO_2 films was retarded in comparison to pure TiO_2.

Gupta et al.[19] synthesized TiO_2 and silver-doped TiO_2 photocatalysts by acid catalyzed sol–gel technique. The antimicrobial performance of TiO_2 and silver-doped TiO_2 NPs was determined against both; Gram-positive bacteria like *S. aureus* and Gram-negative bacteria like *Pseudomonas aeruginosa* and *E. coli*. It was observed that the viability of all the three microorganisms decreased to zero at 60 mg/30 mL culture for Ag-doped TiO_2 nanoparticles.

Xu et al.[56] synthesized the magnetic $Ag_3PO_4/TiO_2/Fe_3O_4$ heterostructured nanocomposite (Fig. 3.4). The as-prepared nanocomposite showed high photocatalytic activity, cycling stability, and durability in the photodegradation of acid orange 7 under visible light. The film prepared from this nanocomposite showed excellent bactericidal activity and recyclability toward *E. coli*. The Ag ions in the nanocomposite proved to be cytotoxic and the high bactericidal efficiency was due to increased photocatalytic activity of nanocomposite. The death of bacteria was caused due to the photogenerated •OH and O_2^- ions at $Ag/Ag_3PO_4/TiO_2$ interfaces, which led to changes in cells.

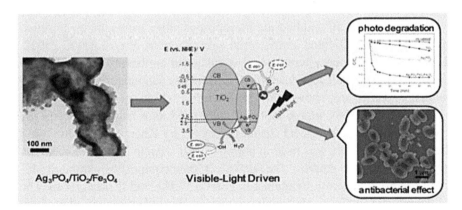

FIGURE 3.4 Antibacterial effect of magnetic $Ag_3PO_4/TiO_2/Fe_3O_4$ heterostructured nanocomposite (Reprinted from Xu, J. W.; Gao, Z. D.; Han, K.; Liu, Y.; Song, Y. Y. *ACS Appl. Mater. Interfaces* **2014**, *6*, 15122–15131. With permission).

Nanofibers having photocatalytic TiO_2 nanoparticle were prepared by electrospinning nylon 6 nanofibers on certain flexible substrates and TiO_2 NPs were electrosprayed on them by Zhang et al.[60] These titania films exhibited efficient photokilling activity against *E. coli* and for photodegradation of methylene blue under UV irradiation. Along with this, soy-protein-containing nanofibers were fabricated by solution blowing, which were then decorated with silver nanoparticles. These nanofibers revealed antibacterial activity against *E. coli* bacteria without exposure to UV light. The so formed material has wide range of applications in water purification and in biotechnology.

Sivakumar et al.[42] prepared nanocomposites of conjugated poly (aniline-*co-o*-anisidine)/zinc oxide [poly(Ani-*co-o*-As)/ZnO] by in situ chemical oxidative polymerization of comonomers using ZnO nanoparticles. They used it as a photocatalyst for the photodegradation of methylene blue. The conductivity measurements exhibited that resulting nanocomposite had higher conductivity in comparison to the pure copolymer. The photocatalytic efficiency was indicated by the degradation of methylene blue dye up to 97% over the nanocomposite catalyst surface under UV-exposure for 3 h. The nanocomposite showed efficient antimicrobial activity.

Mirhoseini and Salabat[32] fabricated an ionic liquid–sensitized TiO_2–loaded poly(methyl methacrylate) (PMMA/TiO_2) films nanocomposite, which acted as an efficient photocatalyst with antibacterial activity against *Klebsiella* spp. The clear zone diameter increased with increasing TiO_2-loaded film until the loading of TiO_2 was done at a value of about 0.01 wt.%. The zone of inhibition was found to decrease on further increasing the TiO_2 doses decrease.

Cendrowski et al.[9] studied the mesoporous silica nanospheres modified by TiO_2 as light activated antibacterial agents against *E. coli* ATCC 25922. The as-prepared nanospheres revealed enhanced antibacterial activity than the commercial catalyst and sample free from the nanomaterials. The silica/titania nanostructures showed low antibacterial activity in dark.

Electrochemically active biofilms were used as a reducing tool in the green synthesis of Ag@CeO_2 nanocomposites by Khan et al.[26] This Ag@CeO_2 nanocomposite was used in antimicrobial, visible light photocatalytic and photoelectrode studies. The Ag@CeO_2 nanocomposites exhibited efficient bactericidal activities against *E. coli* O157:H7, and *P. aeruginosa*. The as-prepared Ag@CeO_2 nanocomposites also showed enhanced photocatalytic degradation of 4-nitrophenol and the dye methylene blue as compared to pure CeO_2. CeO_2 was functionalized using silver NPs, which induced visible light photoactivity by reducing the recombination rate of photogenerated electrons and holes.

Jašková et al.[23] also prepared antimicrobial paints based on the aqueous dispersion of nanoparticles of ZnO and TiO_2. The antimicrobial efficiency and photocatalytic activity of these paints were studied. The best photocatalytic activity was observed in the coating containing the mixture of nano-TiO_2 and nano-ZnO. The agar dilution process was applied to analyze antimicrobial ability. *E. coli*, *P. aeruginosa*, and *S. aureus* were selected as test bacteria and *Penicillium chrysogenum* and *Aspergillus niger* as test molds.

Sol–gel method was used to prepare Au-and Ag-doped TiO_2.[36] Dip coating was done on the soda-lime glass using $TiCl_4$ precursor. Calcinations were carried out at 500°C for half an hour to obtain transparent thin films. These doped TiO_2 sample were used in photodegradation of plasmocorinth B and the antimicrobial activity was observed. It was revealed that the Au-doped TiO_2 has higher antimicrobial activity than the pure sample.

Ghodsi et al.[16] synthesized Fe–TiO_2, Ce–TiO_2, and Ag–TiO_2 thin films using sol–gel dip-coating method. Their antimicrobial activity was studied against the bacteria *S. aureus* by using antibacterial drop test and colony count method. The effect of various factors like thickness and porosity of the films, crystallization, and surface morphology on transition metal-doped TiO_2 was studied. It was found that 70% of cell death was caused in presence of Fe–TiO_2.

Stoyanova et al.[46] studied the photocatalytic and antimicrobial activity of Fe-doped TiO_2. Nonhydrolytic sol–gel method was used to prepare undoped and Fe-doped TiO_2 using titanium tetrachloride, benzyl alcohol, and iron(III) nitrate. The activity of pure and modified samples was tested against *E. coli* under UV-visible light. Ananpattarachai et al.[1] studied the antibacterial activity of N- and Ni-doped TiO_2 on *S. aureus* and *E. coli* bacteria.

TiO_2 is a well-known photocatalysts that has been used to kill cancer cells, bacteria, and viruses under mild-UV irradiation. Silver is also a good antibacterial material as well. The advantage of Ag–TiO_2 nanocomposite can be employed to expand the antibacterial ability of the nanomaterials to a broader range of working conditions. Yu et al.[59] prepared pure TiO_2 and Ag–TiO_2 composite nanofilms on silicon wafer by the sol–gel technique using the spin-coating method. Structural analysis revealed that the Ag NPs were uniformly distributed and firmly attached to the TiO_2 matrix. The antimicrobial activity of the synthesized nanofilms was studied against Gram-negative bacteria (*E. coli* ATCC 29425) in the presence of UV lamp. The Ag–TiO_2 nanocomposite showed greater bactericidal activities compared than the pure TiO_2 both in the dark and under UV irradiation.

Yaithongkum et al.[58] observed that a doping of 0.1–1 mol% Ag on TiO_2/SnO_2/SiO_2 nanocomposite led to change in crystal diameter, morphology, photocatalytic, and fungal growth suppression activities. The nanocomposite powder was prepared by sol–gel method. The *Penicillium expansum* growth was suppressed, when Ag was doped in $TiO_2/SnO_2/SiO_2$ nanocomposite powder under UV illumination. *P. expansum* was completely killed within 1 day of photacatalytic treatment under UV light.

Nanostructured thin films based on neat TiO_2 and TiO_2 modified with vanadium as an additive were prepared using a simple sol–gel process.[38] These nanofilms were deposited on glass and Si substrates. The drop test method was used for the analysis of antifungal and antibacterial activities. The results exhibited that antimicrobial activity of the nanophotocatalyst increases with increasing vanadium concentrations.

Stoyanova et al.[47] synthesized nanosized $ZnTiO_3$ powders using aqueous and nonaqueous sol–gel methods. Zinc acetate and titanium ethoxide were employed as starting materials for the preparation of zinc titanate by aqueous route. Nano-$ZnTiO_3$ powder was also synthesized by a nonaqueous route reaction among $TiCl_4$, $ZnCl_2$, and benzyl alcohol at moderate temperatures. $ZnTiO_3$ powders showed photocatalytic activity in the degradation of malachite green under UV-light illumination. The bactericidal activity of $ZnTiO_3$ was also studied against *E. coli* bacteria.

It was observed that antibacterial effects of nanophotocatalysts remained present even after the UV irradiation was stopped.[7] Resin-based composites having 20% TiO_2 nanoparticles had an efficient antibacterial effect against *E. coli*, *Staphylococcus epidermidis*, *Streptococcus pyogenes*, *Streptococcus mutans*, and *Enterococcus faecalis* for up to after 2-h UV irradiation.

ZnO nanoparticles were synthesized by sol–gel method and characterized by Sunitha and Rao.[49] These nanoparticles were tested for antibacterial activity against the clinical pathogens *E. coli*, *Klebsiella pneumonia*, *P. aeruginosa*, and *S. aureus* using disc well-diffusion method. Positive activity was observed against *P. aeruginosa* and *S. aureus*. The nanoparticles of catalyst were also used for the removal of toxic organic pollutant acid blue 113 under UV irradiation in a batch reactor. The different operational parameters such as the initial concentration of the dye, weight of photocatalyst, and pH on the photocatalytic degradation of the dye were investigated.

Bactericidal effect on Gram-negative bacterium *Vibrio cholerae* 569B was observed in aqueous matrix on carrying out photocatalysis using mediated by Ag@ZnO nanocomposite.[13] They prepared Ag nanoparticles by reducing silver perchlorate followed by precipitation of ZnO. Antibacterial efficiency of nanoparticles was compared with that of pure-ZnO and TiO_2 (Degussa P25). The results showed that nanocomposite system had optimum disinfection ($\approx$98%) at 40–60 min of sunlight irradiation with a catalyst loading of 0.5 mg L^{-1} of the reaction solution. It was concluded that this technique can also be used for water decontamination by reducing bacterial density in aquatic medium.

CuO/Cu(OH)$_2$ nanostructures were synthesized by Azimirad and Safa[4] on immersing copper foil substrate into air, deionized water, and 20-mM NH$_4$OH solution at 25°C. Raman spectroscopy and XRD patterns confirmed the presence of thick layer of the copper oxide nanocrystals at the surface of sample generated in deionized water. The photocatalytic degradation of methylene blue by the CuO/Cu (OH)$_2$ nanostructures was studied. The fungicidal activity of these nanostructured Cu/Cu (OH)$_2$ was also analyzed.

The antifungal and photocatalytic activity of titanium dioxide on number of colony forming units (CFUs) and dry weight of biomass of fungus of *Fusarium oxysporum f. sp. lycopersici Schlecht* was studied by Anbeaki et al.[2] It was observed that TiO$_2$ combined with light resulted into reducing CFUs to 65.66, 12.66, 4, and 4.66 CFUs/0.5 mL after periods of time 30, 60, 90, and 120 min, respectively, as compared to the control in the dark.

Fusarium graminearum is the pathogen causing *Fusarium* head blight in wheat, which could reduce grain quality as well as the yield and generate various mycotoxins contaminating human foods. Zhang et al. [62] prepared a visible light activated palladium-modified nitrogen-doped titanium oxide (TiON/PdO) nanosized photocatalyst as an environmental-friendly fungicide (Fig. 3.5). The nanoparticles were strongly adsorbed onto the macroconidium surface because of the opposite surface charges of TiON/PdO and fungal macroconidium. Photocatalytic disinfection mechanism was due to the cell wall/membrane damage of the fungus resulting from the attack from ROSs.

The antiviral activity of nanosized cuprous iodide (CuI) particles having an average size of 160 nm was examined by Fujimori et al.[13] CuI particles showed aqueous stability and generated hydroxyl radicals, which were probably derived from monovalent copper (Cu$^+$). It was confirmed that CuI particles showed antiviral activity against an influenza A virus of swine origin [pandemic (H1N1) 2009] by plaque titration assay. The virus titer decreased in a dose-dependent manner upon incubation with CuI particles; with the 50% effective concentration being approximately 17 µg mL^1 after exposure for 60 min. Sodium dodecyl sulfate-polyacrylamide gel electrophoresis (SDS–PAGE) analysis confirmed the inactivation of the virus due to the degradation of viral proteins such as hemagglutinin and neuraminidase by CuI. CuI generates hydroxyl radicals in aqueous solution, and radical production was found to be blocked by the radical scavenger *N*-acetylcysteine. These findings indicate that CuI particles exert antiviral activity by generating hydroxyl radicals. Thus, CuI may be a useful material for protecting against viral attacks and may be suitable for applications such as filters, face masks, protective clothing, and kitchen cloths.

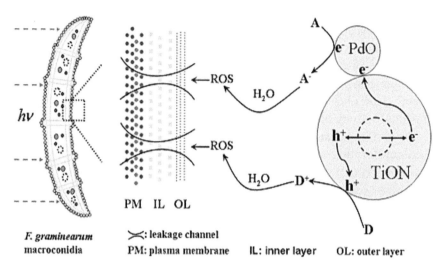

FIGURE 3.5 Antifungal activity and mechanism of palladium-modified nitrogen-doped titanium oxide (Reprinted from Zhang, J.; Liu, Y.; Li, Q.; Zhang, X.; Shang, J. K. *ACS Appl. Mater. Interfaces* **2013**, *5* (21), 10953–10959. With permission).

The photochemical sterilization ability of TiO_2 nanomaterial as an environment-friendly disinfectant against avian influenza (AI) was studied by Cu et al.[11] A neutral and viscous aqueous colloid of 1.6% TiO_2 was prepared from peroxotitanic acid solution using the Ichinose method. It was proved by transmission electron microscopy images that TiO_2 particles were spindle-shaped with an average size of 50 nm. A photocatalytic film of nano-TiO_2 sol was used for inactivating H9N2 AI virus. Such inactivation capabilities were observed under with 365-nm UV light.

A lubricating protective was invented (repair gel), which combined photocatalyst TiO_2 particles with a disinfection material like nanosilver to produce a TiO_2–Ag compound.[48] The compound was mixed with a natural biological organism with a molecular weight below 50 kbp and a concentration of 0.05–5% having small molecules of collagen, chitin, chitosan, hyaluronic acid, vitamin C, and nano-Fe at pH 6.

This mixture had affinity with human skins and was nontoxic to human cells. An additional ingredient like an aromatic agent, a detergent, or a cosmetic agent was added in the manufacturing process to produce an emulsion or a paste for coating or spraying. It was used as a sanitation agent like hand liquid soap, a cosmetic, a shampoo, or a disinfectant. A protective film is formed on the skin as the small sized particles could penetrate the skin cuticle and retained at the derma inside the pores to the skin. The collagen could be absorbed by the skin resulting in tissue repair, lubrication of the skin, and even to stop bleeding. A deeper clean skin can be achieved on applying such a gel, which prevents germs and help in eliminating acnes and pimples. Here, oxidation–reduction mechanism of the photocatalyst TiO_2 has been utilized to achieve long-term disinfection and develop germicidal effect.

Novel modified textile by in situ growth of vertically aligned one-dimensional ZnO nanorods onto textile surfaces was developed by Hatamie et al.[22] These can serve as biosensor, photocatalyst, and antibacterial agent. ZnO nanorods were developed by aqueous chemical growth method. XRD and SEM results showed that these nanorods were dispersed all over the textile surface. These textiles have multifunctional applications like photo-degeradation of organic dyes like methylene blue and Congo red and as an antibacterial agent against *E. coli*. So, the formed textile is quite close to the smart textile used for wearable sensing and antibacterial control without using a deodorant.

Dong et al.[14] prepared Cs/gelatin nano-TiO_2 composite and observed its antibacterial activity. FT-IR, XRD, SEM, and TEM analyses were performed to characterize this nanocomposite. It was found out that the three components, that is, Cs, gelatin, and TiO_2, were bonded through hydrogen bonding.

Phytopathogens cause both qualitative and quantitative loss in all areas of agricultural practices. Phytopathogens are controlled by pesticides, but it is associated with side effects. Most organisms develop resistance to pesticides and high concentration of pesticide is phytotoxic. Wu and Xue combined the effect of nanomaterials and the pesticide. Sunlight was used for photocatalytic degradation of the pesticide. The nanoinorganic material used was nano-CuO and ZnO and thiram as pesticide.[55] Nano-ZnO and the thiram were mixed in a mass concentration ratio of 5:1 to 50:1, while nano-CuO and the thiram were mixed in a mass concentration ratio of 25:1 to 100:1. This mixture was used to prevent and control fungal disease and some pathogenic bacteria. Another benefit was that the pesticidal effect of the thiram was greatly increased, resulting in decreasing the dosage of thiram. The operation and equipment required for the preparation were simple, and the raw material was easily available and low cost also. Continuous photocatalytic degradation or removal of residual pesticide makes it suitable for large-scale industrial production.

Metal oxide NPs like ZnO, CuO, and TiO_2 have been studied for their antibacterial effects, based on their efficiency to generate ROSs by Lucky et al.[30]. The studies revealed that doping metal oxide NPs with transition metals ions, increased their activity in the visible and near-infrared (NIR) range, thereby enhancing their antibacterial activity without causing damage to tissues and cells.

Copper oxide nanoparticles were also prepared by Sivaraj et al.[44] using an Indian medicinal plant *Tabernaemontana divaricate* leaves through a green chemical approach. As-prepared nanoparticles were characterized. The analysis exhibited that these copper oxide nanoparticles were 48 ± 4 nm in size. Antimicrobial activity of these copper oxide nanoparticles were investigated against *E. coli* urinary tract pathogen. Various combinations of antimicrobial agents can be used into packaging materials an active food packaging.[25]

Xue et al.[57] developed a residue-free green nanotechnology, which increased efficiency of the pesticides and also eliminated its residue. A composite antifungal system was prepared by a simple procedure. Observations revealed that 0.25 g L^{-1} ZnO NPs with 0.01 g L^{-1} thiram inhibited the fungal growth. It was also seen that the 0.25 g L^{-1} ZnO NPs completely degraded 0.01 g L^{-1} thiram under simulated sunlight exposure within 6 h. There was electrostatic adsorption of ZnO NPs, which led to the formation of ZnO–thiram antifungal system. Oxidative stress test showed that ZnO-induced oxidative damage is increased by thiram that finally show antifungal effect. Thus, nanotechnology not only controlled the plant disease but also lowered the pesticides concentration used. The photocatalytic degradation of pesticide is also beneficial to the public health as well as environment.

3.5   ANTITUMOR ACTIVITY

Photocatalysts also have a potential to behave as efficient antitumor agents. The nanosized photocatalysts play a significant role in killing carcinogenic tumor cells. Magnetic hyperthermia (MHT) is the generation of heat in the region of the tumor by using magnetic nanoparticles subjected to an alternating magnetic field. MHT has exhibited positive results in preclinical and clinical treatment. Laurent et al. showed the biocompatibility and biomedical potential of superparamagnetic iron oxide NPs. The physicochemical properties and pharmaceutical applications of these nanoparticles were also discussed by them.[27]

Zhang and Sun[61] investigated the photocatalytic killing efficiency of TiO_2 NPs on human colon carcinoma cell line (Ls-174-t). A mechanism was proposed for the action of photoexcited TiO_2 nanoparticles on cancer causing cells. Ls-174-t human colon carcinoma cells were cultured and a GGZ-300-W high-pressure Hg lamp with a maximum UV-A (UVA, 320–400 nm) irradiation peak at 365 nm was used as light source in the photocatalytic killing test. The photocatalytic killing effect was measured using different concentrations of TiO_2 and illumination time. Ls-174-t cells were killed at a much faster rate on adding TiO_2. When 1000 µg mL[1] TiO_2 was used, 44% of the cells were killed after UVA irradiation for 10 min, while 88% were killed after 30 min of UVA exposure.

The apoptotic effect of cerium(IV)-doped titanium dioxide NPs in the presence of visible light was investigated on human hepatoma cell line (Bel 7402) by Wang et al.[53] Impregnation method was used to prepare cerium-doped titanium dioxide nanoparticles. Bel 7402 was cultured in RPMI 1640 medium. A 15-W fluorescent lamp was used as light source in the photocatalytic test. Agarose gel electrophoresis technique was performed to study the apoptotic cells. It was concluded that Ce(IV)-doped TiO_2 nanoparticles induced apoptosis of human hepatoma cells in the presence of visible light. Hackenberg et al.[21] used ZnO NPs to induce photocatalytic cell killing in human head and neck squamous cell carcinoma cell lines in vitro.

TiO_2 and ZnO NPs were used in sunscreen formulation as physical sun blockers[45] and for sunburned skin.[33] The use of deep penetrating NIR light excitable upconversion nanoparticles in combination with photodynamic therapy had exhibited potential in the cure of solid tumors due to its ability to penetrate thick tissues. The surface modification of the as-prepared TiO_2 NPs with polyethylene glycol made them biocompatible, exhibiting better therapeutic efficacy both in vitro and in vivo.

Wason et al.[54] reported that cerium oxide nanoparticles (CONPs) can sensitize pancreatic cancer cells to radiation therapy. CONP treatment prior to radiation therapy caused the cancer cell apoptosis both in culture and in tumors and the inhibition of the pancreatic tumor growth with no harmful effect on the normal tissues. Colon et al.[10] suggested that CeO_2 nanoparticles guard the gastrointestinal epithelium against radiation-induced damage by behaving as free-radical scavengers and enhancing the production of superoxide dismutase 2.

The tumor cells and the normal human cells were treated with nanoceria and irradiated, and then the cell survival rate was determined. Treatment of normal cells guaranteed about 99% protections from radiation-induced cell killing, whereas the same concentration exhibited almost no protection of tumor cells.[50]

CONPs were found to be effective against pathologies related to acute oxidative stress and inflammation[12] and in neurodegradation,[8] which made these NPs favorable for applications in nanobiology and regenerative medicine. Montfort et al.[34] observed a nontoxic and protective effect of cerium oxide NPs in human dermal fibroblasts. Still work is going onto study its effect on important parameters such as cell death, proliferation, and redox state of the cells.

ZnO nanoparticles were prepared by Wahab et al.[52] and used in induction of oxidative stress in Cloudman S91 melanoma cancer cells. Morphology, structure, and optical properties of these NPs were studied. Different doses of ZnO NPs were treated with melanoma cancer cells for 24 h of incubation at 37°C to observe the induction of oxidative stress by as-prepared ZnO nanoparticles in Cloudman S91 melanoma cancer cells. The viability of the cells was determined. The morphology of the cells was analyzed using confocal laser scanning microscopy, which showed that when the time interval was increased, the number of cells decreased.

Zinc oxide nanoparticles (ZnO-NPs) show efficient photocatalytic properties and are, therefore, a major ingredient of sunscreen cosmetics for the degradation of environmental pollutants. Not only this, ZnO-NPs also have a potential to induce tumor-selective cell killing in human squamous cell carcinoma in vitro.[20] The cytotoxic behavior of anticancer drug daunorubicin on leukemia cancer cells in the absence and presence of ZnO-NPs via fluorescence microscopy, UV–Vis absorption spectroscopy, and electrochemical analysis was studied by Guo et al.[18]

Sivaraj et al.[43] prepared copper oxide nanoparticles by biological method using water extract of *Acalyphaindica* leaf and characterized these by UV–visible spectroscopy, XRD, FT-IR, SEM, TEM, and EDX analysis. The as-prepared particles were stable and spherical with particle size in the range 26–30 nm. Copper oxide NPs revealed efficient antibacterial and antifungal effect against *E. coli, Pseudomonas fluorescens*, and *Candida albicans*.

It can be concluded that advance oxidation processes-based disinfection is quick and effective. When applied on a wide scale, these processes are more efficient in improving the sanitary and hygienic conditions, as well as in decreasing food contamination and in lowering the food toxicity. The antimicrobial activity of nanophotocatalyst helps in keeping the building walls clean for long duration, road signs are also protected from getting dirty quickly and mirrors are guarded against fog. It can be observed that nanomaterials play a crucial role in pharmaceuticals (antibacterial, antifungal, antiviral), apart from food hygiene, public health, and crop protection.

KEYWORDS

- **photocatalytic**
- **antimicrobial**
- **nano**
- **antitumor**
- **antifungal**
- **antibacterial**

REFERENCES

1. Ananpattarachai, J.; Boonto, Y.; Kajitvichyanukul, P. Visible Light Photocatalytic Antibacterial Activity of Ni-Doped and N-Doped TiO$_2$ on *Staphylococcus aureus* and *Escherichia coli* Bacteria. *Environ. Sci. Pollut. Res.* **2015**, DOI/10.1007/s11356-015-4775-1.

2. Anbeaki, R. A. A.; Hameed, F. R.; Gassim, F. A. G. Antifungal Activity of Titanium Dioxide Photocatalysis against *Fusarium oxysporum f. sp. lycopersici. Euphrates J. Agri. Sci.* **2009**, *1* (3), 14–26.

3. Armelao, L.; Barreca, D.; Bottaro, G.; Gasparotto, A.; Maccato, C.; Maragno, C.; et al. Photocatalytic and Antibacterial Activity of TiO$_2$ and Au/TiO$_2$ Nanosystems. *Nanotechnology* **2007**, *18* (37), DOI/10.1088/0957-4484/18/37/375709.

4. Azimirad, R.; Safa, S. Photocatalytic and Antifungal Activity of Flower-Like Copper Oxide Nanostructures. *Synth. React. Inorg. Met.-Org. Chem.* **2014**, *44*(6), 798–803.

5. Blanco Gàlvez, J.; Fernàndez Ibànez, P.; Malato-Rodriguez, S. Solar Photocatalytic Detoxification and Disinfection of Water: Recent Overview. *J. Sol. Energy Eng.* **2007**, *129* (1), 4–15.

6. Buzea, C.; Pacheco, I.; Robbie, K. Nanomaterials and Nanoparticles: Sources and Toxicity. *Biointerphases* **2007**, *2*, 17–71.

7. Cai, Y.; Strømme, M.; Welch, K. Photocatalytic Antibacterial Effects are Maintained on Resin-Based TiO$_2$ Nanocomposites after Cessation of UV Irradiation. PLoS One **2013**, *8* (10), DOI/10.1371/journal.pone.0075929.

8. Celardo, I.; Pedersen, J. Z.; Traversa, E.; Ghibelli, L. Pharmacological Potential of Cerium Oxide Nanoparticles. *Nanoscale* **2011**, *3*(4), 1411–1420.

9. Cendrowski, K.; Peruzynska, M.; Markowska-Szczupak, A.; Chen, X.; Wajda, A.; Lapczuk, J., et al. Mesoporous Silica Nanospheres Functionalized by TiO$_2$ as a Photoactive Antibacterial Agent. *J. Nanomed. Nanotechnol.* **2013**, *4* (6), DOI: 10.4172/2157-7439.1000182.

10. Colon, J.; Hsieh, N.; Ferguson, A.; Kupelian, P.; Seal, S.; Jenkins, D. W.; Baker, C. H. Cerium Oxide Nanoparticles Protect Gastrointestinal Epithelium from Radiation-Induced Damage by Reduction of Reactive Oxygen Species and Upregulation of Superoxide Dismutase 2. *Nanomedicine* **2010**, *6*(5), 698–705.

11. Cu, H.; Ji, J.; Gu, W.; Sun, C.; Wu, D.; Yang, T. Photocatalytic Inactivation Efficiency of Anatase Nano-TiO_2 Sol on the H9N2 Avian Influenza Virus. *Photochem. Photobiol.* **2010**, *86* (5), 1135–1139.

12. Das, S.; Dowding, J. M.; Klump, K. E.; McGinnis, J. F.; Self, W.; Seal, S. Cerium Oxide Nanoparticles: Applications and Prospects in Nanomedicine. *Nanomedicine* **2013**, *8* (9), 1483–1508.

13. Das, S.; Sinha, S.; Suar, M.; Yun, S. II; Mishra, A.; Tripathy, S. K. Solar-Photocatalytic Disinfection of *Vibrio cholerae* by Using Ag@ZnO Core–Shell Structure Nanocomposites. *J. Photochem. Photobiol. B* **2015**, *142*, 68–76.

14. Dong, Z. F.; Du, Y. M.; Fan, L. H.; Wen, Y.; Liu, H.; Wang, X. H. Preparation and Properties of Chitosan/Gelatin/Nano-TiO_2 Ternary Composite Films. *J. Funct. Polym.* **2004**, *1*, 61–66.

15. Fujimori, Y.; Sato, T.; Hayata, T.; Nagao, T.; Nakayama, M.; Nakayama, T., et al. Novel Antiviral Characteristics of Nanosized Copper(I) Iodide Particles Showing Inactivation Activity against 2009 Pandemic H1N1 Influenza Virus. *Appl. Environ. Microbiol.* **2012**, *78*(4), 951–955.

16. Ghodsi, F. E.; Dadvar, H.; Khayati, G. Optical, Surface Morphological and Antibacterial Properties of Nanostructured TiO_2: M (M = Fe, Ce, Ag) Thin Films. *Int. J. Multi. Sci. Emerg. Res.* **2014**, *3*, 951–956.

17. Gondal, M. A.; Dastageer, M. A.; Khalil, A. In *Nano-NiO as a Photocatalyst in Antimicrobial Activity of Infected Water Using Laser Induced Photo-Catalysis,* Electronics, Communications and Photonics Conference (SIECPC), 2011 Saudi International, Riyadh, April 24–26, 2011, Riyadh, Saudi Arabia. IEEE; pp 1–5. 10.1109/SIECPC.2011.5876969.

18. Guo, D.; Wu, C.; Jiang, H.; Li, Q.; Wang, X.; Chen, B. Synergistic Cytotoxic Effect of Different Sized ZnO Nanoparticles and Daunorubicin against Leukemia Cancer Cells under UV Irradiation. *J. Photochem. Photobiol. B* **2008**, *93* (3), 119–126.

19. Gupta, K.; Singh, R. P.; Pandey, A.; Pandey, A. Photocatalytic Antibacterial Performance of TiO_2 and Ag-Doped TiO_2 against *S. aureus, P. aeruginosa* and *E. coli. Beilstein J. Nanotechnol.* **2013**, *4*, 345–351.

20. Hackenberg, S.; Scherzed, A.; Harnisch, W.; Froelich, K.; Ginzkey, C.; et al. Antitumor Activity of Photo-Stimulated Zinc Oxide Nanoparticles Combined with Paclitaxel or Cisplatin in HNSCC Cell Lines. *J. Photochem. Photobiol. B* **2012**, *114*, 87–93.

21. Hackenberg, S.; Scherzed, A.; Kessler, M.; Froelich, K.; Ginzkey, C.; et al. Zinc Oxide Nanoparticles Induce Photocatalytic Cell Death in Human Head and Neck Squamous Cell Carcinoma Cell Lines In Vitro. *Int. J. Oncol.* **2010**, *37*(6), 1583–1590.

22. Hatamie, A.; Khan, A.; Golabi, M.; Turner, A. P.; Beni, V.; Mak, W. C.; et al. Zinc Oxide Nanostructure-Modified Textile and Its Application to Biosensing, Photocatalysis, and as Antibacterial Material. *Langmuir* **2015**, *31* (39), 10913–10921.

23. Jašková, V.; Hochmannová, L.; VytLasová, J. TiO_2 and ZnO Nanoparticles in Photocatalytic and Hygienic Coatings. *Int. J. Photoenergy* **2013**, DOI/10.1155/2013/795060.

24. Joost, U.; Juganson, K.; Visnapuu, M.; Mortimer, M.; Kahru, A.; Nõmmiste, E.; et al. Photocatalytic Antibacterial Activity of Nano-TiO_2 (Anatase)-Based Thin Films:

Effects on *Escherichia coli* Cells and Fatty Acids. *J. Photochem. Photobiol. B* **2015**, *142*, 178–185.

25. Kanmani, P.; Rhim, J. W. Nano and Nanocomposite Antimicrobial Materials for Food Packaging Applications. *Prog. Nanomater. Food Packag.* **2014**, 34–48, DOI:10.4155/ EBO.13.303.

26. Khan, M. M.; Ansari, S. A.; Lee, J. H.; Ansari, M. O.; Lee, J.; Cho, M. H. Electrochemically Active Biofilm Assisted Synthesis of Ag@CeO$_2$ Nanocomposites for Antimicrobial Activity, Photocatalysis and Photoelectrodes. *J. Colloid Interface Sci.* **2014**, *431*, 255–263.

27. Laurent, S.; Dutz, S.; Hafeli, U. O.; Mahmoudi, M. Magnetic Fluid Hyperthermia: Focus on Superparamagnetic Iron Oxide Nanoparticles. *Adv. Colloid Interface Sci.* **2011**, *166* (1–2), 8–23.

28. Li, Q.; Mahendra, S.; Lyon, D. Y.; Brunet, L.; Liga, M. V.; Li, D.; Alvarez, P. J. J. Antimicrobial Nanomaterials for Water Disinfection and Microbial Control: Potential Applications and Implications. *Water Res.* **2008**, *42*, 4591–4602.

29. Lipovsky, A.; Gedanken, A.; Lubart, R. Visible Light-Induced Antibacterial Activity of Metaloxide Nanoparticles. *Photomed. Laser Surg.* **2013**, *31* (11), 526–530.

30. Lucky, S. S.; Idris, N. M.; Li, Z.; Huang, K.; Soo, K. C.; Zhang, Y. Titania Coated Upconversion Nanoparticles for Near-Infrared Light Triggered Photodynamic Therapy. *ACS Nano* **2015**, *9* (1), 191–205.

31. Maness, P.; Smolinski, S.; Blake, D. M.; Huang, Z.; Wolfrum, E. J.; Jacoby, W. A. Bactericidal Activity of Photocatalytic TiO$_2$ Reaction: Toward an Understanding of Its Killing Mechanism. *Appl. Environ. Microbiol.* **1999**, *65* (9), 4094–4098.

32. Mirhoseini, F.; Salabat, A. Antibacterial Activity Based Poly(methyl methacrylate) Supported TiO$_2$ Photocatalyst Film Nanocomposite. *Tech. J. Eng. Appl. Sci.* **2015**, *5–1*, 115–118.

33. Monteiro-Riviere, N. A.; Wiench, K.; Landsiedel, R.; Schulte, S.; Inman, A. O.; Riviere, J. E. Safety Evaluation of Sunscreen Formulations Containing Titanium Dioxide and Zinc Oxide Nanoparticles in UVB Sunburned Skin: An In Vitro And In Vivo Study. *Toxicol. Sci.* **2011**, *123* (1), 264–280.

34. Montfort, C. V.; Sarah, L. A.; Hanselmann, T.; Brenneisen, P. Redox-Active Cerium Oxide Nanoparticles Protect Human Dermal Fibroblasts from PQ-Induced Damage. *Redox Biol.* **2015**, *4*, 1–5.

35. Nithya, A.; Jeeva Kumari, H. L.; Rokesh, K.; Ruckmani, K.; Jeganathan, K.; Jothivenkatachalam, K. A Versatile Effect of Chitosan-Silver Nanocomposite for Surface Plasmonic Photocatalytic and Antibacterial Activity. *J. Photochem. Photobiol. B* **2015**, *153*, 412–422.

36. Ogorevc, J. S.; Tratar-Pirc, E.; Matoh, L.; Peter, B. Antibacterial and Photodegradative Properties of Metal Doped TiO$_2$ Thin Films under Visible Light. *Acta Chim. Slov.* **2012**, *59*, 264–272.

37. Pant, H. R.; Pant, B.; Sharma, R. K.; Amarjargal, A.; Kim, H. J.; Park, C. H. Antibacterial and Photocatalytic Properties of Ag/TiO$_2$/ZnO Nano-Flowers Prepared by Facile One-Pot Hydrothermal Process. *Ceram. Int.* **2013**, *39*, 1503–1510.

38. Paul, A. M.; Krishnamoorthy, S.; Anusuya, T.; Sakthivel, P. Visible Photocatalytic Activity of Vanadium Doped with Titanium Dioxide for Biomedical Applications. *Int. J. Chem. Tech. Res.* **2015**, *7* (5), 2125–2129.

39. Rasmussen, J. W.; Martinez, E.; Louka, P.; Wingett. D. G. Zinc Oxide Nanoparticles for Selective Destruction of Tumor Cells and Potential for Drug Delivery Applications. *Expert Opin. Drug Deliv.* **2010**, *7* (9), 1063–1077.

40. Saito, T.; Iwase, T.; Horie, J.; Morioka, T. Mode of Photocatalytic Bactericidal Action of Powdered Semiconductor TiO$_2$ on Mutants Streptococci. *J. Photochem. Photobiol. B* **1992**, *14* (4), 369–379.

41. Sangari, M.; Umadevi, M.; Mayandi, J.; Anitha, K.; Pinheiro, J. P. Photocatalytic and Antimicrobial Activities of Fluorine Doped TiO$_2$-Carbon Nano Cones and Disc Composites. *Mater. Sci. Semicond. Process* **2015**, *31*, 543–550.

42. Sivakumar, K.; Kumar, V. S.; Jae-Jin, S.; Haldorai, Y. Photocatalytic and Antimicrobial Activities of Poly(aniline-*co-o*-anisidine)/Zinc Oxide Nanocomposite. *Asian J. Chem.* **2014**, *26* (2), 600–606.

43. Sivaraj, R.; Rahman, K. S. M.; Rajiv, P.; Narendhran, S.; Venckatesh, R. Biosynthesis and Characterization of *Acalypha indica* Mediated Copper Oxide Nanoparticles and Evaluation of Its Antimicrobial and Anticancer Activity. *Spectrochim. Acta A: Mol. Biomol. Spectrosc.* **2014**, *129*, 255–258.

44. Sivaraj, R.; Rahman, K. S. M.; Rajiv, P.; Salam, H. A.; Venckatesh, R. Biogenic Copper Oxide Nanoparticles Synthesis Using *Tabernaemontana divaricate* Leaf Extract and Its Antibacterial Activity against Urinary Tract Pathogen. *Spectrochim. Acta A: Mol. Biomol. Spectrosc.* **2014**, *133*, 178–181.

45. Smijs, T. G.; Pavel, S. Titanium Dioxide and Zinc Oxide Nanoparticles in Sunscreens: Focus on Their Safety and Effectiveness. *Nanotechnol. Sci. Appl.* **2011**, *4*, 95–112.

46. Stoyanova, A. M.; Hitkova, H. Y.; Ivanova, N. K.; Bachvarova-Nedelcheva, A. D.; Iordanova, R. S.; Sredkova, M. P. Photocatalytic and Antibacterial Activity of Fe-Doped TiO$_2$ Nanoparticles Prepared by Nonhydrolytic Sol–Gel Method. *Chem. Commun.* **2013**, *45*, 497–504.

47. Stoyanova, A.; Hitkova, H.; Bachvarova-Nedelcheva, A.; Iordanova, R.; Ivanova, N.; Sredkova, M. Synthesis, Photocatalytic and Antibacterial Properties of Nanosized ZnTiO$_3$ Powders Obtained by Different Sol–Gel Methods. *Digest J. Nanomater. Biostruct.* **2012**, *7* (2), 777–784.

48. Sun, S. C.; Chiang, C. W. Human Body Affinitive Lubricating Protective and Repair Gel. US Patent 20050249760, Nov 10, 2005.

49. Sunitha, S.; Rao, A. N. Antibacterial and Photocatalytic Activity of ZnO Nanoparticles Synthesized by Sol–Gel Method. *J. Chem. Pharm. Res.* **2015**, *7* (4), 1446–1451.

50. Tarnuzzer, R. W.; Colon, J.; Patil, S.; Seal, S. Vacancy Engineered Ceria Nanostructures for Protection from Radiation-Induced Cellular Damage. *Nano Lett.* **2005**, *5* (12), 2573–2577.

51. Tobaldi, D. M.; Piccirillo, C.; Pullar, R. C.; Gualtieri, A. F.; Seabra, M. P.; Castro, P. M. L.; Labrincha, J. A. Silver-Modified Nano-Titania as an Antibacterial Agent and Photocatalyst. *J. Phys. Chem. C* **2014**, *118* (9), 4751–4766.

52. Wahab, R.; Dwivedi, S.; Umar, A.; Singh, S.; Hwang, I. H.; Shin, H. S.; et al. ZnO Nanoparticles Induce Oxidative Stress in Cloudman S91 Melanoma Cancer Cells. *J. Biomed. Nanotechnol.* **2013**, *9* (3), 441–449.

53. Wang, L.; Mao, J.; Zhang, G. H.; Tu, M. J. Nano-Cerium-Element-Doped Titanium Dioxide Induces Apoptosis of Bel 7402 Human Hepatoma Cells in the Presence of Visible Light. *World J. Gastroenterol.* **2007**, *13* (29), 4011–4014.

54. Wason, M. S.; Colon, J.; Das, S.; Seal, S.; Turkson, J.; Zhao, J.; Baker, C. H. Sensitization of Pancreatic Cancer Cells to Radiation by Cerium Oxide Nanoparticle-Induced ROS Production. *Nanomedicine* **2013**, *9*, 558–569.

55. Wu, Q.; Xue, J. Nano Material Synergistic Pesticide Antibacterial and Continuous Photocatalytic Degradation Residue-Removing Technical Method. Patent CN104,542,707, Apr 29, 2015.

56. Xu, J. W.; Gao, Z. D.; Han, K.; Liu, Y.; Song, Y. Y. Synthesis of Magnetically Separable $Ag_3PO_4/TiO_2/Fe_3O_4$ Heterostructure with Enhanced Photocatalytic Performance under Visible Light for Photoinactivation of Bacteria. *ACS Appl. Mater. Interfaces* **2014**, *6*, 15122–15131.

57. Xue, J.; Luo, Z.; Li, P.; Ding, Y.; Cui, Y.; Wu, Q. A Residue-Free Green Synergistic Antifungal Nanotechnology for Pesticide Thiram by ZnO Nanoparticles. *Sci. Rep.* **2014**, *4*, 5408.

58. Yaithongkum, J.; Kooptarnond, K.; Sikong, L.; Kantachote, D. Photocatalytic Activity against *Penicillium expansum* of Ag-Doped $TiO_2/SnO_2/SiO_2$. *Adv. Mater. Res.* **2011**, *214*, 212–217.

59. Yu, B.; Leung, K. M.; Guo, Q.; Lau, W. M.; Yang, J. Synthesis of Ag-TiO_2 Composite Nano Thin Film for Antimicrobial Application. *Nanotechnology* **2011**, *22*, 115603.

60. Zhang, Y.; Lee, M. W.; An, S.; Sinha-Ray, S.; Khansari, S.; Joshi, B., et al. Antibacterial Activity of Photocatalytic Electrospun Titania Nanofiber Mats and Solution-Blown Soy Protein Nanofiber Mats Decorated with Silver Nanoparticles. *Catal. Commun.* **2013**, *34*, 35–40.

61. Zhang, A. P.; Sun, Y. P. Photocatalytic Killing Effect of TiO_2 Nanoparticles on Ls-174-t Human Colon Carcinoma Cells. *World J. Gastroenterol.* **2004**, *10* (21), 3191–3193.

62. Zhang, J.; Liu, Y.; Li, Q.; Zhang, X.; Shang, J. K. Antifungal Activity and Mechanism of Palladium-Modified Nitrogen-Doped Titanium Oxide Photocatalyst on Agricultural Pathogenic Fungi *Fusarium graminearum*. *ACS Appl. Mater. Interfaces* **2013**, *5* (21), 10953–10959.

CHAPTER 4

NANODEVICES AND ORGANIZATION OF SINGLE ION MAGNETS AND SPIN QUBITS

FRANCISCO TORRENS[1*] and GLORIA CASTELLANO[2]

[1]*Institut Universitari de Ciència Molecular, Universitat de València, Edifici d'Instituts de Paterna, P.O. Box 22085, E-46071 València, Spain*

[2]*Departamento de Ciencias Experimentales y Matemàticas, Facultad de Veterinaria y Ciencias Experimentales, Universidad Catòlica de Valencia San Vicente Màrtir, Guillem de Castro-94, E-46001 València, Spain*

Corresponding author. E-mail: torrens@uv.es

ABSTRACT

Global economic demands and population surges led to dwindling resources and problematic environmental issues. As the climate and its natural resources continue to struggle, it became necessary to research and employ new forms of sustainable technology to help meet the growing demand. The *entanglement* of the quantum states in every pair of correlated particles presents important applications in quantum computing and communication and potential utilization clinically. When photons are created in pairs, they can emerge from different, rather than the same, locations. The discovery, which is in contrast to the general belief that photon pairs must originate from single points in space, could impact the study of quantum physics.

4.1 INTRODUCTION

The *entanglement* (linking actions and states) of the quantum states in every pair of correlated particles presents important applications in quantum computing and communication, and potential utilization clinically.

In earlier publications, fractal hybrid-orbital analysis,[1,2] resonance,[3] molecular diversity,[4] periodic table of the elements,[5,6] law, property, information entropy, molecular classification, simulators,[7-13] labor risk prevention, and preventive healthcare at work with nanomaterials[14-16] were reviewed.

4.2 MOLECULAR DEVICES/MACHINES AND NANOTECHNOLOGY DEVELOPMENT

In everyday life, one makes extensive use of macroscopic devices and machines, which are assemblies of components designed to perform specific functions; for example, hairdryer—it is the result of simpler acts performed by a switch, a heater, and a fan, suitably connected by electric wires and assembled in an appropriate framework.[17] The progress of human civilization was related to the construction of novel devices/machines and miniaturization of their components. The concepts of device and machine were extended to the molecular level. Starting from molecules, the smallest entities of matter that present distinct shapes and properties, chemists developed a *bottom-up* approach to construct nanodevices and machines. A molecular device can be defined as an assembly of a discrete number of molecular components, designed to perform a function under appropriate external stimulation, whereas a molecular machine is a particular type of device where the function is achieved via the mechanical movements of its molecular components. The idea that atoms could be used to construct nanodevices and machines was raised by Feynman.[18] His key sentence follows: "The principles of physics do not speak against the possibility of maneuvering things atom by atom." The construction, futuristic use, and frightening potential of nanomachines were described by Drexler in a book.[19] He claimed the possibility of constructing a general-purpose building nanodevice (*assembler*), which could build almost anything, for example, copies of itself, by atomic-scale precision *pick and place* machine-phase chemistry (mechanosynthesis). His abstract ideas of maneuvering atoms or making molecular mechanosynthesis do not convince chemists, who are well aware of the complexity and subtlety of bond-breaking and -making processes.

In the late 1970s, a new branch of chemistry, *supramolecular chemistry* (SC), emerged from the studies of Lehn, and the idea began to arise that molecules are much more convenient in building blocks than atoms to construct nanodevices and machines.[20] The reasons follow: (1) molecules are stable species, whereas atoms are difficult to handle; (2) nature starts from molecules, not from atoms, to construct the great number and variety of nanodevices and machines that sustain life; (3) most of the laboratory chemical processes deal with molecules, not atoms; (4) molecules are objects that exhibit distinct shapes and carry specific properties (e.g., those that can be manipulated by photochemical, electrochemical inputs) as confirmed by techniques (e.g., single-molecule fluorescence spectroscopy, various types of probe microscopies, capable of *seeing* or *manipulating* single molecules); (5) molecules can self-assemble or be covalently connected to make larger structures. The SC grew rapidly, and the molecular *bottom-up* approach opened virtually unlimited possibilities, concerning design and construction of artificial nanoscale devices and machines, capable of performing specific functions upon stimulation with external energy inputs.

Inspiration to construct molecular devices and machines comes from the outstanding progress of molecular biology, which revealed the secrets of the material base of life. Bottom-up construction of devices and machines as complex as those present in nature is an impossible task. Chemists tried to construct much simpler systems, without mimicking the complexity of the biostructures. The synthetic talent, which was the most distinctive feature of chemists, combined with a device-driven ingenuity evolved from chemists' attention to functions and reactivity, led to achievements; for example, molecular devices capable of behaving as light-harvesting antennae, wires, switches, plug/socket and extension cable systems, memories, logic gates, and molecular machines that perform the function of tweezers, adaptable receptors, propellers, rotors, turnstiles, gyroscopes, gears, brakes, pedals, ratchets, lifts, muscles, valves, processive artificial enzymes, walkers, and catalytic self-propelled micro and nanorods.[21] Two designed and developed systems exploit the action of sunlight to perform their functions. The first behaves as a molecular pump driven by light, which enables the direct conversion of light into mechanical work operating away from thermal equilibrium.[22] The second is a molecular sponge that is wrung out by light.[23] Research on molecular devices and machines presents a great impact for the development of nanotechnology, which can be considered as the ultimate limit to miniaturization. Such a technology should enable to use less materials and resources in constructing the objects people need in their life and solve the four big problems that are faced by a large part of the Earth's population: food, health, energy, and pollution.

4.3　EVALUATION OF LANTHANOID BINDING TAGS AS BIOMOLECULAR HANDLES

Lanthanoid complexes are amongst the most promising compounds in single ion magnetism and as molecular *spin quantum binary-digits* (*bits*) (*qubits*), but their organization remains an open problem. Rosaleny and Gaita-Ariño proposed to combine lanthanoid binding tags (LBTs) with recombinant proteins, as a path for an extremely specific and spatially resolved organization of lanthanoid ions as spin qubits.[24] They developed a new computational subroutine for the freely available code SIMPRE, which allowed an inexpensive estimate of quantum *decoherence* (the noise that lowers the quantum properties of every system) times and qubit–qubit interaction strengths. They used the subroutine to evaluate their proposal theoretically for 63 different systems. They evaluated their behavior as single ion magnets (SIMs) and estimated decoherence caused by the nuclear spin bath and interqubit interaction strength by dipolar coupling. They concluded that Dy^{3+} LBT complexes are expected to behave as SIMs, but Yb^{3+} derivatives should be better spin qubits.

4.4　COHERENCE AND ORGANIZATION IN LANTHANOID COMPLEXES

Molecular magnetism is reaching a degree of development that allows for the rational design of sophisticated systems, among which Gaita-Ariño group focused on those that display single-molecule magnetic behavior, that is, classical memories, and on magnetic molecules that could be used as molecular *spin qubits*, the irreducible components of any quantum technology.[25] Compared with candidates developed from physics, an advantage of molecular spin qubits stems from the power of chemistry for the tailored and inexpensive synthesis of systems for their experimental study; in particular, the so-called lanthanoid-based SIMs, which were for a long time one of the hottest topics in molecular magnetism. They present the potential to be chemically designed, tuning their single-molecule properties and crystalline environment, which allowed the study of the different quantum processes that caused the loss of quantum information, collectively known as *decoherence*, which study in the solid state is necessary to answer fundamental questions and lay the foundations for next-generation quantum technologies. They reviewed the state-of-the-art research in the field and its open problems.

4.5 FINDINGS CHALLENGE UNDERSTANDING OF QUANTUM THEORY

The achievement of optimum conversion efficiency in conventional spontaneous parametric down-conversion requires consideration of quantum processes that entail multisite electrodynamic coupling, actively taking place within the conversion material.[26] The physical mechanism, which operates via virtual photon propagation, provides for photon pairs to be emitted from spatially separated sites of photon interaction; occasionally, pairs were produced in which each photon emerges from a different point in space. The extent of such nonlocalized generation was influenced by individual variations in distance and phase correlation. Mathematical analysis of the global contributions from the mechanism provides a quantitative measure for a degree of positional uncertainty in the origin of downconverted emission.

4.6 DISCUSSION

Global economic demands and population surges led to dwindling resources and problematic environmental issues. As the climate and its natural resources continue to struggle, it became necessary to research and employ new forms of sustainable technology to meet the growing demand. Sustainable nanosystems' development requires emergent research and theoretical concepts in the areas of nanotechnology, photovoltaics, electrochemistry, and materials science and within the physical and environmental sciences. Sustainable nanotechnology is a challenge for researchers, engineers, students, scientists, and academicians.

When photons are created in pairs, they can emerge from different, rather than the same, locations. The discovery, which is in contrast to the general belief that photon pairs must originate from single points in space, could impact the study of quantum physics.

4.7 FINAL REMARKS

From the present discussion, the following final remarks can be drawn:

1. Global economic demands and population surges led to dwindling resources and problematic environmental issues. As the climate and

its natural resources continue to struggle, it became necessary to research and employ new forms of sustainable technology to help meet the growing demand.

2. When photons are created in pairs, they can emerge from different, rather than the same, locations. The discovery, which is in contrast to the general belief that photon pairs must originate from single points in space, could impact the study of quantum physics.

ACKNOWLEDGMENTS

Francisco Torrens belongs to the Institut Universitari de Ciència Molecular, Universitat de València. Gloria Castellano belongs to the Departamento de Ciencias Experimentales y Matemáticas, Facultad de Veterinaria y Ciencias Experimentales, Universidad Católica de Valencia *San Vicente Mártir*. The authors thank support from Generalitat Valenciana (Project No. PROMETEO/2016/094) and Universidad Catolica de Valencia San Vicente Martir (Project No. UCV.PRO.17-18.AIV.03).

KEYWORDS

- molecular device
- molecular machine
- nanotechnology development
- lanthanoid binding tag
- biomolecular handle
- coherence
- lanthanoid complex
- challenge understanding
- quantum theory
- quantum computation
- quantum simulation
- superconducting qubit

REFERENCES

1. Torrens, F. Fractals for Hybrid Orbitals in Protein Models. *Complexity Int.* **2001**, *8*, torren01-1–torren01-13.
2. Torrens, F. Fractal Hybrid-Orbital Analysis of the Protein Tertiary Structure. *Complexity Int.*, Submitted for Publication.
3. Torrens, F.; Castellano, G. Resonance in Interacting Induced-Dipole Polarizing Force Fields: Application to Force-Field Derivatives. *Algorithms* **2009**, *2*, 437–447.
4. Torrens, F.; Castellano, G. Molecular Diversity Classification *via* Information Theory: A Review. *ICST Trans. Complex Syst.* **2012**, *12*(10–12), e4-1–e4-8.
5. Torrens, F.; Castellano, G. Reflections on the Nature of the Periodic Table of the Elements: Implications in Chemical Education. In *Synthetic Organic Chemistry*; Seijas, J. A., Vázquez Tato, M. P., Lin, S. K., Eds.; MDPI: Basel, Switzerland, 2015; Vol. 18; pp 8-1–8-15.
6. Putz, M. V., Ed. The Explicative Handbook of Nanochemistry; Apple Academic–CRC: Waretown, NJ, in press.
7. Torrens, F.; Castellano, G. Reflections on the Cultural History of Nanominiaturization and Quantum Simulators (Computers). In *Sensors and Molecular Recognition*; Laguarda Miró, N., Masot Peris, R., Brun Sánchez, E., Eds.; Universidad Politécnica de Valencia: València, Spain, 2015; Vol. 9; pp 1–7.
8. Torrens, F.; Castellano, G. Ideas in the History of Nano/Miniaturization and (Quantum) Simulators: Feynman, Education and Research Reorientation in Translational Science. In *Synthetic Organic Chemistry*; Seijas, J. A., Vázquez Tato, M. P., Lin, S. K., Eds.; MDPI: Basel, Switzerland, 2016; Vol. 19; pp 1–16.
9. Torrens, F.; Castellano, G. Nanominiaturization and Quantum Computing. In *Sensors and Molecular Recognition*; Costero Nieto, A. M., Parra Álvarez, M., Gaviña Costero, P., Gil Grau, S., Eds.; Universitat de València: València, Spain, 2016; Vol. 10 pp 31-1–31-5.
10. Torrens, F.; Castellano, G. Nanominiaturization, Classical/Quantum Computers/ Simulators, Superconductivity, and Universe. In *Methodologies and Applications for Analytical and Physical Chemistry*; Haghi, A. K., Thomas, S., Palit, S., Main, P., Eds.; Apple Academic–CRC: Waretown, NJ; 2018, p. 27–44.
11. Torrens, F.; Castellano, G. Superconductors, Superconductivity, BCS Theory, and Entangled Photons for Quantum Computing. In *Physical Chemistry for Engineering and Applied Sciences: Theoretical and Methodological Implication*; Haghi, A. K., Aguilar, C. N., Thomas, S., Praveen, K. M., Eds.; Apple Academic–CRC: Waretown, NJ; 2018, p. 379–387.
12. Torrens, F.; Castellano, G. EPR Paradox, Quantum Decoherence, Qubits, Goals, and Opportunities in Quantum Simulation. In *Theoretical Models and Experimental Approaches in Physical Chemistry: Research Methodology and Practical Methods*; Haghi, A. K., Ed.; Apple Academic–CRC: Waretown, NJ; 2018, Vol. 5, p. 317–334.
13. Torrens, F.; Castellano, G. Nanomaterials, Molecular Ion Magnets, Ultrastrong, Spin–Orbit Couplings in Quantum Materials. In *Physical Chemistry for Chemists and Chemical Engineers: Multidisciplinary Research Perspectives*; Vakhrushev, A. V., Haghi, R., de Julián-Ortiz, J. V., Allahyari, E., Eds.; Apple Academic–CRC: Waretown, NJ, in press.
14. Torrens, F.; Castellano, G. In *Book of Abstracts, Certamen Integral de la Prevención y el Bienestar Laboral*, València, Spain, September 28–29, 2016; Generalitat Valenciana–INVASSAT: València, Spain, 2016; p 3.

15. Torrens, F.; Castellano, G. Nanoscience: From a Two-Dimensional to a Three-Dimensional Periodic Table of the Elements. In *Methodologies and Applications for Analytical and Physical Chemistry*; Haghi, A. K., Thomas, S., Palit, S., Main, P., Eds.; Apple Academic–CRC: Waretown, NJ; 2018, p. 3–26.

16. Torrens, F.; Castellano, G. In *Book of Abstracts, Congreso Internacional de Tecnología, Ciencia y Sociedad, València*, Spain, October 19–20, 2017; Universidad Cardenal Herrera CEU: València, Spain, 2017; p 1.

17. Venturi, M. Developing Sustainability: Some Scientific and Ethical Issues. In *Sustainable Nanosystems Development, Properties, and Applications*; Putz, M. V., Mirica M. C., Eds.; IGI Global: Hershey, PA, 2017; pp 657–680.

18. Feynman, R. P. There is Plenty of Room at the Bottom. *Caltech Eng. Sci.* **1960**, *23*, 22–36.

19. Drexler, K. E. *Engines of Creation: The Coming Era of Nanotechnology*; Anchor Press: New York, NY, 1986.

20. Lehn, J. M. *Supramolecular Chemistry: Concepts and Perspectives*; Wiley-VCH: Weinheim, Germany, 1995.

21. Balzani, V.; Credi, A.; Venturi, M. *Molecular Devices and Machines: Concepts and Perspectives for the Nanoworld*; Wiley-VCH: Weinheim, Germany, 2008.

22. Ragazzon, G.; Baroncini, M.; Silvi, S.; Venturi, M.; Credi, A. Light-Powered Autonomous and Directional Molecular Motion of a Dissipative Self-Assembling System. *Nat. Nanotech.* **2015**, *10*, 70–75.

23. Baroncini, M.; d'Agostino, S.; Bergamini, G.; Ceroni, P.; Comotti, A.; Sozzani, P.; Bassanetti, I.; Grepioni, F.; Hernandez, T. M.; Silvi, S.; Venturi, M.; Credi, A. Photoinduced Reversible Switching of Porosity in Molecular Crystals Based on Star-Shaped Azobenzene Tetramers. *Nat. Chem.* **2015**, *7*, 634–640.

24. Rosaleny, L. E.; Gaita-Ariño, A. Theoretical Evaluation of Lanthanide Binding Tags as Biomolecular Handles for the Organization of Single Ion Magnets and Spin Qubits. *Inorg. Chem. Front.* **2016**, *3*, 61–66.

25. Gaita-Ariño, A.; Prima-García, H.; Cardona-Serra, S.; Escalera-Moreno, L.; Rosaleny L. E.; Baldoví, J. J. Coherence and Organisation in Lanthanoid Complexes: From Single Ion Magnets to Spin Qubits. *Inorg. Chem. Front.* **2016**, *3*, 568–577.

26. Forbes, K. A.; Ford, J. S.; Andrews, D. L. Nonlocalized Generation of Correlated Photon Pairs in Degenerate Down-Conversion. *Phys. Rev. Lett.* **2017**, *118*, 133602-1–133602-5.

CHAPTER 5

DEVELOPING SUSTAINABILITY VIA NANOSYSTEMS AND DEVICES: SCIENCE–ETHICS

FRANCISCO TORRENS[1*] and GLORIA CASTELLANO[2]

[1]*Institut Universitari de Ciència Molecular, Universitat de València, Edifici d'Instituts de Paterna, P.O. Box 22085, E-46071 València, Spain*

[2]*Departamento de Ciencias Experimentales y Matemàticas, Facultad de Veterinaria y Ciencias Experimentales, Universidad Catòlica de Valencia San Vicente Màrtir, Guillem de Castro-94, E-46001 València, Spain*

Corresponding author. E-mail: torrens@uv.es

ABSTRACT

Global economic demands and population surges led to dwindling resources and problematic environmental subjects. As the climate and its natural resources continue to struggle, it becomes necessary to research and employ new forms of sustainable technology to help meet the growing demand. This report establishes a dialogue between science and ethics. The aim of this work is to initiate a debate by suggesting a number of questions. The discussion deals with these questions from the perspective of historical research in science. The viewpoint is from the history and sociology of science, making a comparison between *historicism* and *presentism*. There is a boom time for the *circular economy* and creativity. The shape of the future will be circular. However, how will that profit chemistry? Frontier science provides *hope* for many problems but *inequality* persists.

5.1   INTRODUCTION

Venturi group reviewed molecular devices and machines, concepts, and perspectives for the nanoworld.[1] They revised the electrochemistry of functional supramolecular systems.[2]

In earlier publications, fractal hybrid-orbital analysis,[3,4] resonance,[5] molecular diversity,[6] periodic table of the elements,[7,8] law, property, information entropy, molecular classification, simulators,[9–16] labor risk prevention, and preventive healthcare at work with nanomaterials[17–19] were reviewed.

The present report wants to establish a dialogue between science and ethics. The aim of this work is to initiate a debate by suggesting a number of questions (Q), which can arise when addressing ethical subjects of developing sustainability, and providing, when possible, answers (A), hypotheses (H), and/or problems (P). The discussion deals with these questions from the perspective of historical research in science. The viewpoint is from the history and sociology of science, making a comparison between *historicism* and *presentism*. There is a boom time for the *circular economy* and creativity. The shape of the future will be circular. However, how will that profit chemistry? Frontier science provides *hope* for many problems but *inequality* persists.

5.2   SIGNIFICANCE OF MINIATURIZATION IN HISTORY

Communication professionals say that the invention of the *radio* revolutionized the world. However, it was not the radio but the *transistor radio* that did it. For instance, this permitted that a shepherd in the top of a mountain could be in touch with what is going on.

5.3   DEVELOPING SUSTAINABILITY: SOME SCIENTIFIC AND ETHICAL SUBJECTS

Venturi proposed H, Q, and P on developing sustainability, some scientific and ethical issues.[20]

H1. Second law of thermodynamics. In a closed system, limitations exist that cannot be overcome.

Q1. How can one help passengers now traveling in much worse compartments of spaceship Earth?

H2. Modern industrial agriculture is, in fact, the use of land to turn oil into food.

P1. Fossil-fuels extensive use produces harmful effects [e.g., pollution, greenhouse gases (CO_2)].

Q2. (Ciamician, 1912). Is fossil solar energy the only that may be used in modern life civilization?

She reviewed the advantages of producing electricity by *wind*: (1) guaranteed perpetual zero cost of the primary *fuel*; (2) no emissions in the atmosphere or heat to dissipate; (3) a simple technology; (4) short times of construction and wide tunability of installed capacity from a few kilowatts to hundreds of megawatts. She revised the benefits of hydroelectric power: (1) the lowest operating cost and a longer plant lifetime than any other mode of electricity production; (2) easy harvesting of potential energy for peak electricity demand, available within seconds; (3) reliable supply of irrigation and drinking water; (4) protection versus recurrent and, sometimes destructive, floods. She proposed additional Q, A, and H.

Q3. How do light-based technologies promote sustainable development and provide solutions to global challenges in energy, education, agriculture, and health?

Q4. (Ciamician, 1908). What is a reactant of the greatest importance?

A4. (Ciamician, 1908). It is light.

H3. (Ciamician, 1912). Where vegetation is rich, photochemistry is left to plants and, by rational cultivation, solar radiation is used for industrial purposes.

Q5. What is energy?

She reviewed the advantages of organic photovoltaic (PV) devices: (1) low weight and flexibility of PV modules; (2) easy integration into other products; (3) significantly lower manufacturing costs compared to conventional inorganic technologies; (4) manufacturing of devices in a continuous process via state of the art printing tools; (5) short energy payback times, and low environmental impact during manufacturing and operations. She proposed additional Q and H.

Q6. How to make plants bear even more abundant fruit than nature?

H4. Untrue H. The production and use of biofuels is CO_2-neutral.

H5. (Fidel Castro, 2007). Biofuels generation from edible feedstocks is in competition with food production.

H6. (Fidel Castro, 2007). Generation of biofuels from edible feedstocks increases food price.

H7. The finite Earth cannot sustain an endless rise in resource consumption and waste production.

Q7. (Kennedy and Norman, 2005). What do not people know?

H8. (Lehn, 1995). Molecules are more convenient building blocks than atoms to construct nanodevices/machines.

She reviewed the reasons at the basis of *supramolecular chemistry*: (1) molecules are stable species, whereas atoms are difficult to handle; (2) nature starts from molecules, not from atoms, to construct the great number and variety of nanodevices and machines that sustain life; (3) most of the laboratory chemical processes deal with molecules, not atoms; (4) molecules are objects that exhibit distinct shapes and carry specific properties (e.g., those that can be manipulated by photochemical, electrochemical inputs) as confirmed by techniques (e.g., single-molecule fluorescence spectroscopy, various types of probe microscopies, capable of *seeing* or *manipulating* single molecules); (5) molecules can self-assemble or be covalently connected to make larger structures. She proposed additional P, H, and Q.

P2. Big problems that face a large part of Earth's population: food, health, energy, and pollution.

H9. The development of science increases the fragility of the Earth.

H10. (Gould, 1998). Asymmetry principle: The essential human tragedy lies in a great asymmetry in our universe of natural laws. We can only reach our pinnacles by laborious steps, but destruction can occur in a minute fraction of the building time.

H11. (Rees, 2003). No more than 50% probability exists that civilization will survive until the 21st century's end because of bad/incautious use of the most recent developments of science/technology.

Q8. (Greenfield, 2003). What should people do?

Q9. (Greenfield, 2003). What should not people do?

Q10. Should a work on a contagious virus be published in full to aid pandemic preparedness or redacted to prevent misuse by terrorists?

Q11. What can people do with science and technology?

Q12. What can science and technology do of people?

Q13. What could be done in an ecologically *empty world*, when human impact was small?

H12. The larger resources consumption and waste-disposal rates, the more difficult it will be to reach sustainability.

P3. Problem of disparity. The passengers of spaceship Earth travel, indeed, in different *classes*.

H13. The stability of human society decays with rising disparities.

Q14. What will it happen?

Q15. Who is responsible for setting guidelines for defining progress and protecting future-generations interests?

Q16. What is wrong in the social and political organization of people nations and the entire world?

Q17. What is worth making with science?

H14. Establishing equity is not only a moral duty but also a basic need for creating a peaceful world.

H15. (Einstein, 1931). Concern for man himself and his fate must always constitute the chief objective of all technological endeavors.

5.4 CHEMISTRY AND ETHICS

Scheler and Kant considered that knowledge of reality is only possible via the data that experimental sciences supply, and only an aprioristic consideration of ethics would make possible to set out for objective knowledge of the truth that it teaches. It is the way of *positivism*.[21] The value judgments that ethics teaches demand knowledge of man in an integral fashion, the same way as of his nature and own demands that arise from it. In order to be in possession of the truth, an *adaptation* or agreement of reality with the mind that perceive it is essential, *adequatio intelectus et rei*. It is the position of *realism*. This way of seeing the reality happens in the specific human being and occurs in a determined sociocultural context and according to the historical moment that it has to live.

The idea of independence of values in science was not homogeneous throughout history.[22] The neutrality of the values as a dialectic tool to separate science from outside controversies, for example, religious, politics, etc., is not understood from outside a political context[23]; for example, *Darwinism* inspired Spencer's *social Darwinism*[24]; the homage to Darwin (València, 1909) identified *evolutionism* with *socialism* and *collectivism*; *eugenics* was accepted by all *Socialist Parties* and many European states before *Nazism*.

Is there an ethics of chemistry? No, there is a general ethics.

The American Chemical Society informed the *Academic Professional Guidelines* (1991–2016, cf. Table 5.1). However, it presents no practical dimension because it does not explain what to do if the science policy comes into conflict with demands related to the obligation to the boss or employees.

The ethics of chemistry needs, as already happened in the past, to negotiate priorities via a democratic collective deliberation. A general philosophical guide could help to construct an ethics adequate for contemporary research in the nanorevolutionary context.[25]

TABLE 5.1 American Chemical Society (ACS) Academic Professional Guidelines (1991–2016).

Chemical scientists should take personal responsibility for:
Treating coworkers with the respect expected by all professionals
Maintaining high standards of honesty, integrity, ethics, and diligence in the conduct of teaching, research, and all other professional responsibilities
Be concerned with personal health and safety and that of coworkers, consumers, and the community
Utilizing expertise for the good of coworkers, the community, and the world by providing considered comment to the public at large on issues involving the chemical sciences
Establishing and maintaining lines of communication throughout the academic and professional communities
Communicating with scientists and nonscientists accurately, using good oral and written skills. All chemical scientists working in an academic environment in the United States should develop spoken and written English language skills to communicate effectively novel research/educational materials in the language most relevant to a majority of the society and country
Honoring commitments made in the context of fulfilling professional duties, whether to students, colleagues, or employer
Understanding all facets of intellectual property that may be generated from original work
Generating opportunities for appropriate educational and research collaborations
Participating in life-long learning. Chemical scientists should continue their education and professional development, actively participate in appropriate professional societies and interact with other professionals in the field so as to enhance their capabilities
Seeking professional development opportunities to increase mentoring skills

5.5 DISCUSSION

People familiar with matter and how to manipulate it created much of the stuff of civilization (from food to medicines to materials). They may not call themselves chemists, but their experimentation shaped the modern world. Usually, for the sheer joy of knowing, alchemists, chemists, and, later, molecular scientists sought to expand people's understanding of matter from the atomic to the galactic. Chemistry is no longer a self-contained discipline anymore; it becomes nanochemistry, while physics merged with materials science. The natural/artificial dilemma comes from the times of alchemy. The cause is the lack of spreading. The need for communicating is only a secondary effect of creativity. New trends, for example, nanotechnology, etc., show decaying differences between science and humanities, science and technology, physics and chemistry, economy and capitalism, religion and philosophy, etc. This discussion deals with these questions from the perspective of historical research in science. Its viewpoint is from the history and sociology of science, making a comparison between *historicism* and *presentism.*

Global economic demands and population surges led to dwindling resources and problematic environmental subjects. As the climate and its natural resources continue to struggle, it becomes necessary to research and employ new forms of sustainable technology to help meet the growing demand. Sustainable nanosystems development requires emergent research and theoretical concepts in the areas of nanotechnology, PVs, electrochemistry and materials science, and within the physical and environmental sciences. Sustainable nanotechnology is a challenge for researchers, engineers, students, scientists, and academicians. Science is seen from outside as *experimental*, but there is also another *theoretical* science and *classification* is a part of theoretical science. There is a boom time for the *circular economy* and creativity. The shape of the future will be circular. However, how will that profit chemistry? Frontier science provides *hope* for many problems but *inequality* persists.

Chemical science is embedded in cultural values, which matter for the public acceptance of technoscientific innovations. People should take into account the ethical problem that generation of biofuels from edible feedstocks is in competition with food production. Generation of biofuels from edible feedstocks increases food price. Establishing equity is not only a moral duty but also a basic need for creating a peaceful world. To create a critical conscience in the people who must speak of science is still more important than to know science. Know thyself! Although philosophers continue to be necessary, they would perform better their essential work if they would be more interested in the relevant data that scientists unveil. How to translate all the research work in science didactics to the lecture room?

5.6 FINAL REMARKS

From the present discussion, the following final remarks can be drawn.

1. Continuity exists in the relationship between science and society, which will survive the transition from chemistry to nanotechnology. Examination of natural entities, and nanoscale interest, caused a revival of the Faustian ambitions associated with chemistry. Nanotechnology does not only search for imitating nature but also

surpassing it, with scientific visionaries that announce artificial life and self-propagating nanomachines, as precursors of life control by humanity.

2. Chemists were misunderstood as naïf *positivists* when they refused to accept the existence of atoms in the 19th century. After Comte and Mach, chemistry is positivist and not at once. Ostwald and Duhem showed the limit of positivism in chemistry. The discussion of atomism generated by positivism allows exploring the variety of atomisms that existed and exist.

3. Global economic demands and population surges led to dwindling resources and problematic environmental subjects. As the climate and its natural resources continue to struggle, it becomes necessary to research and employ new forms of sustainable technology to help meet the growing demand. There is a boom time for the *circular economy* and creativity. The shape of the future will be circular. However, how will that profit chemistry?

4. Ethics of chemistry needed to negotiate priorities via a democratic collective deliberation. Comparing physics and chemistry, is chemistry the future? New trends (nanotechnology, etc.) show decaying differences between science and humanities, science and technology, physics and chemistry, economy and capitalism, religion and philosophy, etc.

5. Academic professional guidelines present no practical dimension because they do not explain what to do if the science policy comes into conflict with demands related to the obligation to the boss or employees. There should be no indication of prejudice and enmity in a literary work, and researchers should not be afraid to transcend cultural boundaries in search for the truth or to present objectively the view of the *other*.

ACKNOWLEDGMENTS

Francisco Torrens belongs to the Institut Universitari de Ciència Molecular, Universitat de València. Gloria Castellano belongs to the Departamento de Ciencias Experimentales y Matemáticas, Facultad de Veterinaria y Ciencias Experimentales, Universidad Católica de Valencia *San Vicente Mártir*. The authors thank support from Generalitat Valenciana (Project No. PROMETEO/2016/094) and Universidad Catolica de Valencia San Vicente Martir (Project No. UCV.PRO.17-18.AIV.03).

KEYWORDS

- **miniaturization significance**
- **miniaturization history**
- **scientific subject**
- **ethical subject**
- **chemistry**
- **global economy**
- **population surge**

REFERENCES

1. Balzani, V.; Credi, A.; Venturi, M. *Molecular Devices and Machines: Concepts and Perspectives for the Nanoworld*; Wiley-VCH: Weinheim, Germany, 2008.
2. Ceroni, P.; Credi, A.; Venturi, M., Eds. *Electrochemistry of Functional Supramolecular Systems*; Wiley: New York, NY, 2010.
3. Torrens, F. Fractals for Hybrid Orbitals in Protein Models. *Complexity Int.* **2001**, *8*. 01-1–01-13.
4. Torrens, F. Fractal Hybrid-Orbital Analysis of the Protein Tertiary Structure. *Complexity Int.*, Submitted for Publication.
5. Torrens, F.; Castellano, G. Resonance in Interacting Induced-Dipole Polarizing Force Fields: Application to Force-Field Derivatives. *Algorithms* **2009**, *2*, 437–447.
6. Torrens, F.; Castellano, G. Molecular Diversity Classification *via* Information Theory: A Review. *ICST Trans. Complex Syst.* **2012**, *12*(10–12), e4-1–e4-8.
7. Torrens, F.; Castellano, G. Reflections on the Nature of the Periodic Table of the Elements: Implications in Chemical Education. In *Synthetic Organic Chemistry*; Seijas, J. A., Vázquez Tato, M. P., Lin, S. K., Eds.; MDPI: Basel, Switzerland, 2015; Vol. 18, pp 8-1–8-15.
8. Putz, M. V., Ed. *The Explicative Handbook of Nanochemistry*; Apple Academic–CRC: Waretown, NJ, in press.
9. Torrens, F.; Castellano, G. Reflections on the Cultural History of Nanominiaturization and Quantum Simulators (Computers). In *Sensors and Molecular Recognition*; Laguarda Miró, N., Masot Peris, R., Brun Sánchez, E., Eds.; Universidad Politécnica de Valencia: València, Spain, 2015; Vol. 9, pp 1–7.
10. Torrens, F.; Castellano, G. Ideas in the History of Nano/Miniaturization and (Quantum) Simulators: Feynman, Education and Research Reorientation in Translational Science. In *Synthetic Organic Chemistry*; Seijas, J. A., Vázquez Tato, M. P., Lin, S. K., Eds.; MDPI: Basel, Switzerland, 2016; Vol. 19, pp 1–16.
11. Torrens, F.; Castellano, G. Nanominiaturization and Quantum Computing. In *Sensors and Molecular Recognition*; Costero Nieto, A. M., Parra Álvarez, M., Gaviña Costero, P., Gil Grau, S., Eds.; Universitat de València: València, Spain, 2016; Vol. 10, pp 31-1–31-5.

12. Torrens, F.; Castellano, G. Nanominiaturization, Classical/Quantum Computers/ Simulators, Superconductivity, and Universe. In *Methodologies and Applications for Analytical and Physical Chemistry*; Haghi, A. K., Thomas, S., Palit, S., Main, P., Eds.; Apple Academic–CRC: Waretown, NJ; 2018, p 27–44.

13. Torrens, F.; Castellano, G. Superconductors, Superconductivity, BCS Theory and *Entangled* Photons for Quantum Computing. In *Physical Chemistry for Engineering and Applied Sciences: Theoretical and Methodological Implication*; Haghi, A. K., Aguilar, C. N., Thomas, S., Praveen, K. M., Eds.; Apple Academic–CRC: Waretown, NJ, in press.

14. EPR Paradox, Quantum Decoherence, Qubits, Goals, and Opportunities in Quantum Simulation. In Theoretical Models and Experimental Approaches in Physical Chemistry: Research Methodology and Practical Methods; Haghi, A. K., Ed.; Apple Academic– CRC: Waretown, NJ; 2018, Vol. 5, p 317–334.

15. Torrens, F.; Castellano, G. Nanomaterials, Molecular Ion Magnets, Ultrastrong and Spin–Orbit Couplings in Quantum Materials. In *Physical Chemistry for Chemists and Chemical Engineers: Multidisciplinary Research Perspectives*; Vakhrushev, A. V., Haghi, R., de Julián-Ortiz, J. V., Allahyari, E., Eds.; Apple Academic–CRC: Waretown, NJ, in press.

16. Torrens, F.; Castellano, G. Nanodevices and Organization of Single Ion Magnets and Spin Qubits. In *Industrial Chemistry and Chemical Engineering*; Haghi, A. K., Ed.; Apple Academic–CRC: Waretown, NJ, 2018, in press.

17. Torrens, F.; Castellano, G. In *Book of Abstracts*, Certamen Integral de la Prevención y el Bienestar Laboral, València, Spain, September 28–29, 2016; Generalitat Valenciana– INVASSAT: València, Spain, 2016; p 3.

18. Torrens, F.; Castellano, G. Nanoscience: From a Two-Dimensional to a Three-Dimensional Periodic Table of the Elements. In *Methodologies and Applications for Analytical and Physical Chemistry*; Haghi, A. K., Thomas, S., Palit, S., Main, P., Eds.; Apple Academic–CRC: Waretown, NJ, 2018, p 3–26.

19. Torrens, F.; Castellano, G. In *Book of Abstracts*, Congreso Internacional de Tecnología, Ciencia y Sociedad, València, Spain, October 19–20, 2017; Universidad Cardenal Herrera CEU: València, Spain, 2017; p 1.

20. Venturi, M. Developing Sustainability: Some Scientific and Ethical Issues. In *Sustainable Nanosystems Development, Properties, and Applications*; Putz, M. V., Mirica M. C., Eds.; IGI Global: Hershey, PA, 2017; pp 657–680.

21. Zamora Marín, R. Ética en el cuidado del paciente grave y terminal. *Rev. Cubana Salud Pública* **2006**, *32*(4), 4-1–4-7.

22. Arnau, J. Materiales para la Historia de las Matemáticas en la India Antigua y Medieval (hasta el Siglo XII). In *La Circulación del Saber Científico en los Siglos XIX y XX*; Díaz Rojo, J. A., Ed.; Instituto de Historia de la Medicina y de la Ciencia López Piñero: Valencia, Spain, 2011; pp 215–256.

23. Proctor, R. N. *Value-Free Science? Purity and Power in Modern Knowledge*; Harvard University Press: Cambridge, 1991.

24. Spencer, H. *The Principles of Sociology*; Williams and Norgate: London, 1876–1896.

25. Watson, M. When Will Open Science Become Simply Science? *Genome Biol.* **2015**, *16*, 101-1–101-3.

CHAPTER 6

SYNTHESIS, CHARACTERIZATION, AND BIOAPPPLICABILITY OF PROTEIN CONJUGATED NANOMATERIALS

DIVYA MANDIAL and POONAM KHULLAR*

Department of Chemistry, B.B.K. D.A.V. College for Women, Amritsar 143005, Punjab, India

Corresponding author. E-mail: virgo16sep2005@gmail.com

ABSTRACT

Currently, the nanotechnology research is advancing toward the nanomaterial synthesis using the molecules of biological grade. Proteins are the ecofriendly polymers which have led to a new generation synthesis of nanoscale devices with improved properties. The nanodevices designed from bovine serum albumin, soy protein, and Cyt *C* have been projected successfully into living systems to discover their potential in diagnosis and treatment of diseases such as cancer, Alzheimer's disease, diabetes, and chronic lung diseases. Protein nanotechniques aim at providing quick, cost-effective, selective, and pain-free diagnosis of the disease concerned. In coming years, with the evolution of nanomaterials in medical field, dependence on nonbiodegradable hazardous materials is going to get eliminated.

6.1 INTRODUCTION

The growth of nanotechnology in recent years has revolutionized every aspect of human life. Nanotechnology interlinks multiple branches of science such as biotechnology, bioprocess engineering, chemical engineering, genetic engineering, biophysics, optics, photonics, and quantum engineering. Nanoparticles possess improved and new physicochemical properties which make them accessible for environmental use with decreased toxicity levels. NPs have large surface area, high surface energy, reduced imperfections, and spatial confinement as compared to small bulk of material.[1] They show interesting properties of surface plasmon resonance (SPR), light scattering, surface enhanced Rayleigh scattering (SERS).[2] Because of these properties, it is convenient to design a nanodevice for its biological applications, in optoelectronics, and for chemical sensing, as biomarkers, in clinical diagnostics and many others.[3,4]

Nanoparticles can be fabricated either by top-down approach or bottom-up approach. In top-down approach, materials are cut, milled, and shaped in desired order, while in bottom-up approach, particles are self-assembled via chemical and biological methodologies.[5] Nowadays, researchers are exploring cost-effective green protocols for construction of nanodevices which are biocompatible, reusable, viable with high applicability at the same time. Nanotechnology has provided a promising tool for theranosis of most common form of dementia, that is, Alzheimer's disease. It is recognized by problems with memory, thinking, and behavioral changes. The symptoms of the disease develop slowly, get worse with time, and interfere with patient's daily activities. Supraparamagnetic iron oxide NPs (SPION) and amyloid NPs have been found with effective potential for treatment of this disease.[6] Nanotechnology also offers new opportunities to promote efficiency of food production by designing of biosensors for monitoring physical, chemical, and biological properties. These methodologies have efficiency to increase food safety, control pathogens, water treatment, targeted drug delivery of agrochemicals, and improve animal health.[7]

The bioapplicability of these NPs can be enhanced by using biopolymers for designing them. Biopolymers are the polymers synthesized by living organisms and have covalently bonded monomeric units. The three major classes of biopolymers include polynucleotides (RNA and DNA), polypeptides (proteins), and polysaccharides (carbohydrates). Out of these, proteins based biopolymers [bovine serum albumin (BSA), zein, soy protein, wheat gluten, peanut, cotton seed, myofibrils, casein, egg white, Cyt *C*, lysozyme, whey protein, and rice protein] are running successfully for manufacturing of biocompatible nanosystems.[8–10]

Biopolymers can also be combined with synthetic polymers either on their surface or within their interior. These hybrid polymers have increased loading capacity of drug molecules, nucleic acids, magnetic resonance imaging (MRI) contrast agents or catalysts and also their interior provides reaction sites for polymerization reactions. These hybrid polymers have promising applications for bio nanotechnology, biosensing, memory devices, and synthesis of other materials.[11] The biopolymer-based NPs can be characterized by numerous techniques such as UV–visible spectroscopy, fluorescence spectroscopy, transmission electron microscopy (TEM), energy-dispersive X-ray spectroscopy (EDS), Raman spectroscopy, X-ray photo electron spectroscopy, and matrix-assisted laser desorption/ionization mass spectroscopy (MALDI-MS). These biopolymers have not only solved the problem of white pollution but also have decreased the overdependence on nonrenewable sources of energy. The current chapter aims to showcase recent found applications of biopolymers with a marked focus on BSA, soy protein, and Cyt C.

6.2 SYNTHESIS AND CHARACTERIZATION

Protein NPs can be prepared by mixing proteins and metals (Au/Ag/Pd) in their aqueous solutions in the presence or absence of surfactants [sodium dodecyl sulfate (SDS), cetyl trimethyl ammonium bromide (CTAB), sodium tris(2-ethyl hexyl) sulfo-succinate (AOT), dodecyltrimethylammonium bromide (DTAB), 3-(N,N-dimethyltetradecylammonio)propane sulfonate (TPS), 3-(N,N-dimethylhexadecylammonio)propane sulfonate HPS, 3-(N,N-dimethyldodecylammonio)propane sulfonate (DPS) by keeping the reaction at 70°C for 6 h. BSA being a weak reducing agent due to presence to cysteine residue reduces metal to its lower oxidation according to equation[12,13]:

$$Au^{3+} (aq) + BSA \ (aq) + 3e^{-1} \rightarrow Au^0 (S)$$

BSA-stabilized AuNCs can also be synthesized by mixing gold seed NPs with aqueous solution of Au (0.25 mM), HCl (1 M), AgNO$_3$ (3 mM), and ascorbic acid (100 mM). Lastly, BSA solution is added to get greenish-blue-colored NCs. AuNCs can also be synthesized without using BSA as stabilizing agent.[14] NPs prepared by this technique on purifying with distilled water can be characterized by TEM, XRD, EDS, UV–visible spectroscopy. The reaction growth can also be studied w.r.t. time and temperature ranging from 20°C to 70°C by UV–visible measurements as shown in Figure 6.1. TEM studies provide information regarding different shapes obtained by NPs including hexagonal, rombohedral, spherical as shown in Figure 6.2. XRD patterns show prominent growth at {1 1 1} of face centered cubic (FCC) geometry. SEM analysis can be done by taking photomicrographs in bright field scanning/imaging mode with a spot size of ~1 nm and 12 cm of camera length. Bright and dark SEM images confirm the presence of BSA as a gel support for the NPs. EDS analysis shows the emission only due to gold as shown in Figure 6.3.

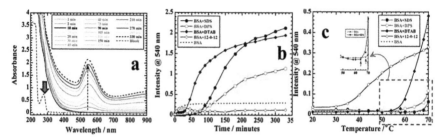

FIGURE 6.1 **(See color insert.)** (a) UV–visible scans of BSA+SDS+HAuCl$_4$ mixture with (BSA/residue)/(SDS) mole ratio = 88 at 70°C. The block arrow shows the absorbance due to tryptophan residues, while the dotted indicates the increasing absorbance of AuNPs due to SPR with time. Blank means when no HAuCl$_4$ is added. (b) and (c) Plots of intensity at 540 nm versus reaction time and temperature, respectively, for different mixtures in the presence and absence of surfactants. (Reproduced with permission from Reference [12]. © 2012 American Chemical Society.)

Folic acid conjugated BSA NPs and soy protein NPs with or without metal can be prepared by desolvation technique followed by redistribution of purified NPs in water. The FA solution activated with EDC N-[3-(dimethyl amino)propyl-N-ethyl carboimide] is added dropwise to NPs to get FA-BSA NPs.[15,16] Finally, organoseleno compounds (PSeD) can be loaded on these NPs to record their potential application as radiosensitizer.[17] Atomic force microscopic image of a single NP demonstrates smooth surface of FS-BSA NPs. The zeta potential as recorded from DLS technique shows the potential of plain, FA, and PSeD loaded BSA NPs to be −51.8, −31.6, and −22.5 mV, respectively.

The encapsulation of curcumin on FA-soy protein isolate (SPI) NPs in the presence of ethanol as evaluated by SEM analysis shows spherical particles with smooth surfaces in case of FA-SPI NPs and dense clusters on incorporating curcumin. On evaporating ethanol, NPs separate from each other as shown in Figure 6.4. The DLS data shows varied size ranging from 150 to 250 nm. The XRD analysis shows highly crystalline nature of curcumin as exhibited by peaks shown in Figure 6.5.[18,19,48]

BSA a zwitterionic protein at its isoelectric point (PI = 4.7) can bind simultaneously with Ag$^+$ and Au$^-$ ions. Due to this versatile character, BSA coated Au–Ag alloy NPs can be synthesized without using any external stabilizing agent. The alloy NPs are preferred over individual metals due to their advanced optoelectronic and catalytic properties. The alloy formation is supported by the linear trend followed by plasmon absorption maximum with respect to mixture of metals in alloy NPs. The reduction of metal mixture by using sodium citrate is assumed to be possibly catalyzed by Au, due to the inability of citrate to reduce Ag alone. Best foaming templates for the synthesis of NPs are SDS, AOT, and CTAB.

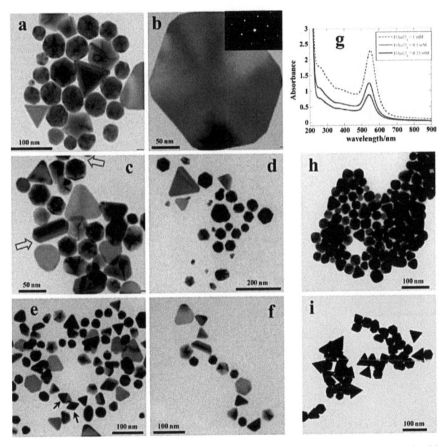

FIGURE 6.2 (See color insert.) (a) TEM images of purified AuNPs prepared with BSA+SDS+HAuCl$_4$ mixture at 70°C. Part (b) shows a single hexagonal NP with diffraction image in inset. (c) and (d) TEM images of purified AuNPs prepared with BSA+DPS+HAuCl$_4$ and BSA+DTAB+HAuCl$_4$ mixtures, respectively, at 70°C. Empty block arrows in (c) indicate the presence of thin BSA coating around each NP. (e) and (f) TEM images of purified AuNPs prepared with BSA+TPS+HAuCl$_4$ at 70°C. A pearl necklace arrangement of BSA conjugated NPs in (f) is due to the fibrillation of BSA. (g) UV–visible scans of different samples of as-prepared 16-2-16+HAuCl$_4$ mixture at 70C with different concentrations of gold salt. Parts (h) and (i) show their respective TEM images with HAuCl$_4$ = 0.25 and 1 mM, respectively. (Reproduced with permission from Reference [12]. © 2012 American Chemical Society.)

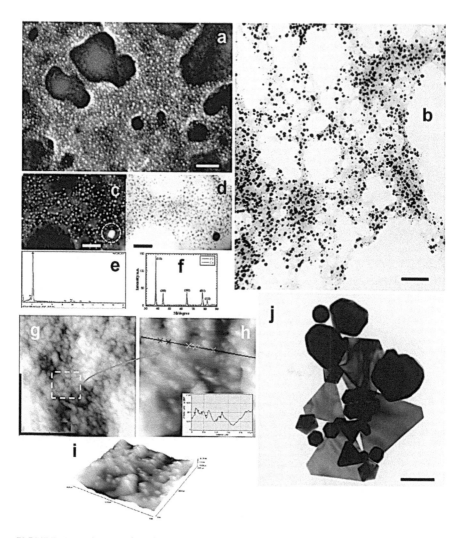

FIGURE 6.3 (See color insert.) (a) FESEM (scale bar = 200 nm) and (b) TEM micrographs of AuNPs as bright spots on unfolded BSA as a soft template of a sample with Au/BSA–mole ratio 167 synthesized at 80°C (scale bar = 100 nm). (c) and (d) Bright field (scale bar = 250 nm) and dark field images (scale bar = 250nm), respectively, differentiating between AuNPs and unfolded BSA. (e) and (f) EDS spectrum and XRD patterns of AuNPs, respectively. (g) AFM height image of unfolded BSA as a soft template bearing AuNPs (scale bar = 870 nm). (h) Close-up image showing the line analysis of various AuNPs (scale bar = 100 nm). (i) Topography of the BSA film bearing AuNPs showing peaks and valleys. (j) TEM micrograph of large-plate-like NPs of a sample with Au/BSA–mole ratio 17 prepared at 80°C (scale bar = 100 nm). (Reproduced with permission from Reference [13]. © 2011 American Chemical Society.)

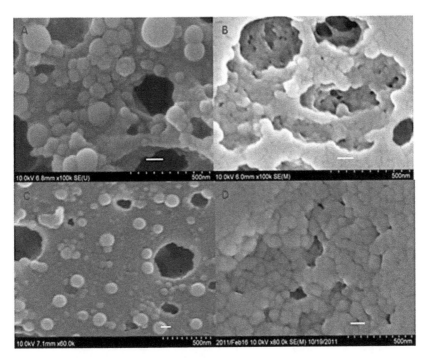

FIGURE 6.4 SEM images of (A) FA-SPI (F20) nanoparticles desolvated by 80% (v/v) ethanol, (B) curcumin-loaded FA-SPI (F20) nanoparticles, (C) same as (B), with ethanol evaporated, and (D) nanoparticles formed with control SPI. The white bars in the micrographs represent for 100 nm. (Reproduced with permission from Reference [48].)

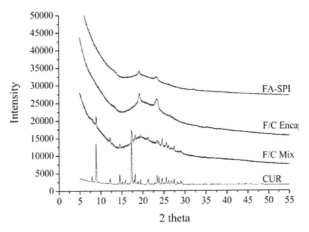

FIGURE 6.5 XRD patterns of FA-SPI, curcumin (CUR), their physical mixtures (F/C mix, curcumin/protein = 1:20 and FA-SPI encapsulated curcumin (F/C Encap, curcumin/ protein—1:20). (Reproduced with permission from Reference [48]. © 2013 American Chemical Society.)

Alloy NPs can be prepared in a column with a sintered frit at the bottom, the protein along within mixture of metals, that is, Ag and Au are inserted in column and foam is generated by passing N_2 gas through the frit. Excess liquid can be drained out and the remaining foam is treated with hydrazine hydrate vapors. The progress of the reaction can be determined by change in color of the foam. The experiment can be repeated by changing the ratio of Ag and Au metal. The resultants NPs as evaluate from TEM determines polydispersed nature of NPs with an average size of 30–60 nm for AuNPs, 20–80 nm for AgNPs, and 5–15 for Au–Ag alloy NPs. The selected area electron diffraction (SAED) for AuNPs can be indexed to 111, 200, 220, 311, and for AgNPs can be indexed to 111, 200, 220, 311 Braggs reflections from FCC structure of Au, Ag as shown in Figure 6.6. SAED pattern for alloy NPs confirms its crystalline nature as shown in Figure 6.7. The UV–visible analysis shows the absorbance of Au and AgNPs at 570 nm and 429 nm, respectively. The alloy NPs with different ratios of Au and Ag have peak in between the two as shown in Figure 6.8.

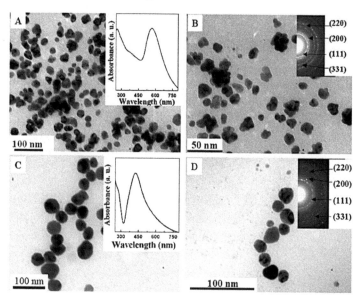

FIGURE 6.6 (A) and (B) Representative TEM micrographs obtained from drop-coated films of Au nanoparticles prepared in the BSA foam. The inset in (A) shows the UV–vis absorption spectrum of AuNPs. The inset in (B) shows the SAED pattern recorded from the Au nanoparticles. (C) and (D) Representative TEM micrographs obtained from a drop-coated film of AgNPs prepared in the BSA foam. The inset in (C) shows the UV–vis absorption spectrum of the AgNP sol. The inset in (D) corresponds to the SAED of the silver nanoparticles. (Reproduced with permission from Reference [20]. © 2005 Royal Society of Chemistry.)

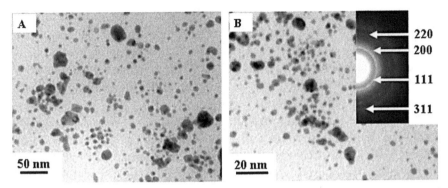

FIGURE 6.7 (A) and (B) Representative TEM micrographs obtained from the drop-coated films of Au–Ag alloy nanoparticles obtained from the 3Au:1Ag mixture. The SAED pattern of the Au–Ag nanoparticles is shown in the inset of (B). (Reproduced with permission from Reference [20]. © 2005 Royal Society of Chemistry.)

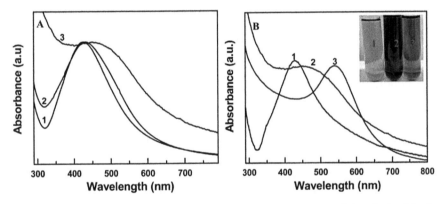

FIGURE 6.8 (A) UV–vis absorption spectra recorded from nanoparticles obtained in the BSA foam experiment from (1) 1Au:3Ag mixture, (2) 1Au:1Ag mixture, and (3) 3Au:1Ag mixture. (B) UV–vis absorption spectra recorded from (1) BSA-capped AgNP solution, (2) BSA-capped Au–Ag alloy NP solution obtained from 3Au:1Ag mixture, and (3) BSA-capped AuNP solution. Photograph of the three test tubes with aqueous solutions of Ag, Au–Ag alloy, and AuNPs, respectively, is shown in the inset of (B). (Reproduced with permission from Reference [20]. © 2005 Royal Society of Chemistry.)

BSA due to its biostability is used for in situ synthesis of metals alloys. Metal BSA, in its dry foam, is at its isoelectric point, where both C and N terminals are exposed by binding with oppositely, charged metal ions, that is, Ag binds to COO^- and $AuCl_4^-$ binds to NH_3^+ thus leading to alloy NPs formation simultaneously. Thus, the existing method can be preferred over the classical synthesis method of individual metal NPs. Fourier-transform infrared spectroscopy studies show that NH_3^+ binds to $AuCl_4^-$ ion leading to formation of $NH^{3+}–AuCl_4^-$ complex; therefore, no peak of NH bending vibration at 1494 cm^{-1} is observed in case of BSA complexed with Au. Similarly, the decrease in intensity of peak above 3200 cm^{-1} shows the complexation of carbohydrate group with Ag^+ ion, while in case of alloys NPs, the bands overlap show no clear peaks.

Au Cyt C nanostructures with required number ratio of protein to Au can be synthesized by dropwise addition of stock solution of Cyt C to gold colloidal sol. For 5700:1 Cyt C/Au ratio, 89 µl of Cyt C is added to 711 µl of colloidal gold solution. Acetonitrile assists solvent precipitation of Cyt C in the presence of excipient methyl-β-cyclodextrin-succinimidyl-3(2-pyridyldithio) propionate linker can be used to modify NPs surface. These NPs can be dissolved in acetonitrile to form thiol bond with linker. The NPs can be purified to remove unreacted reagents for their active targeting on cancerous cells.[21] These NPs show characteristics absorbance at 80 and 365 nm in UV–visible spectroscopic analysis. SEM images determine the morphology after coating process. DLS analysis shows increased zeta potential of Cyt C after NPs formation as well as NPs coating as shown in Figure 6.9. The increased positive charge is attributed to higher amount of Cyt C in the resulting system. The resulting system has high colloidal stability due to micellar like NPs, and this may also increase circulation time in blood systems.[22]

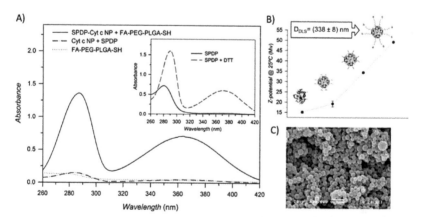

FIGURE 6.9 Characterization of the modification of protein-based NPs with FA-PEG-PLGA-SH. (A) UV–vis spectra of the supernatant recovery after the coating process. The release of the pyridine-2-thione group produced an absorbance increase at around 280 and 365 nm. (B) Z-potential of Cyt C in water before and after nanoprecipitation and after each coating step. The particle diameter of the resulted nanoparticle was measured with DLS. (C) SEM image of the resulting Cyt C-based NPs coated with FA-PEG-PLGA. (Reproduced with permission from Reference [21]. © 2016 American Chemical Society.)

6.3 APPLICATIONS OF PROTEIN CONJUGATED NANOMATERIALS

BSA is a single polypeptide chain consisting of 583 amino acid residues. Its secondary structure is highly α-helical and tertiary structure consists of 3 homologous domains, 2 tryptophans at positions 134, 213 with cysteine residues forming disulfide bonds.[23] BSA is reported to combine with metal NPs without loss of its biological activity hence is a favorite biopolymer for researchers to work with.[24] BSA has numerous applications in various branches due to its nontoxic, biodegradable nature, and it also stabilizes NPs.

BSA@AuNCs have potential nanoplatform for SERS, CT imaging, photothermal therapy of cancer.[25] BSA@AuNPs are promising sensitizers for glioblastoma tumor radiation therapy with least toxicity as recorded in vivo and in vitro conditions.[26,27] BSA in combination with organoseleno compounds and folic acid can fabricate a cancer targeted nanosystem to improve cancer radiotherapy.[28] It also increases the bioavailability of tea polyophenol NPs which act as valuable radioprotectors.[29] BSA-coated NPs have been found to be accepted preferentially by cancerous cells than normal cells for targeted drug delivery.[30–32] Therefore, they are recognized as best drug releasing tools. Microwave synthesized BSA@AuNCs can detect toxic metals such as lead and mercury from environmental samples. The technique provides low coat and high sensitivity for better results.[33,34] Pt@BSA nanocomposites can sense the glucose levels in human blood serum samples; thus, diabetes can also be diagnosed.[35] Au@BSA-modified electrodes can be employed for enantioselective recognition of propranolol (PRO) enantiomers.[36] This provides an alternate methodology to discriminate between chiral molecules.

BSA is a powerful chiral selector recorded in recent era. As we know, most of the biomolecules possess a determinant property of chirality, and this property finds widespread usage for the recognition and separation of chiral molecules such as PRO [1-isopropylamino-3-(1-napthyloxy)-2-propanolol]. PRO is an essential β-adrenergic blocking agent used in the treatment of hypertension, cardiac arrhythmia, and angina pectoris. It exists as S-PRO and R-PRO forms. The former is 100 times better isomer than the later. The two isomers can be differentiated by constructing a simple biosensor based on Au@BSA nanocomposite coated on gold electrode. Au@BSA nanocomposite when fabricated on sensitive electrochemical sensors using green protocol is effective to replace laborious, high cost and complicated techniques to distinguish the enantiomers. Differential pulse voltammetry (DPV) shows almost similar amperometic currents of two enantiomers of PRO on named gold electrode and BSA coated electrode. But different peak currents were observed for two enantiomers on Au@BSA-modified biosensor with lower peak current of S-PRO than R-PRO as shown in Figure 6.10. The average peak current recorded was about 30.5 μA with a relative standard deviation of 2.43%. Likewise, other proteins also interact with NPs to give high relevance for chiral recognition of pharmacological drugs.[36]

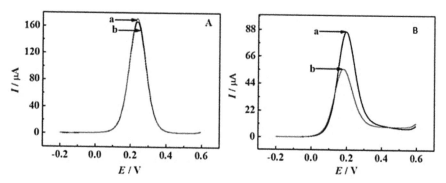

FIGURE 6.10 DPV curves of (A) the bare gold electrode, (B) Au@BSA-modified interface after interaction with 5 mM (a) R-PRO, (b) S-PRO in 5 mM [Fe(CN)6]4_/3_ solution (pH 7.4). (Reproduced with permission from Reference [36]. © 2016 Royal Society of Chemistry.)

Au@BSA NPs have been recorded as promising radiation therapy sensitizing agents for cancer treatment. The radio sensitizer effects of AuNPs have been reported in various types of cancers, including cervix cancer, prostate cancer, colon cancer, and breast cancer.[37–41] Least cytotoxicity, good biocompatibility of Au@BSA NPs makes them eligible to be used for radiation therapy for glioblastoma without detrimental effect on other organs and tissues. Tumor cells, when incubated with Au@BSA NPs, tend to accumulate on the surface of cells through EPR (enhanced permeability and retention effect) as shown in Figure 6.11. The retended NPs exhibit strong absorption of radiations followed by generating secondary electrons under ionizing radiations.

BSA-AuNPs can reduce the rapid growth of tumor cells compared with radiations alone. The radiosensitivity of tumor cells in the presence of BSA-AuNPs can be characterized by breaking of DNA double strands and cell apoptosis. The DNA damage is induced by combined photoelectric effect and compton scattering. Increased ratio of cell apoptosis, 48 h after irradiation is shown in Figure 6.12. The data show significant radio sensitization when BSA-AuNPs are accumulated on tumor cells by EPR. The in vivo radiation therapy carried out on a mouse suffering from tumor shows the reduction in weight of increasing tumor.[26]

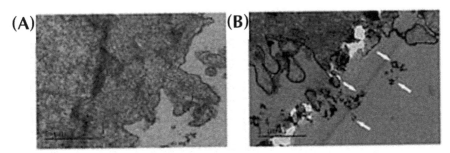

FIGURE 6.11 (A) TEM image of U87 cells without adding BSA-AuNPs (36 μg mL⁻¹) as control. (B) TEM image of U87 cells incubated with BSA-AuNPs (36 μg mL⁻¹) for 6 h. (Reproduced with permission from Reference [26]. © 2015 Royal Society of Chemistry.)

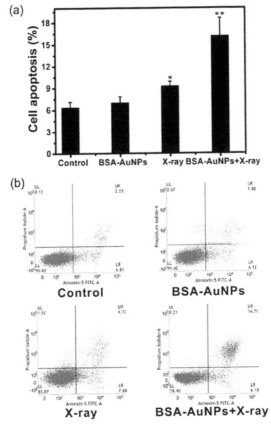

FIGURE 6.12 BSA-AuNPs increased the apoptosis of U87 cells. (a) Apoptosis ratio of U87 cells. Cells were irradiated after treatment with 36 mg mL⁻¹ BSA-AuNPs or vehicle control. * $p < 0.05$ when compared with control, ** $p < 0.05$ when compared with the group of X-ray. (b) Cell apoptosis of BSA-AuNPs. (Reproduced with permission from Reference [26]. © 2015 Royal Society of Chemistry.)

BSA-FA conjugated AuNSs (BSA-FA-AuNSs) have been reported for efficiently targeted photothermal ablation of cancer cells. BSA due to its nontoxicity, good biocompatibility, biodegradability, and stability have been found to be preferentially taken up by cancer cells compared to normal cells, which signifies its potential application for diagnosis and treatment of cancer. When modified with folic acid (FA) as receptor, BSA-coated AuNPs can be effectively used for molecular imaging, thermal therapy, and controlled drug delivery as represented in Scheme 6.1. The photothermal performance of BSA-AuNSs and BSA-FA-AuNSs displayed an increase in the temperature of aqueous solutions from 26°C to 69°C in 300 s as shown in Figure 6.13a. The temperature change is directly related to concentration of Au as shown in Figure 6.13b. The temperature increase shows no drastic change for BSA-AuNSs and BSA-FA-AuNSs even after five repeated laser on off cycles in which sample is irradiated by 805 nm laser for 5 min and then cooled to room temperature as shown in Figure 6.13c.[42]

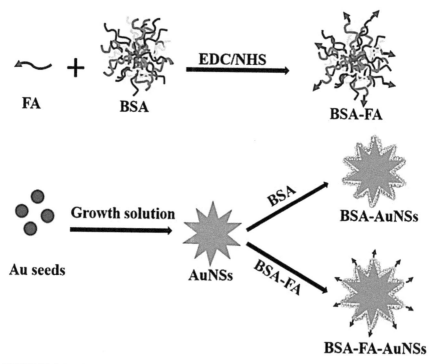

SCHEME 6.1 Schematic of the synthesis of the BSA-FA conjugate, BSA-AuNSs, and BSA-FA-AuNSs. (Reproduced with permission from Reference [42]. © 2015 Royal Society of Chemistry.)

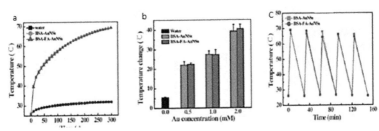

FIGURE 6.13 (See color insert.) (a) Temperature increase in water, an aqueous solution of BSA-AuNSs or BSA-FA-AuNSa; (b) temperature change in an aqueous solution of BSA-AuNSs or BSA-FA-AuNSs at different Au concentrations; (c) temperature change in an aqueous solution of BSA-AuNSs or BSA-FA-AuNSs over five laser on–off cycles. (Reproduced with permission from Reference [42]. © 2015 Royal Society of Chemistry.)

Alb-GNS (albumin gold nanostars) manifest clearly photothermal cytotoxicity on cancerous cells due to excellent photothermal effects of AuNSs.[14] Coating of albumin over AuNSs makes nanostarts nontoxic for its biomedical application. Alb-GNS when treated with cancerous cells did not show generation of any reactive oxygen species as shown in Figure 6.14(a), which confirms its biodegradability. Cancerous cells loaded with Alb-GNS when subjected to laser radiations for 10 min showed marked cells death than the cells just treated with laser beam only as shown in Figure 6.14(c).

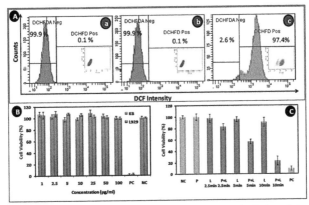

FIGURE 6.14 (See color insert.) (A) Flow cytogram depicting intracellular ROS generation by DCFH-DA assay in KB cells treated with (a) media alone (negative control), (b) Alb-GNS (100 mg mL^{-1}), and (c) 30 mM H$_2$O$_2$ (positive control); (B) graphical representation of cell viability tests on KB and L929 cells treated with Alb-GNS at varying concentration for 24 h; (C) graphical representation of cell viability test on KB cells treated with 100 mg mL^{-1} of Alb-GNS (P) and laser (L) for different time period. NC and PC represent negative control (untreated cells) and positive control (cells treated with 1% Triton X-100), respectively. (Reproduced with permission from Reference [25]. © 2016 Royal Society of Chemistry.)

BSA@AuNPs are best suited for drug delivery due to their least hemolytic and cytotoxic responses as shown in Scheme 6.2.[12] The unfolding of BSA can be carried out by using toxic/zwitterionic surfactants due to their strong electrostatic interactions with BSA. The cytotoxic response of BSA-coated AuNPs as recorded using MTT assay by subjecting glioma cells to BSA-coated AuNPs shows no cytotoxic effect. While cytotoxic effects can be observed in the presence of surfactant-coated AuNPs with 20% cytotoxicity at 50 µg mL^{-1} concentration of AuNPs. Likewise, BSA-coated AuNPs show negligible hemolysis in comparison to surfactant-coated AuNPs as depicted in Figure 6.15. Thus, AuNPs can be introduced into blood stream by complexing with BSA without any damage to the cells which confirm biomedical use of BSA-coated AuNPs.

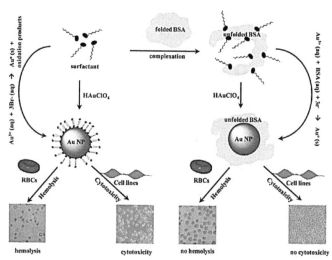

SCHEME 6.2 Schematic representation of the proposed mechanism of the synthesis of BSA and 16-2-16-coated AuNPs. (Reproduced with permission from Reference [12]. © 2012 American Chemical Society.)

Brain gliomas are responsible for 40% of all brain tumors. Brain tumors can be destroyed with surgeries followed by radiation therapy. This radiation therapy is not applicable every time as it can be intolerant for normal tissues at high dosage. Therefore, targeted therapy may be the best way out to treat brain gliomas. In connection to this, nanocarriers are best tool to develop a drug delivery system that even penetrates, blood–brain barrier to destruct tumor of targeted area. BSA-NPs modified with Lactoferrin (Lf) and mPEG2000 loading doxorubicin have been designed which are further treated for potential cytotoxicity on primary brain capillary endothelial cells and glioma cells as shown in Scheme 6.3. Initially, Doxorubicin-BSA NPs are synthesized by a desolvation technique. Then, those NPs are treated with mPEG-2000. Finally, P_{2000}-NPs are incubated with LF to achieve Lf-NPs. DOX laded NPs can be studied for in vitro and in vivo release of DOX to targeted cells. Both investigations exhibit the improved long circulation effect of Lf-NPs with less zeta potential as shown in Figure 6.16.[27]

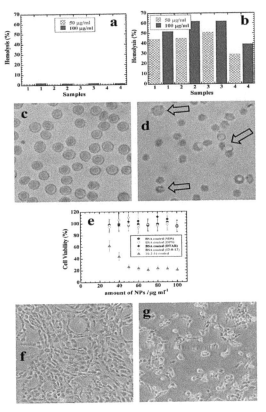

FIGURE 6.15 **(See color insert.)** (a) Percentage hemolysis of purified BSA-coated AuNPs prepared with different mixtures, BSA+SDS+HAuCl$_4$ (1); BSA+DPS+HAuCl$_4$ (2); BSA+DTAB+HAuCl$_4$ (3); BSA+12-0-12+HAuCl$_4$ (4), with doses of 50 and 100 µg mL^{-1}. (b) Percentage hemolysis of purified 16-2-16-coated AuNPs prepared with 16-2-16+HAuCl$_4$ mixtures at 70°C at fixed HAuCl$_4$ = 0.25 mM and with different concentrations of 16-2-16 = 2 mM (1); 4 mM (2); 8 mM (3); and at HAuCl$_4$ = 1 mM and 16-2-16 = 4 mM (4), with doses of 50 and 100 µg mL^{-1}. (c) and (d) Bright field optical microscopic images of RBCs without and with 16-2-16-coated AuNPs, respectively, of "sample 2" of (b). Empty block arrows in (d) indicate broken cells with released contents. (e) Percentage cell viability of glioma cell lines with different amounts of BSA and 16-2-16-coated NPs. (f) and (g) Bright field optical microscopic images of glioma cell lines without and with 16-2-16-coated AuNPs, respectively, of "sample 2" of (b). (Reproduced with permission from Reference [12]. © 2012 American Chemical Society.)

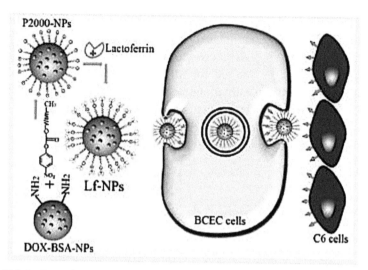

SCHEME 6.3 Lactoferrin modified P$_{2000}$-NPs targeting endothelial (BCECs) and C6 cells. (Reproduced with permission from Reference [27]. © 2014 American Chemical Society.)

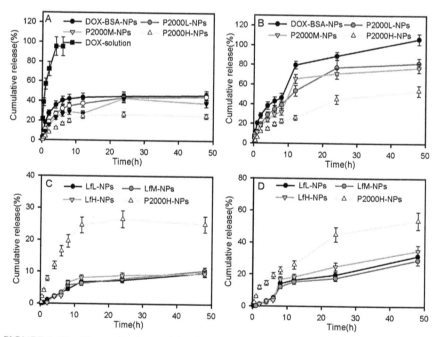

FIGURE 6.16 (See color insert.) In vitro release curve of DOX from DOX solution and DOX-loaded NPs: (A) and (C) in saline and (B) and (D) in saline with trypsin. Data were presented as the mean ± SD ($n = 3$). (Reproduced with permission from Reference [27]. © 2014 American Chemical Society.)

The in vivo tissue distribution in glioma model rats release longer systemic circulation of P_{2000}-NPs and LF-NPs than DOX-BSA NPs. In addition, it also shows less accumulation of DOX in heart and kidney which decreases cardiac and renal toxicity of DOX as shown in Figure 6.17.

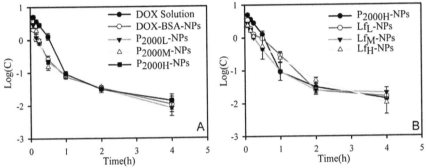

FIGURE 6.17 **(See color insert.)** Plasma concentration–time curves of DOX in rats after intravenous administration of (A) DOX solution, DOX-BSA-NPs, P_{2000}L-NPs, P_{2000}M-NPs, and P_{2000}H-NPs and (B) P_{2000}H-NPs, LfL-NPs, LfM-NPs, and LfH-NPs. Data were presented as the mean ± SD ($n = 6$). (Reproduced with permission from Reference [27]. © 2014 American Chemical Society.)

Recently, chemical therapeutics such as cisplatin and paclitaxel have been developed to be used as radiosensitizers for treatment of lung, cervical, gastric, head, and neck cancers. It is viable to use a nanosystem to construct a nanosensitizer for enhanced tumor targeted drug delivery and drug localization. BSA NPs fabricated with folate (FA) ligand are effective carriers of organic selenadiazole compounds, phenylbenzo(1,2,5)selenadiazole (PseD) which have excellent chemopreventive and chemotheraneutic effects on cancer treatment with low toxicity. PSeD due to its metalloid properties such as SPR effect, heavy metal effect, and high refractive index is considered to be a promising radio sensitizer. Colonogenic assay exhibits the appealing effect of combined FA-BSA NPs and X-rays in inhibition of cells survival to 4.3% in comparison to single FA-BSA NPs and X-rays treatments which decreases HeLa cells colony formation to 67.1% and 55.7%, respectively. The (3-(4,5-dimethylthiazol-2-yl)-2,5-diphenyltetrazolium bromide) MTT assay further demonstrates the radiosensitization caused by FA-BSA NPs. Fluorescence image also displays increased level of DNA fragmentation (green fluorescence) and nucleus condensation (blue fluorescence) in comparison to single treatment alone as shown in Figure 6.18.[28] The signaling pathway of FA-BSA NPs and radiations in combination is shown in Figure 6.19. The presence of FA-BSA NPs inhibits the X-rays caused VEGF/VEGFR2 and

XRCC-1 overexpression indicating that FA-BSA NPs successfully reversed X-rays resistance in HeLa cells.

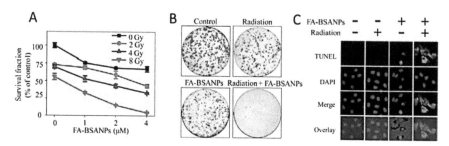

FIGURE 6.18 (See color insert.) FA-BSA NPs in combination with X-ray radiation increased G2/M phase arrest and apoptosis in HeLa cells. (A) Combined treatment of FA-BSA NPs and X-ray radiation decreased HeLa cells survival detected by clonogenic assay. HeLa cells were pretreated FA-BSA NPs at different concentrations for 2 h and then were irradiated by X-ray radiation at different dosages. Values expressed were means ± SD of triplicate. (B) Colony formation of HeLa cells under the cotreatment of FA-BSA NPs and radiation (8 Gy). (C) Fluorescence images of DNA fragmentation and nuclear condensation after exposure to FA-BSA NPs or/and X-ray radiation. Cells pretreated with FA-BSA NPs (4 μM) and X-ray radiation (8 Gy) were stained with TUNEL working buffer and DAPI for DNA fragmentation and nucleus visualization, respectively. (Reproduced with permission from Reference [28]. © 2014 American Chemical Society.)

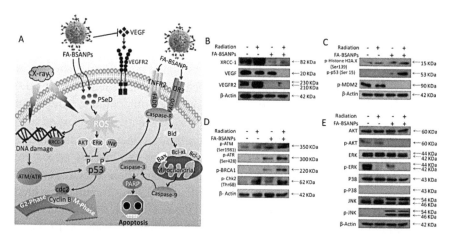

FIGURE 6.19 Signaling pathways triggered by FA-BSA NPs and X-ray radiation. (A) Overview of G2/M phase arrest and apoptosis signaling pathway caused by FA-BSA NPs and X-ray radiation in combination in HeLa cells. Western blot analysis for the expression of (B) XRCC-1, VEGF, VEGFR2; (C) p-Histone, p-p53, p-MDM2; (D) p-ATM, p-ATR, p-BRCA1, p-Chk2; (E) AKT, p-AKT, ERK, p-ERK, P38, p-P38, JNK, p-JNK. β-Actin was used as the loading control. (Reproduced with permission from Reference [28]. © 2014 American Chemical Society.)

BSA provides the matrix to increase the bioavailability of tea polyphenols (TPs) which are effective in reducing radiation-induced oxidative damage to normal cells caused during radio therapy by restoring the redox status through NrF_2–ERK (transcription factor governing antioxidant response elements–extracellular-signal-regulated kinase) pathway. BSA due to its remarkable efficacy is approved by US Food and Drug Administration (FDA) for formulation and delivery. TP nanoparticles synthesized using BSA as matrix and chitosan as covering shell are investigation to reduce the radiation induced lethality when introduced orally in mice. The treatment of TPNPs, BSA tea polyphenols NPs, chitosan-coated BSATP NPs increased the survival rate from 6 to 25, 30, 40%, respectively, due to preservation and enhanced delivery of TPNPs in the gastric cell as shown in Figure 6.20.[29] The number of spleen colonies increase, when pretreated with chitosan-coated BSATP NPs proving its biological efficacy. The radiation-induced damage caused to DNA and membranes is also reduced when pretreated with chitosan-coated BSATP NPs. The free radical mediated oxidative damage caused to normal cell is reduced by TPs by scavenging them resulting in protection of hematopoietic system.

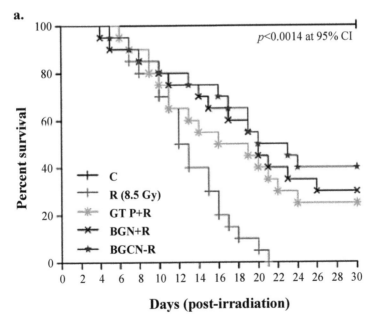

FIGURE 6.20 **(See color insert.)** Effect of GTP, BGN, and BGCN on radiation-induced damage to hematopoietic system and mortality. (Reproduced with permission from Reference [29]. © 2016 American Chemical Society.)

BSA/SPION exhibit great potential in MRI as the liver-specific biocompatible agents. BSA/SPION hybrid nanoclusters of uniform size ~86 nm provide large imaging time-window and gradual excretion, as compared to the toxic contrast agents such as gadolinium diethylenetriaminepentaacetic acid (Gd-DTPA). BSA/SPION hybrid NCs also exhibit good colloidal stability, high relaxivity. The schematic illustration of one-pot facile SPION-clustering process is shown in Scheme 6.4. The in vivo MRI studies conducting by injecting [2.5 mg (Fe) kg^{-1} body weight] of BSA/SPION shows the MR signal enhancement in liver and prolonged up to 48 h as depicted in Figure 6.21. The ICP (inductively coupled plasma) analysis shows the uptake, retention, and clearance route of BSA/SPION hybrid NCs via hepatobiliary (Hb) processing with excretion into bile over a period of 14 days with no in vivo toxicity to normal cells.[43]

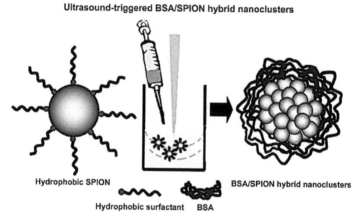

SCHEME 6.4 Schematic diagram showing the formation of BSA/SPION hybrid nanoclusters under ultrasonication in a one-pot facile approach. (Reproduced with permission from Reference [43]. © 2012 American Chemical Society.)

Pt@BSA microspheres with improved electrocatalytic activity are a low cost, highly sensitive approach in clinical diagnostic of glucose in blood. Covalent adsorption of glucose oxidase (GOD) on Pt@BSA nanocomposite makes it an enzyme labeled new generation glucose sensor. BSA layer acts as a protective film to maintain water permeability and to restore the bioactivity of anchored enzyme. The two highly bound flavin adenine dinucleotide (FA) cofactors embedded in GOD acts as active redox center to electrocatalyze reduction of dissolved oxygen as shown in Scheme 6.5.[35] A well-defined reduction peak is observed at around −0.20 V as the result of direct electron

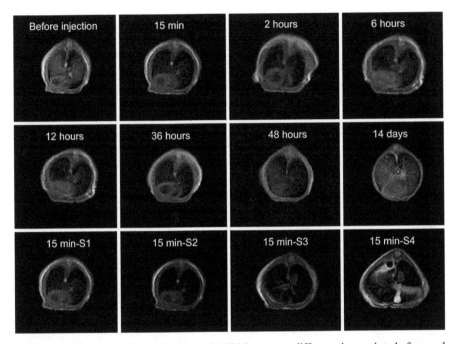

FIGURE 6.21 Mouse liver T2-weighted MRI images at different time points before and after administration of BSA/SPION hybrid nanoclusters, and axial slices of T2-weighted MR images of BSA/SPION hybrid nanoclusters in liver at 15 min postinjection. All images were obtained at a dose of 2.5 mg (Fe) kg¹ body weight. (Reproduced with permission from Reference [43]. © 2012 American Chemical Society.)

transfer from GOD in presence of dissolved oxygen as shown in Figure 6.22. The reduction of dissolved oxygen on electrode surface is catalyzed by GOD according to following equations:

$$\text{GOD}\left(\text{FAD}\right) + 2e^- + 2H^+ \rightleftharpoons \text{GOD}\left(\text{FADH}_2\right)$$
$$\text{GOD}\left(\text{FADH}_2\right) + O_2 \rightarrow \text{GOD}\left(\text{FAD}\right) + H_2O_2$$

When glucose is added to oxidized product, it results in decrease of reduction peak current due to restrained activity of enzyme catalyst as shown in following reaction:

$$\text{Glucose} + \text{GOD}\left(\text{FAD}\right) \rightarrow \text{Gluconolactone} + \text{GOD}\left(\text{FADH}_2\right)$$

The reduction peak current decreases with increasing glucose concentration, making it an efficient biosensor for determination of glucose. DPV investigation determines the performance of biosensor giving a linear response range from 0.05 to 12.05 mM with optimal detection limit (DL) of 0.015 mM. The practical clinical application on human blood serum samples shows good detection of glucose with analytical recoveries from 97.5 to 104.0% demonstrating its practical application in human blood samples.

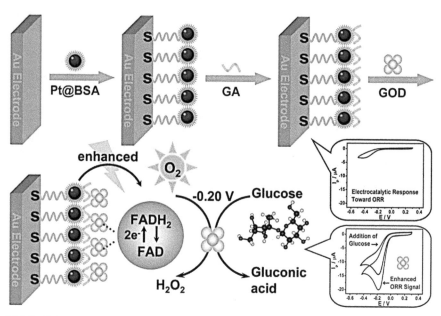

SCHEME 6.5   Schematic illustration of the fabrication process and sensing mechanism of the Pt@BSA-based electrochemical glucose biosensor. (Reproduced with permission from Reference [35]. © 2014 American Chemical Society.)

Microwave synthesized BSA-Au nanoclusters have been found for determination of toxic ions such as Hg^{2+}, Pb^{2+}, and melamine from environmental water samples, sea water, and milk, respectively.[33,34] According to US Environmental Protection Agency, maximum limit of Hg in drinking water should be 10 nM (2.0 ppb). Likewise, FDA has limited Pb concentration of 2.5 pm (518 μg^{-1}) for products to be consumed by children and melamine concentration of 2.5 pm (20 μM) for human consumption. Higher level of these toxin metals can have effects severe effects on human health and can even prove to be fatal. There are powerful techniques that exist to determine the ultrabase amounts of these toxic metals such as ICP mass spectrometry and X-ray fluorescence spectrometry, atomic absorption spectroscopy, and anodic stripping voltammetry. However, these techniques are quite expensive and require careful handling. On the other hand, BSA@AuNCs offers sensitive, selective, and reliable analysis of these toxic metals. The sensing potential of BSA@AuNCs is based on the $d^{10}-d^{10}$ metallophilic interaction between Au^+ ($4f^{14}\,5d^{10}$) and Hg^{2+} ($4f^{14}\,5d^{10}$) ions which results in fluorescence quenching. The addition of $NaBH_4$ weakens quenching process as $NaBH_4$ is a strong reducing agent and reduces Hg^{2+} to Hg^0 which ends in destroyed interaction between Au^+ and Hg^{2+} as shown in Figure 6.23. Sensing of Pb^{2+} by BSA@AuNCs is based on interaction between amino and of BSA and Pb^{2+} that leads to fluorescence quenching as shown Figure 6.24. Moreover, TEM images, EDS spectrum, and Dynamic Light Scattering results favor the interaction of BSA@AuNCs with Pb^{2+}. The presence of melamine inhibits the fluorescence quenching caused by Hg^{2+} at optimum concentration of Hg^{2+} as 20 μm and pH 7.0, respectively. The addition of melamine leads to its co-ordination with Hg^{2+} thus preventing the fluorescence quenching caused by Hg^{2+} as shown in Figure 6.25. Thus, these microwave assisted synthesis can find good application for synthesis of HSA, trypsin, lysozyme-stabilized AuNCs which may be used for detecting antibodies and receptors.

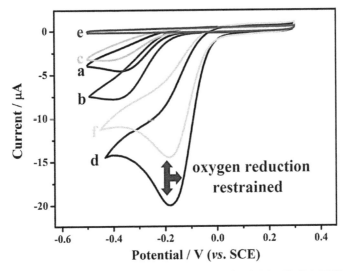

FIGURE 6.22 Cyclic voltammograms in air-saturated 0.1 M, pH 7.4 PBS at (a) an electrochemically pretreated bare Au electrode, (b) after formation of Pt@BSA layer, (c) after cross-linking reaction with GA, after covalent immobilization of GOD in the (d) presence and (e) absence of dissolved oxygen, and (f) with the addition of 6.55 mM glucose to air-saturated buffer (scan rate: 50 mV s^{-1}). (Reproduced with permission from Reference [35]. © 2014 American Chemical Society.)

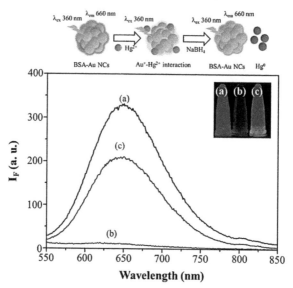

FIGURE 6.23 Schematic of Hg²⁺ sensing based on the fluorescence quenching of BSA@AuNCs that resulted from the metallophilic interaction between Hg²⁺ and Au⁺ ions. Fluorescence spectra of BSA@AuNCs (a) in the absence and (b) presence of Hg²⁺ ions (1.0 mM) and (c) in the presence of Hg²⁺ ions (1.0 mM) and NaBH4 (1.0 M). (Reproduced with permission from Reference [33]. © 2015 Royal Society of Chemistry.)

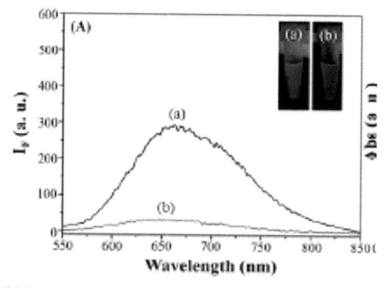

FIGURE 6.24 Fluorescence of MW_BSA@AuNCs. (Reproduced with permission from Reference [34]. © 2016 Royal Society of Chemistry.)

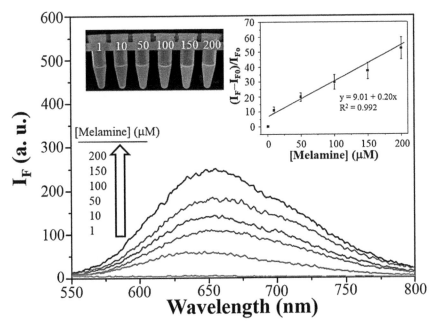

FIGURE 6.25 Fluorescence spectra for the responses of MW_BSA@AuNCs–Hg²⁺ for different concentrations of melamine. Inset: values of (IF _ IF0)/IF0 and photographs of the MW_BSA@AuNC–Hg²⁺ probes with different concentrations of melamine (lex: 365 nm), where IF and IF0 represent the fluorescence at 660 nm of MW_BSA@AuNCs–Hg²⁺ probe in the presence and absence of melamine, respectively. The error bars represent the standard deviations from three repeated experiments. (Reproduced with permission from Reference [34]. © 2016 Royal Society of Chemistry.)

BSA is efficient in forming biodegradable protein films on aggregation in unfolded state. Zein protein films of average thickness 0.05 mm can be synthesized out of BSA conjugated AuNPs using glycerol as plasticizer as shown in Scheme 6.6.[13] Apart from being biodegradable, these films have better tensile strength, flexibility. The formation of BSA@AuNPs depends on concentration, pH, temperature, and physical state of BSA, and it is recorded that Au/BSA mole ratio of 167 is most suitable for the formation of soft protein film in colloidal state as shown in Figure 6.26a and b. The isotropic nature of protein films is the measure of its brightness and color coordinates. A lower measure of brightness indicates the uniform distribution of BSA@AuNPs. A mole ratio of 167 produces remarkably strong and flexible films than in absence of BSA@AuNPs as indicated in Figure 6.26d and e.

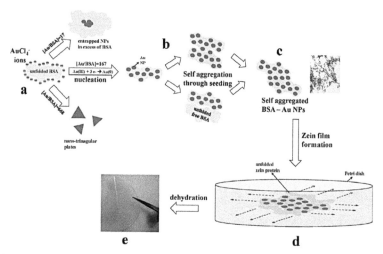

SCHEME 6.6 Representation of the synthesis of BSA conjugated NPs under different experimental conditions and their use in the zein film formation. (Reproduced with permission from Reference [13]. © 2011 American Chemical Society.)

Ag due to its antibacterial properties is of much interest in the medical field. Negatively charged coating on BSA, a protein in serum can be used to release sufficient amount of Ag in the solution. Ag in its 0 oxidation state is ineffective antibacterial agent, while its ion is remarkably reactive. The release of Ag from its metallic form is thus of huge importance in biological systems. Positively charged lysozyme suppresses the release of Ag, while the negatively charged BSA enhances the released Ag ions at neutral pH conditions. The absorption of BSA on AgNPs can be studied by using QCM-D (quartz crystal microbalance) in combination with ellipsometry.[44]

QCM-D studies show the rapid shifting down of frequency change (Δf) and rapid shifting up of (ΔD) dissipation energy due immediate absorption of BSA on Ag surface. Afterwards there occurs linear increase of Δf without any increase in ΔD which is attributed to the release of Ag form underlying surface. More Ag is released from the surface due to the presence of BSA in the solution. The adsorption of BSA on Ag surface occurs in the presence of $NaNO_3$ which screens the charges of two surfaces with efficiency. This reduces the repulsion between BSA and Ag. The adsorption may be also due to binding of sulfur of disulfide bridges of BSA to silver. BSA adsorbs as hexagonal closed packed (HCP) monolayers with side on orientation of Ag surface. Silver ions are released from Ag-coated crystals in the presence of $NaNO_3$ both with and without BSA. In case of nonprotein solution, Ag is predominantly released during initial 2 h, while it proceeds up to 24 h in case of BSA containing solutions. But release of Ag is more in case of BSA than a nonprotein solution. As compared to BSA, the presence of poly glutamic acid sodium salt (PGA) does not influence the release of Ag into the solution due to overall negatively charged surface which is attributed to the presence of glutamic acid. The adsorption of negatively charged BSA on silver results in the increment of counter ions such as H^+ and Na^+ at the interface of two layers. This lowers the pH and enhances the release of Ag ions. Thus, the amount of Ag released increases with significant increase in concentration of BSA. This increment can be recorded by means of graphite furnace atomic absorption spectroscopy (GF-AAS0.

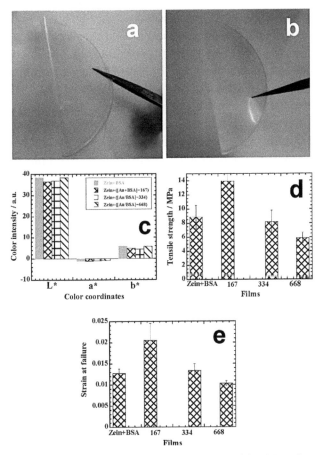

FIGURE 6.26   Photos of biodegradable protein films made with BSA conjugated NPs with Au/BSA mole ratios of (a) 167 and (b) 668. (c) Histogram of color coordinates for the films made with different mole ratios. (d) and (e) Histograms of tensile strength and strain at failure versus different mole ratios, respectively. (Reproduced with permission from Reference [13]. © 2011 American Chemical Society.)

Another biopolymer, that is, soy protein or its isolate with 90% protein content is most widely produced plant protein in the world. It has high content of hydrophobic amino acids with significant amount of polar and charged residue.[45,46] It is a complete protein with all essential amino acids and is a cheap source of dietary protein therefore considerable for all vegetarians. It has been among those proteins which are used to synthesize drug loaded NPs.[47] In conjugation with folic acid, soy protein NPs (SP-NPs) are efficiently taken up by tumor cells for enhanced delivery of antitumor drugs.[48] SP opens up the accessibility of folic acid in targeted cells. SP synthesized NPs help to improve bioavailability of curcumin; a natural polyphenolic compound is used as antioxidant, anti-inflammatory, anticancer, antiallergic and has other healing properties but is poorly soluble in water. SP-NP not only improves solubility of curcumin but also makes it resistant to the enzymes present in the gastrointestinal track and enhances its absorbance.[49,50] Renewable and biodegradable nanocomposites have been designed by crosslinking jute fabric with soyflour.[51] These nanocomposites are inexpensive with improved physical, thermal, mechanical properties.[52] Considering the serious environmental pollution caused by the heat and harmful gases evolved on burning, soy protein based protein films have replace nonbiodegradable petroleum derived plastics for food pacakaging.[53]

NPs with zeta potential of -36 mV can be successfully synthesized from soy protein. These NPs can be encapsulated by drugs to be delivered into targeted area. Curcumin as a model drug can be encapsulated with efficiency of 97.22% on soy protein NPs for its targeted delivery.[47] The SPI NPs synthesized by modified desolvation process can be loaded with curcumin, and its releasing rates can be compared at different curcumin/protein ratio.[53,54] Therefore, the release of curcumin in Phosphate buffer saline/Tween systems follow a biphasic trend. The biphasic trend observed for 1/100, 2/100, 5/100 c/p ratio determined the release of 50% encapsulated curcumin in first 1.5 h with sustained release between 2 and 8 h which shows increasing trend with time as shown in Figure 6.27. The release of curcumin is attributed to burst effect due to swelling and rupture of protein matrix. Similar NPs can be synthesized from water soluble and water insoluble proteins such as BSA and zein.[55,56]

SP finds good application to synthesize biobased polymer nanocomposite in renewable resource based industries.[51] Modified soy protein along with bamboo, palm tree fibers can also be used to fabricate biodegradable nanocomposites.[57,58] The thermomechanical properties can be improved by using gluteraldehyde along with cellulose whiskers and nanoclay as nanofillers, as recorded by thermogravimitric analysis[59-65] TEM micrographs denote the jute-based cross-linked SF composite matrix as shown in Figure 6.28. Increased limiting oxygen index value is exhibited by increasing interfacial interaction between nanofillers and matrix. Thus, cellulose whiskers in addition with nanoclay significantly improve the characteristic properties of bionanocomposites.

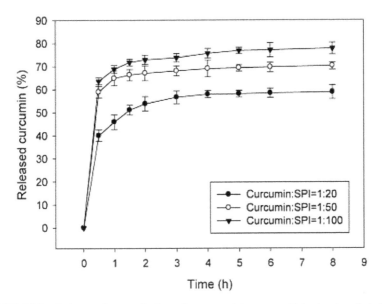

FIGURE 6.27 Release of curcumin from the soy protein nanoparticles in phosphate buffer saline with Tween 20. (Reproduced with permission from Reference [47].)

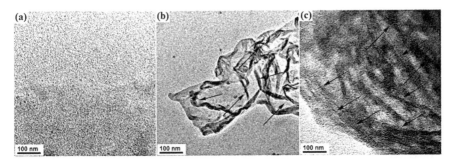

FIGURE 6.28 TEM micrographs of (a) SF/J/G50, (b) SF/J/G50/C5, and (c) SF/J/G50/C5/M5. (Reproduced with permission from Reference [51]. © 2013 American Chemical Society.)

Curcumin is a bioactive compound extracted from curcuma longa. It indulges in various biological activities such as anti-inflammation, antitumor, antiviral, antioxidant, anti-HIV, and many other clinical applications.[66] Its water insoluble property limits its oral intake in the form of polymeric liposome, micelles, dendrimers, and nanoparticles.[67] The entry of many of its form is resisted in the blood stream; thus, it is necessary to design oral nanovectors to determine its practical application both in vitro and in vivo. Oral bioavailability of curcumin can be improved by fabricating it with soybean derived Bowman–Birk inhibitor (BBI) protein. BBI is a stable, nonallergenic protein which is resistant to pH of digestive enzymes in gastric track and gets readily absorbed. It is also known to prevent carcinogenesis. In vitro bioaccessibility of curcumin fabricated with BBI NPs and sodium caseinate (SC) NPs can be compared, in which Cur-BBI-NPs are digested without any significant loss of curcumin with remaining amount of 51.23% curcumin in case of Cur-BBI-NPs in aqueous phase as shown in Figure 6.29. The in vivo bioavailability of curcumin determines rapid metabolism of curcumin in case of Cur-SC-NPs due to free curcumin and sustained metabolism in case of Cur-BBI-NPs due to protection of curcumin by BBI. The relevant pharmacokinetic parameters including C_{max}, T_{max}, and area under the plasma concentration–time curve from 0 h to ∞ ($AUC_{0-\infty}$) confirm the sustained plasma concentration of curcumin with delayed C_{max} at 2 h in case of Cur-BBI-NPs and C_{max} at 1 h in case of Cur-SC-NPs. The $AUC_{0-\infty}$ for curcumin was significantly higher in case of Cur-BBI-NPs with bioaccessibility of 3.11 as compared to Cur-SC-NPs as shown in Figure 6.30.[49]

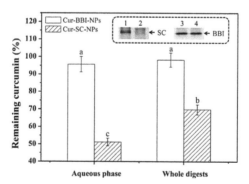

FIGURE 6.29 Percentage of curcumin remaining in the aqueous phase or the whole digests of Cur-BBI-NPs and Cur-SC-NPs after the whole simulated digestion of 180 min under darkness. Means of columns marked with different letters indicate a significant difference between each other ($p < 0.05$). (Inset) Lanes 1 and 2 are SDS–PAGE patterns of SC before and after in vitro digestion, respectively; lanes 3 and 4 are SDS–PAGE patterns of BBI before and after in vitro digestion, respectively. (Reproduced with permission from Reference [49]. © 2017 American Chemical Society.)

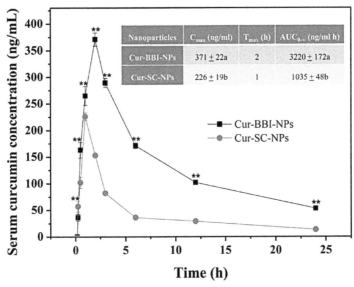

FIGURE 6.30 (See color insert.) Bioavailability of Cur-BBI-NPs and Cur-SC-NPs. All values are presented as the mean ± SD, $n = 6$.** Indicates an extremely significant difference between two groups ($p < 0.01$). (Inset) Pharmacokinetics parameters of two curcumin nanoparticles. AUC, area under the plasma concentration–time curve from 0 h to ∞; C_{max}, peak concentration; T_{max}, time to reach peak concentration. Means of symbols marked with different letters indicate an extremely significant difference between two groups ($p < 0.01$). (Reproduced with permission from Reference [49]. © 2017 American Chemical Society.)

Folic acid, a diet derived micronutrient, finds good application as target-specific ligand for tumor cells.[68] Many of biopolymers such as chitosan, human serum albumin, BSA, and polylysine have been conjugated with FA to be used as anticancer drugs.[69–72] Fluent expression of GP38, glycosyl[73]-phosphatidylinositol anchored glycoprotein, a FA receptor protein shows the applicability of FA in treatment of a variety of carcinomas including ovarian, prostate, breast, colon, and lung cancer. SP when used for synthesis of NPs conjugated with FA forms the NPs of average size 150–170 nm with zeta potential ranging from −36 to −42 mv. These SP-NPs are successfully loaded with FA with encapsulation and loading efficiencies of 72.7% and 5.4%, respectively. SP-FA-NPs can be loaded with curcumin, an antitumor bioactive compound to study its cellular uptake by tumor cells.[48] It is recorded that SP-FA-NPs release curcumin faster (29%) than SP-NPs (24%) in PBS–Tween 20 buffer during the first hour and at the end of eighth hour (58%), and 39% of curcumin was released[74,75] as shown in Figure 6.31. This is attributed to higher encapsulating and loading efficiency of SP-FA-NPs.

The cellular uptake comparison of SP-NPs and SP-FA-NPs by Caco-2 cells shows the uptake up to 93% by SP-FA-NPs than that for control SP-FA-NPs. Therefore, are attractive candidates for encapsulated and target-specific delivery of drugs.

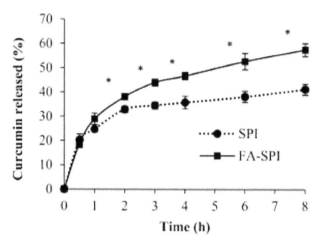

FIGURE 6.31   Release of curcumin from SPI and FA-SPI nanoparticles in Tween 20/PBS buffer. t Test was performed in each pair of values obtained SPI and FA-SPI nanoparticles corresponding to a same time. An asterisk (*) indicates significant ($p < 0.05$) difference. (Reproduced with permission from Reference [48]. © 2013 American Chemical Society.)

Bioaccessibility of water insoluble curcumin can be improved by complexing it with proteins including BSA, α_{S1}-casein, β-lactoglobulin, SPI, and zein protein.[76–81] The stability depends on hydrogen bonds and hydrophobic interactions.[82–90] When curcumin is heated with β-lactoglobulin NPs, its water solubility increases by 20,000 fold. SPI on nanocomplexation with curcumin improves its stability (up to 90%) and bioaccessibility and also increases the nutritional value of soy protein. On increasing the temperature from 75°C to 95°C, protein denaturation of β-conglycinin and glycinin, the components of SPI increases and the surface hydrophobicity (H_0) increases from 2248 to 5173. The bioaccessibility of curcumin depends on the state of curcumin, whether free or bound, and also on presence of protease enzyme. In case of free and bound curcumin, 88% and 30% of curcumin degrades, respectively, after 3 h as shown in Figure 6.32B.[50] In presence of protease, the amount of curcumin transferred to aqueous phase of the digests is more than in its absence as shown in Figure 6.32A. The in vitro digestibility of curcumin in NPs can be explained by SDS–PAGE and trichloroacetic acid

(TCA) soluble nitrogen method. The effect of complexation with curcumin on digestibility of protein in simulated gastric fluid (SGF) and simulated intestinal fluid (SIF) in heated and heated environment is as shown in Figures 6.7 and 6.8. In TCA analysis, TCA soluble nitrogen is released during SGF digestion which determines the improved protein digestion on complexation with curcumin as shown in Figure 6.33.

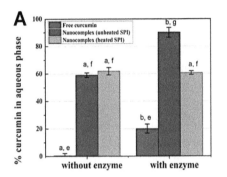
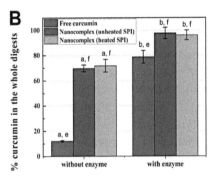

FIGURE 6.32 (See color insert.) (A) Percentage of curcumin remaining in the aqueous phase for free curcumin and curcumin nanocomplexes with unheated and heated (at 95°C) SPI after the whole simulated digestion of 180 min in the absence or presence of proteases. (B) Percentage of curcumin in the whole digests of free curcumin and curcumin nanocomplexes with unheated and heated SPI after the digestion of 180 min in the absence or presence of proteases. (Reproduced with permission from Reference [50]. © 2015 American Chemical Society.)

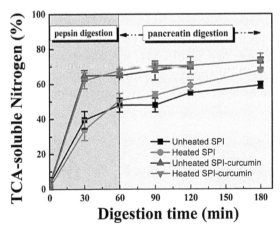

FIGURE 6.33 (See color insert.) Release kinetics of TCA-soluble nitrogen during the sequential SGF and SIF digestion of unheated and heated SPI and its nanocomplexes with curcumin. Each data point is the mean and standard deviation of triplicate measurements on separate samples. (Reproduced with permission from Reference [50]. © 2015 American Chemical Society.)

Another bio-based polymer, that is, Cyt *C* is a small water soluble mitochondrial protein (MW = 12 kDa) that is responsible for cellular respiration. It facilitates electron transport between complexes III (coenzyme Q–Cyt *C* reductase) and IV (Cyt *C* oxidase).[91] It also acts as antioxidative enzyme by limiting the production of O_2^- and H_2O_2 in mitochondria. It also plays a key role in cell apoptosis.[92–95] Encapsulated Cyt *C* by polymeric NPs when delivered to cancer cells cause programmed cell death.[96,97] Involvement of FA as folate receptor targeting amphiphilic copolymer enhances the targeting action of Cyt *C* NPs on tumor cells as compared to healthy cells.[21] Cyt *C* encapsulated by NPs also design aerogels for development of biosensors and other bioanalytical devices.[98]

Cyt *C*, an apoptic protein, is responsible for apoptosis of various cancer cells when delivered into cytoplasm. However, due to poor stability, use of Cyt *C* in therapeutics is a major challenge. In some cases, biodegradable polymer such as poly(lactic-*co*-glycolic acid) (PLGA) has been used to coat protein NPs to improve its delivery to targeted site.[99,100] FDA has approved only two nanoformulations, namely, Abraxane and Rexin, for passive and passive targeting. Folate receptor targeting amphiphilic copolymer, folate-poly(ethylene glycol)-PLGA, when coated on Cyt *C* NPs forms a smart release nanocarrier which is stable in extracellular environment and targets only intracellular reducing environment of cell.[21] The nanodevice consists of micellar like system in which hydrophobic part of polymer (PLGA) makes it water soluble thus making it a suitable candidate for intravenous administration. FA receptor is a well-known tumor marker and also mediates the synthesis of essential nucleotide bases.[101–103] The micellar system increases the surface charges thus increasing the stability of Cyt *C* NPs in extracellular nonreducing medium. In cytosol, Cyt *C* is released from micellar like NP and interacts with Apaf-1 (apoptic protease activating factor) thus causing apoptosis as shown in Scheme 6.7. The in vivo application can be determined by

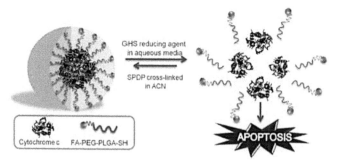

SCHEME 6.7   Scheme of Cyt *C* release in the intracellular reducing environment to induce apoptosis. (Reproduced with permission from Reference [21]. © 2016 American Chemical Society.)

incubating HeLa cells with FS-PEG-PLGA-coated Cyt *C* NPs which drastically reduces the cell viability up to 20% after 6 h when concentration of Cyt
C is 50 μg mL^{-1}. Folate-decorated NPs can also be used to treat malignant
glioma, primary tumors within central nervous system (CNS). The confocal
fluorescence in the cytoplasm of tumor cells as compared to a healthy cell is
shown in Figure 6.34.

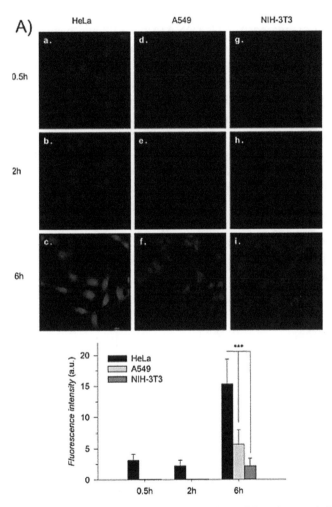

FIGURE 6.34   In vitro and ex vivo internalization of the folate-decorated Cyt *C*-based
NPs observed by confocal microscopy after different incubation times. (A) A folate-positive
carcinoma HeLa cell line (a–c) showed better uptake behavior in comparison with folate-
deficient carcinoma A549 (d–f) and normal NIH-3T3 (g–i) cell lines. (Reproduced with
permission from Reference [21]. © 2016 American Chemical Society.)

Cyt C can be detected by fluorometry, spectrophotometry, mass spectrometry, chemiluminescence, electrophoresis, high performance liquid chromatography (HPLC). But due to complexity and low selectivity of these methods, they are not preferred for Cyt C analysis. Recently, aptamer-based biosensors (aptasensors) have received attention for selective and sensitive detection of Cyt C. They possess advantage of easy synthesis, stability, reusability, and general availability for most of the proteins. Such two fluorescent active systems, Hb-stabilized AuNCs (Hb/AuNCs) and aptamer-stabilized AgNCs (DNA/AgNCs), have been synthesized for responsive determination of Cyt C.[96] The analysis is based on quenching observed on transfer of fluorescence resonance energy (FRET) from HB/AuNCs to Cyt C and photoinduced electron transfer from DNA/AgNCs to aptamer Cyt C complex. FRET process enhances with increasing concentration of Cyt C as shown in Figure 6.35A. Stern volmer plot between F_0/F versus Cyt C concentration for quantitative analysis of Hb/AuNCs quenching shows good linearity for 0–10 μM of Cyt C with correlation coefficient of 0.998 (R^2), K_{sv} value of 9.6×10^4 M^{-1} [$F_0/F = 1 + K_{sv}$ (Cyt C)], and DL of 14.3 μM as shown in Figure 6.35B. Likewise, DNA/AgNCs quenching also show good linear growth ranging from 0 to 1.0 μM. The in vivo detection of Cyt C in human embryonic kidney cells can be determined from quenched fluorescence of DNA/AgNCs on increasing the cell lysate as shown in Figure 6.36. Therefore, DNA/AgNCs find good potential for future biological research.

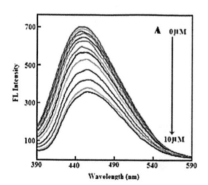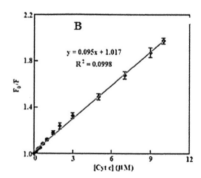

FIGURE 6.35 **(See color insert.)** (A) Fluorescence emission spectra upon addition of different concentrations of Cyt C (0, 0.16, 0.32, 0.48, 0.70, 1.0, 1.5, 2.0, 3.0, 5.0, 7.0, 9.0, and 10 μM). The spectra were recorded at time intervals of 5 min. (B) Stern–Volmer plot of fluorescence quenching of the Hb/AuNCs by Cyt C. All experiments were carried out in 10 mM PBS of pH 7.4 with Hb/AuNCs concentration of 2.2 μM. Error bars represented as $\pm 3\sigma$. (Reproduced with permission from Reference [96]. © 2016 American Chemical Society.)

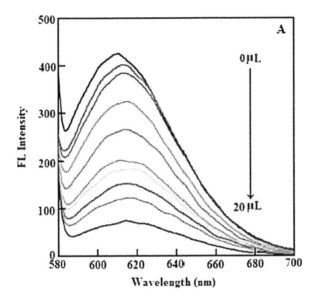

FIGURE 6.36 **(See color insert.)** Fluorescence emission spectra of the Cyt C-2/AgNCs in the presence of variable concentrations of cancer cell extracts treated with (a) hypotonic buffer and (b) with CCLR buffer (both in the presence of 0, 3.0, 5.0, 7.0, 9.0, 11.0, 13.0, 15.0, 17.0, and 20 μL aliquots taken from initial 10,000 cells mL^{-1} solution). The spectra were recorded at time intervals of 20 min. (Reproduced with permission from Reference [96]. © 2016 American Chemical Society.)

Cyt C can be encapsulated in silica gels to construct bioaerogels. Aerogels are high surface area, high porosity devices with sol–gel-derived materials which contain a pore solid 3D nanoscale mesh network, important for surface reactions.[104,105] Bioaerogels have been synthesized using enzymes, bacteria, oliginucleotide, antibodies, and biomembranes.[106–110] These bioaerogels can be designed with and without involvement of metal such as Au and Ag. In the presence of metal, a multilayered superstructures of proteins Au~Cyt C are formed which are further encapsulated in aerogels.[111] These exhibit a rapid gas phase recognition property for nitric oxide (NO), a toxic pollutant affecting nervous, cardiovascular, and immune systems.[112] Au~Cyt C aerogels are highly responsive (150 s) as compared to their counterpart xerogels (200 s). The structural stability of Cyt C is retained when it is encapsulated within aerogels in the absence of metal and displays its soret band at 409 nm. Cyt C–SiO$_2$ aerogels have attractive feature of less reversible ligand (NO) binding capability as compared to Au~Cyt C–SiO$_2$ aerogels thus designing a better biosensor.

Cyt *C* can also be encapsulated by polymeric NPs of hyperbranched poly-hydroxyl (HBPH) polymer for targeted drug delivery of Cyt *C* to cancerous cells.[97] The polymeric NPs have amphiphilic nanocavities to bind success-fully to hydrophobic Cyt *C* protein. The positively charged lysine residue on proteins surface interacts electrostatically with HBPH nanocavities. Cyt *C*-HBPH NPs, when incubated with MCF7 breast carcinomas cells, cause oxidation of 3,3',5,5'-tetramethylbenzidine (TMB) by Cyt *C*'s peroxidase activity, these results in shifting of absorbance at 625 nm as compared to untreated MCF7 cells. To explore the potential application in targeted drug delivery, DiI encapsulating folate conjugated HBPH NPs were incubated with lung cancerous cells (A549). The enhanced fluorescence in cytoplasm of A549 cells determines its applicability in therapeutics. Also, A549 cells when treated with Fol-Cyt *C*/DiI-HBPH NPs lead to cell apoptosis. This is attributed to observed heterogeneous chromatin condensation and nuclear fragmentation.

6.4 CONCLUSION

Overdependence on petrochemical products has created an environ-mental pressure which can only be faded by opting green approaches. Biopolymers are the topic of interest due to their renewable, nontoxicity, and biocompatible nature. They also have additional advantages of water solubility and bioaccessibility for biological applications. Therefore, they have been used in variety of applications, like therapeutic aids, medicines, coatings, biosensing, clinical diagnostics, radiosensitizing, renewable resource-based industries, food products, and packing mate-rials. Biopolymers have significant roles to play in the advancement of green nanotechnology. NPs synthesized using biopolymers have shown least cytotoxicity but enhanced targeted actions. Biopolymer@ metal NPs have been successfully reported to treat a variety of cancers such as cevix cancer, prostate cancer, brain gliomas, neck cancer, lung cancer, colon cancer, breast cancer, and other primary tumors. In addi-tion, folate receptor mediated targeted delivery is the most efficient form of drug delivery. Bioavailability of water insoluble curcumin has also been enhanced. An attempt was made to quote applications of BSA, soy protein, Cyt *C* through this chapter.

KEYWORDS

- proteins
- biopolymers
- applications
- BSA
- soy protein
- Cyt *C*

REFERENCES

1. Akhtar, M. S.; Panwar, J.; Yun, Y. S. Biogenic Synthesis of Metallic Nanoparticles by Plant Extracts. *ACS Sustainable Chem. Eng.* **2013**, *1*, 591–602.
2. Jain, P. K.; Huang, X.; El-Sayed, I. H.; El-Sayed, M. A. Review of Some Interesting Surface Plasmon Resonance-Enhanced Properties of Noble Metal Nanoparticles and Their Applications to Biosystems. *Plasmonics* **2007**, *2*, 107–118.
3. Wong, T. S.; Schwaneberg, U. Protein Engineering in Bioelectrocatalysis. *Curr. Opin. Biotechnol.* **2003**, *14*, 590–596.
4. Ramanavicius, A.; Kausaite, A.; Ramanaviciene, A. Biofuel Cell Based on Direct Bioelectrocatalysis. *Biosens. Bioelectron.* **2005**, *20*, 1962–1967.
5. Thakkar, K. N.; Mhatre, S. S.; Parikh, R. Y. Biological Synthesis of Metallic Nanoparticles. *Nanomedicine* **2010**, *6*, 257–262.
6. Amiri, H.; Saeidi, K.; Borhani, P.; Manafirad, A.; Ghavami, M.; Zerbi, V. Alzheimer's Disease: Pathophysiology and Applications of Magnetic Nanoparticles as MRI Theranostic Agents. *ACS Chem. Neurosci.* **2013**, *4*, 1417–1429.
7. Rodrigues, S. M.; Demokritou, P.; Dokoozlian, N.; Hendren, C. O.; Karn, B.; Mauter, M. S.; Sadik, O. A.; Safarpour, M.; Unrine, J. M.; Viers, J.; Welle, P.; White, J. C.; Wiesner, M. R.; Lowry, G. V. Nanotechnology for Sustainable Food Production: Promising Opportunities and Scientific Challenges. *Environ. Sci. Nano* **2017**, *4*, 767.
8. Hengartner, M. O. The Biochemistry of Apoptosis. *Nature* **2000**, *407* (6805), 770–776.
9. Cao, X.; Bennett, R. L.; May, W. S. c-Myc and Caspase-2 are Involved in Activating Bax During Cytotoxic Drug-Induced Apoptosis. *J. Biol. Chem.* **2008**, *283* (21), 14490–14496.
10. Jurgensmeier, J. M.; Xie, Z.; Deveraux, Q.; Ellerby, L.; Bredesen, D.; Reed, J. C. Bax Directly Induces Release of Cytochrome *C* from Isolated Mitochondria. *Proc. Natl. Acad. Sci. U.S.A.* **1998**, *95* (9), 4997–5002.
11. Rother, M.; Nussbaumer, M. G.; Renggli, K.; Bruns, N. Protein Cages and Synthetic Polymers: A Fruitful Symbiosis for Drug Delivery Applications, Bionanotechnology and Materials Science. *Chem. Soc. Rev.* **2016**, *45* (22), 6213–6249.
12. Khullar, P.; Singh, V.; Mahal, A.; Dave, P. N.; Thakur, S.; Kaur, G.; Singh, J.; Kamboj, S. S.; Bakshi, M. S. Bovine Serum Albumin Bioconjugated Gold Nanoparticles:

Synthesis, Hemolysis, and Cytotoxicity Toward Cancer Cell Lines. *J. Phys. Chem. C* **2012**, *116* (15), 8834–8843.

13. Bakshi, M. S.; Kaur, H.; Khullar, P.; Banipal, T. S.; Kaur, G.; Singh, N. Protein Films of Bovine Serum Albumen Conjugated Gold Nanoparticles: A Synthetic Route from Bioconjugated Nanoparticles to Biodegradable Protein Films. *J. Phys. Chem. C* **2011**, *115* (7), 2982–2992.

14. Yuan, H.; Khoury, C. G.; Hwang, H.; Wilson, C. M.; Grant, G. A.; Dinh, V. Gold Nanostars: Surfactant-Free Synthesis, 3D Modeling, and Two-Photon Photoluminescence Imaging. *Nanotechnology* **2012**, *23*, 75102–75111.

15. Du, C.; Deng, L.; Shan, S.; Wan, J.; Cao, J.; Tian, S. A pH-Sensitive Doxorubicin Prodrug Based on Folate-Conjugated BSA for Tumor-Targeted Drug Delivery. *Biomaterials* **2013**, *34*, 3087–3097.

16. Teng, Z.; Luo, Y.; Wang, T.; Zhang, B.; Wang, Q. Development and Application of Nanoparticles Synthesized with Folic Acid Conjugated Soy Protein. *J. Agric. Food Chem.* **2013**, *61*, 2556–2564.

17. Zhang, B.; Li, Q.; Yin, P.; Rui, Y.; Qiu, Y.; Wang, Y.; Shi, D. Ultrasound-Triggered BSA/SPION Hybrid Nanoclusters for Liver-Specific Magnetic Resonance Imaging. *ACS Appl. Mater. Interfaces* **2012**, *4*, 6479–6486.

18. Teng, Z.; Luo, Y.; Wang, T.; Zhang, B.; Wang, Q. Development and Application of Nanoparticles Synthesized with Folic Acid Conjugated Soy Protein. *J. Agric. Food Chem.* **2013**, *61*, 2556–2564.

19. Patel, A.; Hu, Y. C.; Tiwari, J. K.; Velikov, K. P. Synthesis and Characterisation of Zein-Curcumin Colloidal Particles. *Soft Matter* **2010**, *6*, 6192–6199.

20. Singh, A. V.; Bandgar, B. M.; Kasture, M.; Prasad, B. L. V.; Sastry, M. Synthesis of Gold, Silver and Their Alloy Nanoparticles Using Bovine Serum Albumin as Foaming and Stabilizing Agent. *J. Mater. Chem.* **2005**, *15*, 5115–5121.

21. Cruz, M. M.; Montañez, A. C.; Figueroa, C. M.; Robles, T. G.; Davila, J.; Inyushin, M.; Rosas, S. A. L.; Molina, A. M.; Perez, L. M.; Kucheryavykh, L. Y.; Tinoco, A. D.; Griebenow, K. Combining Stimulus-Triggered Release and Active Targeting Strategies Improves Cytotoxicity of Cytochrome *C* Nanoparticles in Tumor Cells. *Mol. Pharmaceutics* **2016**, *13*, 2844–2854.

22. Fang, J.; Nakamura, H.; Maeda, H. The EPR Effect. Unique Features of Tumor Blood Vessels for Drug Delivery, Factors Involved, and Limitations and Augmentation of the Effect. *Adv. Drug Delivery Rev.* **2011**, *63*, 136–151.

23. Sen, T.; Haldar, K. K.; Patra, A. Au Nanoparticle-Based Surface Energy Transfer Probe for Conformational Changes of BSA Protein. *J. Phys. Chem. C* **2008**, *112*, 17945–17951.

24. Feng, H.; Wang, H.; Zhang, Y.; Yan, B. N.; Shen, G. L.; Yu, R. Q. A Direct Electrochemical Biosensing Platform Constructing by Incorporating Carbon Nanotubes and AuNPs onto Redox Poly(thioninr) Film. *Anal. Sci.* **2007**, *23*, 235.

25. Sasidharan, S.; Bahadur, D.; Srivastava, R. Albumin Stabilized Gold Nanostars: A Biocompatible Nanoplatform for SERS, CT Imaging and Photothermal Therapy of Cancer. *RSC Adv.* **2016**, *6*, 84025–84034.

26. Chen, N.; Yang, W.; Bao, Y.; Xu, H.; Qin, S.; Tu, Y. BSA Capped Au Nanoparticle as an Efficient Sensitizer for Glioblastoma Tumor Radiation Therapy. *RSC Adv.* **2015**, *5*, 40514–40520.

27. Su, Z.; Xing, L.; Chen, Y.; Xu, Y.; Yang, F.; Zhang, C.; Ping, Q.; Xiao, Y. Lactoferrin-Modified Poly(ethylene glycol)-Grafted BSA Nanoparticles as a Dual-Targeting Carrier for Treating Brain Gliomas. *Mol. Pharmaceutics* **2014**, *11*, 1823–1834.

28. Huang, Y.; Luo, Y.; Zheng, W.; Chen, T. Rational Design of Cancer-Targeted BSA Protein Nanoparticles as Radiosensitizer to Overcome Cancer Radioresistance. *ACS Appl. Mater. Interfaces* **2014**, *6*, 19217–19228.

29. Kumar, S.; Meena, R.; Rajamani, P. Fabrication of BSA–Green Tea Polyphenols–Chitosan Nanoparticles and Their Role in Radioprotection: A Molecular and Biochemical Approach. *J. Agric. Food Chem.* **2016**, *64*, 6024–6034.

30. Murawala, P.; Tirmale, A.; Shiras, A.; Prasad, B. In Situ Synthesized BSA Capped Gold Nanoparticles: Effective Carrier of Anticancer Drug Methotrexate to MCF-7 Breast Cancer Cells. *Mater. Sci. Eng., C* 2014, *34*, 158–167.

31. Alkilany, A. M.; Nagaria, P. K.; Hexel, C. R.; Shaw, T. J.; Murphy, C. J.; Wyatt, M. D. Cellular Uptake and Cytotoxicity of Gold Nanorods: Molecular Origin of Cytotoxicity and Surface Effects. *Small* 2009, *5*, 701–708.

32. Khandelia, R.; Jaiswal, A.; Ghosh, S. S.; Chattopadhyay, A. Gold Nanoparticle-Protein Agglomerates as Versatile Nanocarriers for Drug Delivery. *Small* 2013, *9*, 3494–3505.

33. Hsu, N. Y.; Lin, Y. W. Microwave-Assisted Synthesis of Bovine Serum Albumin–Gold Nanoclusters and Their Fluorescence-Quenched Sensing of Hg^{2+} Ions. *New J. Chem.* **2016**, *40*, 1155–1161.

34. Lee, C. Y.; Hsu, N. Y.; Wu, M. Y.; Lin, Y. W. Microwave-Assisted Synthesis of BSA-Stabilised Gold Nanoclusters for the Sensitive and Selective Detection of Lead(II) and Melamine in Aqueous Solution. *RSC Adv.* **2016**, *6*, 79020–79027.

35. Hu, C.; Yang, D. P.; Zhu, F.; Jiang, F.; Shen, S.; Zhang, J. Enzyme-Labeled Pt@BSA Nanocomposite as a Facile Electrochemical Biosensing Interface for Sensitive Glucose Determination. *ACS Appl. Mater. Interfaces* **2014**, *6*, 4170–4178.

36. Xuan, C.; Xia, Q.; Xu, J.; Wang, Q. A Biosensing Interface Based on Au@BSA Nanocomposite for Chiral Recognition of Propranolol. *Anal. Methods* **2016**, *8*, 3564–3569.

37. Berbeco, R. I.; Korideck, H.; Ngwa, W. R. DNA Damage Enhancement from Gold Nanoparticles for Clinical MV Photon Beams. *Radiat. Res.* **2012**, *178*, 604–608.

38. Butterworth, K. T.; Coulter, J. A.; Jain, S. Evaluation of Cytotoxicity and Radiation Enhancement Using 1.9 nm Gold Particles: Potential Application for Cancer Therapy. *Nanotechnology* 2010, *21*, 295101–295118.

39. Butterworth, K. T.; Coulter, J. A.; Jain, S. Cell-Specific Radiosensitization by Gold Nanoparticles at Megavoltage Radiation Energies. *Biol. Phys.* 2011, *79*, 531 539.

40. Liu, C. L.; Wang, C. H.; Chen, S. T.; Chen, H. H.; Leng, W. Development of Bacteriochlorophyll *a*-Based Near-Infrared Photosensitizers Conjugated to Gold Nanoparticles for Photodynamic Therapy of Cancer. *Phys. Med. Biol.* 2010, *55*, 931–945.

41. Chattopadhyay, N.; Cai, Z.; Kwon, Y. L.; Lechtman, E. Molecularly Targeted Gold Nanoparticles Enhance the Radiation Response of Breast Cancer Cells and Tumor Xenografts to X-Radiation. *J. Breast Cancer Res. Treat.* 2013, *137*, 81–91.

42. Li, J.; Cai, R.; Kawazoea, N.; Chen, G. Facile Preparation of Albumin-Stabilized Gold Nanostars for the Targeted Photothermal Ablation of Cancer Cells. *J. Mater. Chem. B* **2015**, *3*, 5806–5814.

43. Zhang, B.; Li, Q.; Yin, Q.; Rui, Y. Ultrasound-Triggered BSA/SPION Hybrid Nanoclusters for Liver-Specific Magnetic Resonance Imaging. *ACS Appl. Mater. Interfaces* **2012**, *4*, 6479–6486.

44. Wang, X.; Herting, G.; Wallindera, I.; Blomberg, E. Adsorption of Bovine Serum Albumin on Silver Surfaces Enhances the Release of Silver at pH Neutral Conditions. *Phys. Chem. Chem. Phys.* **2015**, *17*, 18524–18534.

45. Netto, F. M.; Galeazzi, M. A. M. Production and Characterization of Enzymatic Hydrolysate from Soy Protein Isolate. *Food Sci. Technol.* **1998**, *31*, 624–631.
46. Riche, M.; Williams, T. N. Apparent Digestible Protein, Energy and Amino Acid Availability of Three Plant Proteins in *Florida pompano, Trachinotus carolinus* L. in Seawater and Low-Salinity Water. *Aquacult. Nutr.* **2010**, *16*, 223–230.
47. Teng, Z.; Luo, Y.; Wang, Q. Nanoparticles Synthesized from Soy Protein: Preparation, Characterization, and Application for Nutraceutical Encapsulation. *J. Agric. Food Chem.* **2012**, *60*, 2712–2720.
48. Teng, Z.; Luo, Y.; Wang, Q. Development and Application of Nanoparticles Synthesized with Folic Acid Conjugated Soy Protein. *J. Agric. Food Chem.* **2013**, *61*, 2556–2564.
49. Liu, C.; Cheng, F.; Yang, Q. Fabrication of a Soybean Bowman–Birk Inhibitor (BBI) Nanodelivery Carrier to Improve Bioavailability of Curcumin. *J. Agric. Food Chem.* **2017**, *65*, 2426–2434.
50. Chen, F. P.; Li, B. S.; Tang, C. H. Nanocomplexation between Curcumin and Soy Protein Isolate: Influence on Curcumin Stability/Bioaccessibility and In Vitro Protein Digestibility. *J. Agric. Food Chem.* **2015**, *63*, 3559–3569.
51. Iman, M.; Bania, K. K.; Maji, T. K. Green Jute-Based Cross-Linked Soy Flour Nanocomposites Reinforced with Cellulose Whiskers and Nanoclay. *Eng. Chem. Res.* **2013**, *52*, 6969–6983.
52. (a) Plackett, D.; Andersen, T. L.; Pedersen, W. B.; Nielsen, L. Biodegradable Composites Based on L-Polylactide and Jute Fibres. *Compos. Sci. Technol.* **2003**, *63*, 1287. (b) Kafi, A. A.; Magniez, K.; Fox, B. L. A Surface–Property Relationship of Atmospheric Plasma Treated Jute Composites. *Compos. Sci. Technol.* **2011**, *71*, 1692. (c) Zaman, H. U.; Khan, M. A.; Khan, R. A. Comparative Experimental Measurements of Jute Fiber/Polypropylene and Coir Fiber/Polypropylene Composites as Ionizing Radiation. *Polym. Compos.* **2012**, *33*, 1077. (d) Roy, A.; Chakraborty, S.; Kundu, S. P.; Basak, R. K.; Majumder, S. B.; Adhikari, B. Improvement in Mechanical Properties of Jute Fibres through Mild Alkali Treatment as Demonstrated by Utilisation of the Weibull Distribution Model. *Bioresour. Technol.* **2012**, *107*, 222. (e) Ray, D.; Das, M.; Mitra, D. A Comparative Study of the Stress-Relaxation Behavior of Untreated and Alkali-Treated Jute Fibers. *J. Appl. Polym. Sci.* **2012**, *123*, 1348. (f) Zhuang, R. C.; Doan, T. T. L.; Liu, J. W.; Zhang, J.; Gao, S. L.; Mäder, E. Multi-Functional Multi-Walled Carbon Nanotube-Jute Fibres and Composites. *Carbon* **2011**, *49*, 2683.
53. Song, F.; Tang, D. L.; Wang, X. L.; Wang, Y. Z. Biodegradable Soy Protein Isolate-Based Materials: A Review. *Biomacromolecules* **2011**, *12*, 3369–3380.
54. Weber, C.; Coester, C.; Kreuter, J.; Langer, K. Desolvation Process and Surface Characterisation of Protein Nanoparticles. *Int. J. Pharm.* **2000**, *194* (1), 91–102.
55. Jithan, A.; Madhavi, K.; Prabhakar, K.; Madhavi, M. Preparation and Characterization of Albumin Nanoparticles Encapsulating Curcumin Intended for the Treatment of Breast Cancer. *Int. J. Pharm. Invest.* **2011**, *1* (2), 119–125.
56. Patel, A.; Hu, Y. C.; Tiwari, J. K.; Velikov, K. P. Synthesis and Characterisation of Zein–Curcumin Colloidal Particles. *Soft Matter* **2010**, *6* (24), 6192–6199.
57. (a) Islam, M. R.; Beg, M. D. H.; Gupta, A.; Mina, M. F. Optimal Performances of Ultrasound Treated Kenaf Fiber Reinforced Recycled Polypropylene Composites as Demonstrated by Response Surface Method. *J. Appl. Polym. Sci.* **2013**, *128*, 2847. (b) Liang, K.; Gao, Q.; Shi, S. Q. Kenaf Fiber/Soy Protein Based Biocomposites Modified with Poly(carboxylic acid) Resin. *J. Appl. Polym. Sci.* **2013**, *128*, 1213. (c) Phong, N. T.;

Gabr, M. H.; Okubo, K.; Chuong, B.; Fujii, T. Enhancement of Mechanical Properties of Carbon Fabric/Epoxy Composites Using Micro/Nano-Sized Bamboo Fibrils. *Mater. Des.* **2013**, *47*, 624. (d) Zainuddin, S. Y. Z.; Ahmad, I.; Kargarzadeh, H.; Abdullah, I.; Dufresne, A. Potential of Using Multiscale Kenaf Fibers as Reinforcing Filler in Cassava Starch–Kenaf Biocomposites. *Carbohydr. Polym.* **2013**, *92*, 2299. (e) Saadaoui, N.; Rouilly, A.; Fares, K.; Rigal, L. Characterization of Date Palm Lignocellulosic By-Products and Self-Bonded Composite Materials Obtained Thereof. *Mater. Des.* **2013**, *50*, 302. (f) Bamufleh, H. S.; Alhamed, Y. A.; Daous, M. A. Furfural from Midribs of Date-Palm Trees by Sulfuric Acid Hydrolysis. *Ind. Crops Prod.* **2013**, *42*, 421.

58. Huang, X.; Netravali, A. Biodegradable Green Composites Made Using Bamboo Micro/Nano-Fibrils and Chemically Modified Soy Protein Resin. *Compos. Sci. Technol.* **2009**, *69*, 1009.

59. Hussain, F.; Hojjati, M.; Okamoto, M.; Gorga, R. E. Review Article: Polymer-Matrix Nanocomposites, Processing, Manufacturing, and Application: An Overview. *J. Compos. Mater.* **2006**, *40*, 1511.

60. Samir, M. A. S. A.; Alloin, F.; Dufresne, A. Review of Recent Research into Cellulosic Whiskers, their Properties and their Application in Nanocomposite Field. *Biomacromolecules* **2005**, *6*, 612.

61. Tang, L.; Weder, C. Cellulose Whisker/Epoxy Resin Nanocomposites. *ACS Appl. Mater. Interfaces* **2010**, *2*, 1073.

62. Samir, M. A. S. A.; Alloin, F.; Sanchez, J. Y.; Dufresne, A. Cellulose Nanocrystals Reinforced Poly(oxyethylene). *Polymer* **2004**, *45*, 4149.

63. Anglès, M. N.; Dufresne, A. Plasticized Starch/Tunicin Whiskers Nanocomposites. 1. Structural Analysis. *Macromolecules* **2000**, *33*, 8344.

64. Goesmann, H.; Feldmann, C. Nanoparticulate Functional Materials. *Angew. Chem. Int. Ed.* **2010**, *49*, 1362.

65. Kashiwagi, T.; Du, F.; Douglas, J. F.; Winey, K. I.; Harris, R. H., Jr.; Shields, J. R. Nanoparticle Networks Reduce the Flammability of Polymer Nanocomposites. *Nat. Mater.* **2005**, *4*, 928.

66. Cui, J.; Yu, B.; Zhao, Y.; Zhu, W.; Li, H.; Lou, H.; Zhai, G. Enhancement of Oral Absorption of Curcumin by Self-Microemulsifying Drug Delivery Systems. *Int. J. Pharm.* **2009**, *371*, 148–155.

67. Roger, E.; Lagarce, F.; Garcion, E.; Benoit, J. P. Biopharmaceutical Parameters to Consider in order to Alter the Fate of Nanocarriers after Oral Delivery. *Nanomedicine* **2010**, *5*, 287–306.

68. Sudimack, J.; Lee, R. J. Targeted Drug Delivery via the Folate Receptor. *Adv. Drug Delivery Rev.* **2000**, *41*, 147–162.

69. Yang, S. J.; Lin, F. H.; Tsai, K. C.; Wei, M. F.; Tsai, H. M.; Wong, J. M.; Shieh, M. J. Folic Acid-Conjugated Chitosan Nanoparticles Enhanced Protoporphyrin IX Accumulation in Colorectal Cancer Cells. *Bioconjugate Chem.* **2010**, *21*, 679–689.

70. Li, Q.; Liu, C.; Zhao, X.; Zu, Y.; Wang, Y.; Zhang, B.; Zhao, D.; Zhao, Q.; Su, L.; Gao, Y.; Sun, B. Preparation, Characterization and Targeting of Micronized 10-Hydroxycamptothecin-Loaded Folate-Conjugated Human Serum Albumin Nanoparticles to Cancer Cells. *Int. J. Nanomed.* **2011**, *6*, 397–405.

71. Zhao, D.; Zhao, X.; Zu, Y.; Li, J.; Zhang, Y.; Jiang, R.; Zhang, Z. Preparation, Characterization, and In Vitro Targeted Delivery of Folate Decorated Paclitaxel-Loaded Bovine Serum Albumin Nanoparticles. *Int. J. Nanomed.* **2010**, *5*, 669–677.

72. Mislick, K. A.; Baldeschwieler, J. D.; Kayyem, J. F.; Meade, T. J. Transfection of Folate-Polylysine DNA Complexes Evidence for Lysosomal Delivery. *Bioconjugate Chem.* **1995**, *6*, 512–515.

73. MaHam, A.; Tang, Z. W.; Wu, H.; Wang, J.; Lin, Y. H. Proteinbased Nanomedicine Platforms for Drug Delivery. *Small* **2009**, *5*, 1706–1721.

74. Hu, B.; Pan, C. L.; Sun, Y.; Hou, Z. Y.; Ye, H.; Hu, B.; Zeng, X. X. Optimization of Fabrication Parameters to Produce Chitosan Tripolyphosphate Nanoparticles for Delivery of Tea Catechins. *J. Agric. Food Chem.* **2008**, *56*, 7451–7458.

75. Teng, Z.; Luo, Y. C.; Wang, Q. Nanoparticles Synthesized from Soy Protein: Preparation, Characterization, and Application for Nutraceutical Encapsulation. *J. Agric. Food Chem.* **2012**, *60*, 2712–2720.

76. Wang, Y. J.; Pan, M. H.; Cheng, A. L.; Lin, L. I.; Ho, Y. S.; Hsiech, C. Y.; Li, J. K. Stability of Curcumin in Buffer Solutions and Characterization of its Degradation Products. *J. Pharm. Biomed. Anal.* **1997**, *15*, 1867–1876.

77. Leung, M. H. M.; Kee, T. W. Effective Stabilization of Curcumin by Association to Plasma Proteins: Human Serum Albumin and Fibrinogen. *Langmuir* **2009**, *25*, 5773–5777.

78. Yang, M.; Wu, Y.; Li, J.; Zhou, H.; Wang, X. Binding of Curcumin with Bovine Serum Albumin in the Presence of ι-Carrageenan and Implications on the Stability and Antioxidant Activity of Curcumin. *J. Agric. Food Chem.* **2013**, *61*, 7150–7155.

79. Sneharani, A. H.; Singh, S. A.; Rao, A. G. A. Interaction of αs1-Casein with Curcumin and its Biological Implications. *J. Agric. Food Chem.* **2009**, *57*, 10386–10391.

80. Sneharani, A. H.; Karakkat, J. V.; Singh, S. A.; Rao, A. G. A. Interaction of Curcumin with β-Lactoglobulin Stability, Spectroscopic Analysis, and Molecular Modeling of the Complex. *J. Agric. Food Chem.* **2010**, *58*, 11130–11139.

81. Tapal, A.; Tiku, P. K. Complexation of Curcumin with Soy Protein Isolate and Its Implications on Solubility and Stability of Curcumin. *Food Chem.* **2012**, *130*, 960–965.

82. Sahu, A.; Kasoju, N.; Bora, U. Fluorescence Study of the Curcumin–Casein Micelle Complexation and Its Application as a Drug Nanocarrier to Cancer Cells. *Biomacromolecules* **2008**, *9*, 2905–2912.

83. Sneharani, A. H.; Karakkat, J. V.; Singh, S. A.; Rao, A. G. A. Interaction of Curcumin with β-Lactoglobulin Stability, Spectroscopic Analysis, and Molecular Modeling of the Complex. *J. Agric. Food Chem.* **2010**, *58*, 11130–11139.

84. Tapal, A.; Tiku, P. K. Complexation of Curcumin with Soy Protein Isolate and Its Implications on Solubility and Stability of Curcumin. *Food Chem.* **2012**, *130*, 960–965.

85. Esmaili, M.; Ghaffari, M.; Moosavi-Movahedi, Z.; Atri, M. S.; Sharifizadeh, A.; Farhadi, M.; Yousefi, R.; Chobert, J.-M.; Haertlé, T.; Moosavi-Movahedi, A. A. Beta Casein-Micelle as a Nano Vehicle for Solubility Enhancement of Curcumin; Food Industry Application. *LWT Food Sci. Technol.* **2011**, *44*, 2166–2172.

86. Yazdi, S. R.; Corredig, M. Heating of Milk Alters the Binding of Curcumin to Casein Micelles. A Fluorescence Spectroscopy Study. *Food Chem.* **2012**, *132*, 1143–1149.

87. Bourassa, P.; Kanakis, C. D.; Pollissiou, M. G.; Tajmir-Riahi, H. A. Resveratrol, Genistein, and Curcumin Bind Bovine Serum Albumin. *J. Phys. Chem. B* **2010**, *114*, 3348–3354.

88. Kiminaga, Y.; Nagatsu, A.; Akiyama, T.; Sugimoto, N.; Yamazaki, T.; Maitani, T.; Mizukami, H. Production of Unnatural Glucosides of Curcumin with Drastically Enhanced Water Solubility by Cell Suspension Cultures of *Catharanthus roseus*. *FEBS Lett.* **2003**, *555*, 311–316.

89. Pan, K.; Zhong, Q.; Baek, S. J. Enhanced Dispersibility and Bioactivity of Curcumin by Encapsulation in Casein Nanocapsules. *J. Agric. Food Chem.* **2013**, *61*, 6036–6043.

90. Sahoo, B. K.; Ghosh, K. S.; Dasgupta, S. Investigating the Binding of Curcumin Derivatives to Bovine Serum Albumin. *Biophys. Chem.* **2008**, *132*, 81–88.

91. Green, D. R.; Reed, J. C. Mitochondria and Apoptosis. *Science* **1998**, *281* (5381), 1309–1312.

92. Slowing, I. I.; Trewyn, B. G.; Lin, V. S. Y. Mesoporous Silica Nanoparticles for Intracellular Delivery of Membrane-Impermeable Proteins. *J. Am. Chem. Soc.* **2007**, *129* (28), 8845–8849.

93. Kam, N. W.; Dai, H. Carbon Nanotubes as Intracellular Protein Transporters: Generality and Biological Functionality. *J. Am. Chem. Soc.* **2005**, *127* (16), 6021–6026.

94. Hengartner, M. O. The Biochemistry of Apoptosis. *Nature* **2000**, *407* (6805), 770–776.

95. Cao, X.; Bennett, R. L.; May, W. S. c-Myc and Caspase-2 are Involved in Activating Bax During Cytotoxic Drug-Induced Apoptosis. *J. Biol. Chem.* **2008**, *283* (21), 14490–14496.

96. Shamsipur, M.; Molaabasi, F.; Hosseinkhani, S.; Rahmati, F. Detection of Early Stage Apoptotic Cells Based on Label-Free Cytochrome *C* Assay Using Bioconjugated Metal Nanoclusters as Fluorescent Probes. *Anal. Chem.* **2016**, *88*, 2188–2197.

97. Santra, S.; Kaittanis, C.; Perez, J. M. Cytochrome *C* Encapsulating Theranostic Nanoparticles: A Novel Bifunctional System for Targeted Delivery of Therapeutic Membrane-Impermeable Proteins to Tumors and Imaging of Cancer Therapy. *Mol. Pharmaceutics* **2010**, *7*, 1209–1222.

98. Leatherman, A. S. H.; Iftikhar, M.; Ndoi, A.; Scappaticci, S. J.; Lisi, G. P.; Buzard, K. L.; Garvey, E. M. Simplified Procedure for Encapsulating Cytochrome *C* in Silica Aerogel Nanoarchitectures while Retaining Gas-Phase Bioactivity. *Langmuir* **2012**, *28*, 14756–14765.

99. Ye, Q.; Asherman, J.; Stevenson, M.; Brownson, E.; Katre, N. V. DepoFoam Technology: A Vehicle for Controlled Delivery of Protein and Peptide Drugs. *J. Controlled Release* **2000**, *64*, 155–166.

100. Morales-Cruz, M.; Figueroa, C. M.; Gonzalez-Robles, T.; Delgado, Y.; Molina, A.; Mendez, J.; Morales, M.; Griebenow, K. Activation of Caspase-Dependent Apoptosis by Intracellular Delivery of Cytochrome *C*-Based Nanoparticles. *J. Nanobiotechnol.* **2014**, *12*, 33.

101. Danhier, F.; Feron, O.; Preat, V. To Exploit the Tumor Microenvironment: Passive and Active Tumor Targeting of Nanocarriers for Anti-Cancer Drug Delivery. *J. Controlled Release* **2010**, *148*, 135–146.

102. Cho, K.; Wang, X.; Nie, S.; Chen, Z. G.; Shin, D. M. Therapeutic Nanoparticles for Drug Delivery in Cancer. *Clin. Cancer Res.* **2008**, *14*, 1310–1316.

103. Jaracz, S.; Chen, J.; Kuznetsova, L. V.; Ojima, I. Recent Advances in Tumor-Targeting Anticancer Drug Conjugates. *Bioorg. Med. Chem.* **2005**, *13*, 5043–5054.

104. Fricke, J., Ed. Aerogels—Proceedings of the First International Symposium, Wurzburg, FRG, Sept 23–25, 1985; Springer-Verlag: Berlin, 1986.

105. Hüsing, N.; Schubert, U. Aerogels Airy Materials: Chemistry, Structure, and Properties. *Angew. Chem. Int. Ed.* **1998**, *37*, 22–45.

106. Basso, A.; De Martin, L.; Ebert, C.; Gardossi, L.; Tomat, A.; Casarci, M.; Rosi, O. L. A Novel Support for Enzyme Adsorption: Properties and Applications of Aerogels in Low Water Media. *Tetrahedron Lett.* **2000**, *41*, 8627–8630.

107. Power, M.; Hosticka, B.; Black, E.; Daitch, C.; Norris, P. Aerogels as Biosensors: Viral Particle Detection by Bacteria Immobilized on Large Pore Aerogel. *J. Non-Cryst. Solids* **2001**, *285*, 303–308.

108. Li, Y. K.; Yang, D.-K.; Chen, Y.-C.; Su, H.-J.; Wu, J.-C.; Chen-Yang, Y. W. A Novel Three-Dimensional Aerogel Biochip for Molecular Recognition of Nucleotide Acids. *Acta Biomater.* **2010**, *6*, 1462–1470.

109. Li, Y. K.; Chen, Y.-C.; Jiang, K.-J.; Wu, J.-C.; Chen-Yang, Y. W. Three-Dimensional Arrayed Amino Aerogel Biochips for Molecular Recognition of Antigens. *Biomaterials* **2011**, *32*, 7347–7354.

110. Weng, K. C.; Stalgren, J. J. R.; Duval, D. J.; Risbud, S. H.; Frank, C. W. Fluid Biomembranes Supported on Nanoporous Aerogel/Xerogel Substrates. *Langmuir* **2004**, *20*, 7232–7239.

111. Wallace, J. M.; Rice, J. K.; Pietron, J. J.; Stroud, R. M.; Long, J. W.; Rolison, D. R. Silica Nanoarchitectures Incorporating Self-Organized Protein Superstructures with Gas-Phase Bioactivity. *Nano Lett.* **2003**, *3*, 1463–1467.

112. Culotta, E.; Koshland, D. E. NO News is Good-News. *Science* **1992**, *258*, 1862–1865.

PART III
Key Issues in Industrial Chemistry and Chemical Engineering

CHAPTER 7

A COMPREHENSIVE REVIEW OF THYMOQUINONE EXTRACTION FROM *Nigella sativa* AND ITS CHARACTERIZATION

ALI YALÇIN and MEHMET GÖNEN*

Department of Chemical Engineering, Engineering Faculty, Süleyman Demirel University, E-13 Blok, Batı Yerleşkesi, Isparta 32260, Turkey

Corresponding author. E-mail: mehmetgonen@sdu.edu.tr

ABSTRACT

The oil of *Nigella sativa* is widely used in the medical, food, and pharmaceutical industries due to the presence of thymoquinone (TQ). Seeds of *N. sativa* are traditionally consumed in the preparation of bakery products, as well. The properties of oil obtained from *N. sativa* are mostly related to the amount of TQ molecule. In this chapter, separation methods were compared for showing differences in the extraction efficiency and in the oil properties. The concentration of TQ obtained by supercritical CO_2 extraction is much higher than one obtained by cold press, soxhlet extraction, and microwave extraction. As CO_2 has low critical temperature, it is advised as a green solvent for the extraction of temperature sensitive molecules.

7.1 INTRODUCTION

The use of plants as a source of healing and for the treatment of most diseases has been known since ancient times. The knowledge of this fact by human beings has led to an increase in researches on plants, especially for recent years. One of the most important plants used for the treatment and prevention of diseases is black seed (*Nigella sativa*). It is cultivated in different parts of the world such as Mediterranean countries, Southern Europe, and North Africa.[16,62]

Black seeds of *N. sativa* are commonly used in bakery products especially in Middle East and Mediterranean countries. They are widely used for the treatment of diseases such as headache, asthma,[34] cancer (kidney, skin, liver, lung),[25] bronchitis, hypertension[16] cough, rheumatism, and fever.[63] It has been reported that crude oil obtained from black seeds has antioxidant properties,[34] anti-inflammatory,[50] antibacterial activity,[40] antitumor,[12] antidiabetic,[63] anticarcinogenic[31] on the immune system. *N. sativa* seeds usually consist of moisture 5.52–7.43%, protein 20–27%, ash 3.77–4.92%, carbohydrate 23.5–33.2%, and crude oil 34.49–38.72%.[9] The crude oil obtained from black seeds includes volatile oils in the range of 0.4–0.44%.[60] According to the recent study, the thymoquinone (TQ) content of volatile oil obtained from black seeds varies from 27.8% to 57.0%.[31]

TQ, also known as 2-isopropyl-5-methyl-1,4-benzoquinone, is an essential component in volatile oil.[14,60] It is a substance having an aromatic ketone structure which can also be synthesized by thymol or carvacrol oxidation.[33,53] Due to its strong natural antioxidant properties, it has recently been preferred in the formulation pharmaceutical, food, and cosmetic products rather than the synthetic antioxidants of butylated hydroxytoluene, butylated hydroxyanisole, and tertiary butyl hydroquinone.[52]

The crude oil could be obtained through the processing of *N. sativa* seeds by supercritical fluid extraction,[50,58] solvent extraction,[14] microwave-assisted extraction,[24] and cold press extraction.[1] Supercritical CO_2 (SC-CO_2) extraction technique emerges as an environment friendly and free of solvent residues in the product. Thus, it is mostly preferred for the separation and synthesis of drug raw materials.

In this review, extraction techniques for oil separation from *N. sativa* seeds were summarized and discussed. The use of TQ and its derivatives for the treatment of diseases were investigated based on the recent studies.

7.1.1 PHYSICOCHEMICAL PROPERTIES OIL OBTAINED FROM Nigella sativa

Composition of *N. sativa* seeds having moisture, oil, protein, carbohydrate, and ash are given in Table 7.1. The chemical composition of *N. sativa* varies because of different cultivated regions, maturity stage, and storage circumstances, as seen in Table 7.1. For instance, according to Table 7.1, oil content of the *N. sativa* grown in Iran has higher than the one grown in Egypt,

Malaysia, and Pakistan. This situation can be observed in the subregions of the country of plantation.[1,9,34,56] The bitterness of the oil obtained from *N. sativa* is related to the presence of some components such as carvacrol and thymol.

TABLE 7.1   The Chemical Composition of the *Nigella sativa* Seeds.

Moisture	Oil	Protein	Carbohydrate	Ash	Region	References
4.08	40.35	22.60	32.70	4.41	Iran	[9]
7.00	34.80	20.80	33.70	3.70	Alexandria/Egypt	[1]
4.99	31.72	23.07	34.91	5.29	Tehran/Iran	[51]
6.67	32.26	19.19	35.04	6.82	West Malaysia	[34]
6.46	31.16	22.80	–	4.20	Chakwal/Pakistan	[56]

The physicochemical properties of oil obtained from *N. sativa* seeds, such as iodine value (IV), free fatty acids (FFAs) content, refractive index, saponification value (SV), viscosity, and peroxide value (PV), are shown in Table 7.2. The physicochemical properties of the oil vary according to the extraction methods as seen in Table 7.2. The reasons behind that variation might be the parameters in the extraction process, which influences both quality and quantity of the oil extracted. For instance, processing temperature may influence both oil composition and PV. Elevated temperatures may alter structural changes such as increase in FFAs. It was pointed out that the amount of components such as vitamin E in olive oil decreases with rising temperature. Same phenomenon occurs for TQ which presents in volatile oils.

TABLE 7.2   Physicochemical Properties of Oil Obtained from *Nigella sativa* Seed by Different Extraction Methods.

Physiochemical properties	Supercritical CO_2 extraction		Solvent extraction		Cold pressing		Microwave extraction
FFA (%)	5.98	5.83	22.70	6.40	11.00	0.90	9.51
SV (mg of KOH/g of oil)	243.52	194.00	211.00	190.00	192.00	–	–
IV (g of I_2/100 g of oil)	115.10	127.00	119.00	121.00	115.00	128.00	–
PV (meq O_2/kg of oil)	3.40	5.62	5.65	7.30	13.50	3.40	21.45
Refractive index	1.47	–	1.47	–	1.47	1.47	1.47
Viscosity (mPa s)	6.26	–	–	–	–	–	–
References	[34, 51]		[9, 51]		[1, 14]		[26]

FFA, free fatty acid; SV, saponification value; IV, iodine value; PV, peroxide value.

7.1.2 CHEMICAL PROPERTIES OF TQ

TQ is abundant component in volatile oil of herbs, such as *N. sativa*,[13] *Monarda fistulosa*,[53] *Monarda didyma*,[53] *Satureja montana*,[17,64] and *Origanum vulgare*.[33] Recently, there has been increased research on TQ which presents in the oil extracted from *N. sativa*. It was reported that *N. sativa* seeds contain TQ in the volatile oil determined by nuclear magnetic resonance analysis. Isopropyl group in the TQ structure was detected from signals at 1.93 ppm, 1.94 ppm, and 2.12 ppm with the coupling constant of 5.3 Hz, and the chemical structure of TQ is shown in Figure 7.1. In addition, the crystal structure of TQ was determined by X-ray diffraction.[47] The physical properties of TQ are given in Table 7.3.

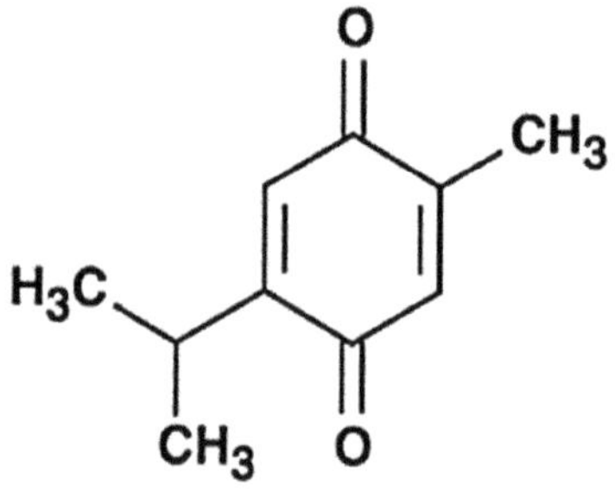

FIGURE 7.1   The chemical structure of thymoquinone.

The separation TQ from *N. sativa* oil could be performed by extraction. In this method, the oil sample extracted from *N. sativa* is taken in a volumetric flask with 10 mL of methanol and is sonicated for 20 min. Methanol layer is removed from oil. Liquid obtained is loaded on silica gel column and eluted with petroleum ether. The fastest moving yellow color spot is concentrated and is subjected to high performance liquid chromatography (HPLC) analysis.[35]

TABLE 7.3   The Physical Properties of the Thymoquinone.[37]

Molecular formula	$C_{10}H_{12}O_2$
Molecular weight	164.20 g/mol
State	Solid
Melting point	44–45°C
Boiling point	230–232°C
Appearance	Golden brown to dark brown flaky crystals

Coenzyme Q10 is a compound that chemically resembles TQ. It has antioxidant properties like TQ that can be used as a nutritional supplement as well as cosmetic industry. It has been widely used in the treatment of many diseases such as ischemic heart disease, diastolic dysfunction of the left ventricle, and congestive heart failure.[19] The chemical structure of the Coenzyme Q10, which has remarkable similarity with TQ, is given in Figure 7.2. This molecule is mainly responsible for the electron transfer in mitochondria of both plant and mammalian cells.

Coenzyme Q10 is a member of class of compounds that are characterized by their quinone moiety and lipophilic tail which could be in different composition and length as shown in Figure 7.2. The quinone group consists of benzene ring having two conjugated oxygen atoms, two methoxy groups, and a methyl group. The hydrophobic tail is made of isoprenoid chain with different lengths ($n = 6$–10). The characteristic quinone group can accept electrons from different biological sources and is converted into semiquinone group by a one electron transfer or to the more stable hydroquinone by two electrons. This feature makes Coenzyme Q10 cellular electron transfer molecule in the respiratory chain of mitochondria.

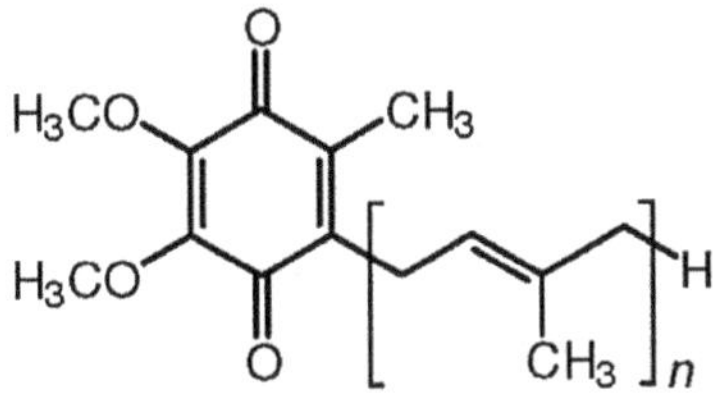

FIGURE 7.2   The chemical structure of the Coenzyme Q10.

7.2   EXTRACTION METHODS

Active components from plants and their seeds are obtained by conventional techniques, such as Soxhlet extraction, cold press, and supercritical fluid extraction. Those techniques are generally used in the separation of oil from *N. sativa* seeds. TQ content of oil obtained from *N. sativa* by different extraction methods are shown in Table 7.4. TQ concentration in oil varies according to the extraction method used and extraction conditions as seen in Table 7.4. For instance, extraction method and conditions are very important in the determination of oil properties.

TABLE 7.4 Thymoquinone Concentration of Oil from *Nigella sativa* by Different Extraction Method.

Extraction technique	Extraction parameters	Thymoquinone content	References
Supercritical carbon dioxide extraction	350 bar, at 50°C	4.07 mg/g oil	[51]
Cold pressing	30°C, without heating	1.56 mg/mL oil	[34]
Supercritical carbon dioxide extraction	150 bar, at 40°C	4.09 mg/mL oil	[50]
Microwave extraction	10 min, 1 atm, at 45°C	5.65 µg/g oil	[26]
Soxhlet extraction	40–60°C	1.06 mg/g oil	[51]
Supercritical carbon dioxide extraction	–	4.48 mg/g oil	[58]
Supercritical carbon dioxide extraction	600 bar, at 40°C	6.63 mg/mL oil	[34]
Cold pressing	At 25°C, without heating	14.40 µg/g oil	[26]
Cold pressing	Mechanical pressing without heating	8.73 mg/g oil	[30]
Supercritical carbon dioxide extraction	150 bar, at 60°C	20.80 mg/mL oil	[50]
Microwave extraction	10 min, 1 atm, 50 mL distilled water	0.49 mg/g black seed	[4]
Soxhlet extraction	*n*-hexane	6.20 µg/g oil	[26]

As shown in Table 7.4, the content of TQ in volatile oil obtained from *N. sativa* was found to be 4.07 mg/g oil, 4.09 mg/mL oil, 4.48 mg/g oil, 6.63 mg/mL oil, and 20.80 mg/mL in supercritical fluid extractions. The issue in comparison of those values is units, some of them were reported as mg/g oil, and the others were reported as mg/mL oil. When density of oil is considered, it would be easy to compare those values. The concentration of TQ in volatile oil obtained from *N. sativa* by using cold pressing, microwave extraction, and solvent extraction is generally less than that of obtained by using supercritical fluid extraction. It can be concluded that the extraction method is an important parameter for obtaining TQ as a valuable component.[34]

7.2.1 SOLVENT EXTRACTION

Solvent extraction is a separation technique that is commonly used due to its simplicity and fastness. It is widely accepted method to extract substances which present as trace amount in the bulk materials.[27] Although the solvent extraction is inexpensive, there are some issues which hinder its use such as toxicity of solvent and solvent losses in recovery step.[22] In this method, chemical stability of solvent is very important, and it is expected to be stable during extraction and recovery steps.[57] Commonly used organic solvents are methanol,[57] ethanol,[41] hexane,[34] and chloroform.[31] The crude oil extraction efficiency from *N. sativa* was calculated as 37% and 27% (by wt.) by using hexane extraction and cold pressing, respectively. The composition of crude oil obtained plants grown in Moroccan was similar to that of one from other Mediterranean countries.[14] A crude oil from *N. sativa* was obtained by SC-CO$_2$ extraction in which ethyl alcohol was used as cosolvent. Optimum conditions for the highest oil yield of 23.20% were determined as 350 bar, 60°C, and 120 min. On the other hand, the maximum TQ concentration was determined as 4.09 mg/mL under the following conditions: 150 bar, 40°C, 120 min.[50] The most significant disadvantage of solvent extraction is the remaining of solvent residues in the crude components after the extraction. So, the presence of solvent has negative effects on human health.[10]

7.2.2   COLD PRESS EXTRACTION

Cold press is a technique used to separate liquids contained in a solid matrix without heat and chemical treatments.[38] Recently, consumers demand natural and extracted products with free of solvents. Consumption of natural products and oils obtained by cold press method improves human health and prevents certain diseases.[48] TQ concentration of oil extracted using supercritical fluid and cold press method was found to be 6.63 mg/mL and 1.56 mg/mL, respectively.[34] The concentration of TQ was also determined as 14.40-µg/g, 5.65-µg/g, and 6.20-µg/g oil cold press, microwave-assisted extraction and solvent extraction, respectively.[26] The extraction temperature affects the amount of oil obtained from *N. sativa* in the cold press. The most important reason behind this variation is the viscosity of the oil. The effect of temperature on viscosity of crude oil obtained from *N. sativa* was investigated at different temperatures (50, 60, 70, 80, 90, and 100°C) in cold press technique. The viscosity values were found to be 64.53 ± 0.31, 63.80 ± 0.61, 66.23 ± 1.29, 66.47 ± 0.81, 69.53 ± 0.31, and 71.47 ± 0.45 (mPa s), respectively.[67] In fact, it is anticipated that viscosity values would decrease with the increase of temperature. As seen from the above values, viscosity increases with rising temperature. There might be two reasons behind this viscosity increase. The first one is the removal of volatile components in the oil extracted, and the second one is the increase of longer chains alkyl-groups that was polymerized at higher temperatures. Therefore, the change in viscosity with temperature can cause significant differences in the yield of *N. sativa* oil obtained by cold pressing. The temperature should not be above 60°C, as the lowest viscosity was obtained at that temperature during the cold press.

7.2.3 SUPERCRITICAL FLUID EXTRACTION

If liquid and gas phases are beyond critical temperature and critical pressure, this single phase exhibits properties of both liquid and gases and fluid in this region called supercritical fluid. Above these critical points, the coexistence of liquid and gas phases exhibits a characteristic property which is different than that of liquids or gases under standard conditions. Figure 7.3 shows the relationship between pressure and temperature including supercritical region. In most of the studies in the literature show that about 98% of the supercritical fluid extraction is carried out by using CO_2.[57] The use of carbon dioxide as a supercritical fluid has several advantages, such as solvent recovery,[3] easy separation from the extracted material, not flammable, abundant, low cost, and requiring low energy in the region close to the critical point.[57] As carbon dioxide has lower critical temperature (31.1°C), it is the only solvent for the extraction of materials which decomposes at higher temperatures.

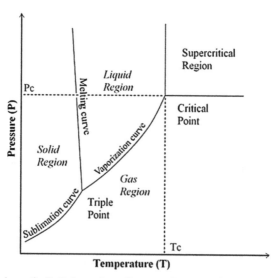

FIGURE 7.3   Schematic *P–T* plane phase diagram of a pure substance.

Optimum parameters affecting supercritical fluid extraction is of great importance in terms of the quantity and quality of the desired substance. Parameters such as particle size and homogeneity, sample pre-drying, different solvent and co-solvent, temperature, pressure, flow rate, and outlet valve temperature influence properties of the final product.[57] Extracts obtained through supercritical fluid have high purity and no solvent residue.[3]

Increasing the pressure in supercritical fluid extraction raises the solvation power of the fluid and the solubility of the components, which induces a higher yield.[50] The density of the fluid decreases with increasing temperature, at constant pressure in this process. Therefore, a decrease in density affects adversely the solvation power of the fluid.[29] But, while the density of the fluid is decreasing with temperature, the solubility of substance is increased because of the increase of the solute's vapor pressure.[28] It was reported that the power of the solvent is increased when the extraction pressure is increased from 10,000 to 14,000 kPa,[28] and the extraction yield from black poplar by supercritical fluid has increased with rising temperature from 40°C to 80°C.[49] In another work carried out using ethyl acetate and ethanol as co-solvents, the extraction yield was found to be 14.54% and 16.36% for ethyl acetate and ethanol, respectively.[8]

7.2.4 CHARACTERIZATION EXPERIMENTS

7.2.4.1 HPLC ANALYSIS

Chromatography is a separation technique which widely used in the quantification of components of liquid. Liquid chromatography plays an important role in the development of HPLC which is a suitable method of analysis for quantification of high molecular weight and high boiling point components.[46] The operating rule of the HPLC device is based on the principle that the sample molecules to be analyzed move at different rates through column and change the retention time. The mobile phase forms a gradient elution by carrying the stationary phase. This gradient separates the analyte mixtures through the column. The retention times identified by the detector determine the contents and the quantity of components in the analyzed sample.[32]

The liquid phase dissolved in a suitable solvent is pumped to the column packed at a constant rate. Before entering the column, the sample is injected into the carrier stream. When the sample components reach the column, they are retained on the basis of physicochemical interactions between the stationary phase and analyte molecules. The mobile phase flows at a steady-state and then components elute in operating conditions. The concentration of the components is determined with the help of the detector.[46]

Aqueous solubility and degradation kinetics of the TQ was investigated using ultraviolet–visible spectroscopy. The peak at 254–257 nm is known as the characteristic peak of quinone group.[45] The concentration of TQ obtained from *N. sativa* seeds by cold pressing was determined as 14.40 µg/g by using HPLC.

7.2.4.2   GC–MS ANALYSIS

Gas chromatography (GC) is a widely used technique for determining the number of components and their concentration in a mixture. However, it is inadequate to determine the chemical structure and nature of these components. That is, a spectroscopic detection system is required for accurate determination. The mass spectrometric (MS) detector is widely used to provide information on the molecular weight, elemental composition, functional groups, the spatial isomerism, and geometry of the molecule.[54] GC–MS device is based on the principle in which samples carried by the inert gas have different retention times through the column.[20]

During the transfer into the GC, a liquid sample in a solution is volatilized by exposing to high temperature (200–300°C). The evaporated gas is mixed with the carrier gas and enters the separation section. Analyte molecules are partitioned between the carrier gas stream (mobile phase) and the stationary phase. At the end of the separation section, the molecules reach a detection system that determines the amount of molecules having same properties.[54]

Raw oil was obtained from *N. sativa* seeds by SC-CO$_2$ extraction. The yield of oil extraction was determined by using GC–MS. The peak area and retention time of TQ found to be 16.80 and 11.32, respectively.

The concentration of TQ in the extracted oil from *N. sativa* seeds was determined by GC–MS. Oils obtained from six different samples of *N. sativa* seeds have been characterized. All samples contain one major component (TQ), its concentration vary between 27.8 and 57.0%.[7] It has been reported that TQ is mostly present in volatile oil of *N. sativa* as understood from GC–MS analysis.[43]

7.2.4.3   FTIR ANALYSIS

Fourier-transform infrared spectroscopy (FTIR) provides the ability to monitor the vibrations of the functional groups which characterize molecular structure and govern the course of chemical reactions.[18] According to the results of TQ analysis by FTIR, stretching bands at wavenumber 2967 cm^{-1} and 1650 cm^{-1} represent C–H stretching vibration of aliphatic groups and the stretching vibration of the carbonyl group of a cyclohexadiene in the TQ, respectively. Since the strong carboxylic stretching band of TQ ranges 1640–1675 cm^{-1} frequency, the C=C stretching band cannot be clearly identified. The band observed at a wavelength of 3040 cm^{-1} designates stretching observed in the C=C–H groups.[37]

7.2.5 APPLICATIONS OF TQ

TQ is widely used as an active component to treat wide range of diseases.[13] Recently, many studies have been performed related to effects of TQ on anticancer,[44] antifungal,[13] antimicrobial,[42] antioxidant,[61] anti-inflammatory,[59] antitumor,[66] anticonvulsant,[13] antiaggregation,[21] antifibrotic,[15] and antibacterial.[13] Some of those studies were summarized below.

7.2.5.1 ANTIOXIDANT PROPERTIES

Antioxidants are compounds which slow or retard the oxidation rate of chemicals that prone to oxidation.[6] They are used to inhibit the formation of free radicals resulting from the oxidation.[65] Natural antioxidants that extracted from various plants are used instead of synthetic antioxidants which adversely affect human health.[6,61]

Oxidation reactions cause the initiation and development of Parkinson's disease. Rotenone is a compound which induces to these oxidation reactions. The formation of free radicals is reduced because of antioxidant properties of TQ. Cell damage was prevented by decreasing the formation of those free radicals. In the treatment of Parkinson's disease, rotenone was injected to male Wistar rats. It has been observed that TQ prevents the formation of rotenone-induced defects. These results have led to the conclusion that TQ is associated by neuroprotective and antioxidant effects.[11]

It has been carried out a research related to the protective effect of TQ on survival, mesenteric artery blood flow, vascular reactivity, oxidative, and inflammatory injuries in a murine sepsis model induced by cecal ligation and puncture. Wistar rats separated in four groups were injected with TQ at different rates for 3 days. Tissue samples were taken and analyzed for histopathological and biochemical examinations. The histopathological protective effect of TQ was observed on organ damage. TQ has been shown to have therapeutic effects on sepsis due to the contractile function of the aorta and its anti-inflammatory and antioxidative effects.[36] The effects of TQ were investigated in acetaminophen induced renal toxicity. TQ has been shown to be an important component in the prevention of acetaminophen-induced nephrotoxicity in rats due to its antioxidative effects.[2]

7.2.5.2   ANTICANCER EFFECTS

It is a known fact that TQ has significant effects on chemotherapeutic anti-tumor and chemopreventive. Many studies have reported that TQ inhibits the growth of cancer cells. The interaction of TQ with G-quadruplex DNA was investigated because of the anticancer effect of TQ. It has been reported that TQ acts as a G-quadruplex DNA stabilizer and inhibits cancer's proliferation and telomerase enzyme.[44]

There are studies in the literature showing that TQ provides antiprolif-eration to cancer cells. Asthma is a chronic obstructive disease infected to humans by air. The effect of TQ on the inflammation, vascular, and neoan-giogenesis caused by ovalbumin in asthmatic mice were investigated. At the end of the conducted studies, it is emphasized that TQ is an important component in ameliorating asthma.[55]

The effect of *N. sativa* essential oil on breast cancer cells has been investigated. As a result, it has been found that this essential oil significantly reduces the viability of breast cancer cells. Findings obtained suggest that essential oil including TQ is effective on breast cancer.[39]

7.2.5.3   ANTIMICROBIAL EFFECTS

Microbial resistance development is an important issue, and antimicrobial agents are used to increase this microbial resistance. There are some complica-tions for patients exposed to skin infections. The oil extracted from *N. sativa* seeds has been used for the treatment of skin infections. Antimicrobial effect of oil obtained from the *N. sativa* was also investigated against skin pustules infection. It has been reported that oil has antimicrobial effects against staphy-lococcal infections, and no side effects were observed.[42] The cytotoxic and the antibacterial activities of *N. sativa* essential oil and TQ were investigated against various cariogenic bacteria such as *Streptococcus mutans*, *Strepto-coccus mitis*, *Streptococcus constellatus*, and *Gemella haemolysans*. It was determined that the essential oil containing TQ and pure TQ have antibacterial activity against those bacteria.[23]

7.2.5.4 ANTI-INFLAMMATORY EFFECTS OF TQ

TQ in the volatile oil of *N. sativa* is known to have therapeutic effects on chronic inflammatory. Some studies have been conducted related to its anti-inflammatory effects on disease such as asthma, arthritis, and neurodegeneration. Anti-inflammatory effect of TQ was investigated in lipopolysaccharide-stimulated BV-2 murine microglia cells. It was detected that TQ retarded the formation of inflammation-mediated neurodegenerative disorders.[59] In a recent study, it was detected that TQ inhibits granules exocytosis and *N*-formyl-methionyl-leucyl-phenylalanine-induced superoxide production in neutrophils. It is reported that TQ possesses promising due to anti-inflammatory therapeutic potential, since it exhibits anti-inflammatory properties.[5]

7.3 CONCLUSION

The oil of *N. sativa* has been widely used in the medical, food, and pharmaceutical industries due to the presence of TQ. The properties of *N. sativa* oil are mostly related to the presence of TQ molecule. The mechanism in prevention of those diseases is unknown. Extraction method and extraction parameters affect the concentration of TQ in the oil obtained from *N. sativa* seeds. The concentration of TQ obtained by SC-CO_2 extraction is much higher than that obtained by cold press, Soxhlet extraction, and microwave extraction. SC-CO_2 extraction is mostly preferred technique in wide range of industrial separations as there is no solvent residue in the product and being an environmentally-friendly method.

KEYWORDS

- thymoquinone
- *Nigella sativa*
- supercritical CO_2
- extraction

REFERENCES

1. Atta, M. B. Some Characteristics of Nigella (*Nigella sativa* L.) Seed Cultivated in Egypt and Its Lipid Profile. *Food Chem.* **2003**, *83*, 63–68.
2. Aycan, I. O.; Tokgoz, O.; Tufek, A.; Alabalik, U.; Evliyaoglu, O.; Turgut, H.; Celik, F.; Guzel, A. The Use of Thymoquinone in Nephrotoxicity Related to Acetaminophen. *Int. J. Surg.* **2015**, *13*, 33–37.
3. Barroso, P. T.; de Carvalho, P. P.; Rocha, T. B.; Pessoa, F. L.; Azevedo, D. A.; Mendes, M. F. Evaluation of the Composition of *Carica papaya* L. Seed Oil Extracted with Supercritical CO$_2$. *Biotechnol. Rep.* **2016**, *11*, 110–116.
4. Benkaci-Ali, F.; Baaliouamer, A.; Meklati, B. Y.; Chemat, F. Chemical Composition of Seed Essential Oils from Algerian *Nigella sativa* Extracted by Microwave and Hydro-distillation. *Flavour Fragrance J.* **2007**, *22* (2), 148–153.
5. Boudiaf, K.; Hurtado-Nedelec, M.; Belambri, S. A.; Marie, J. C.; Derradji, Y.; Benbou-betra, M.; El-Benna, J.; Dang, P. M. Thymoquinone Strongly Inhibits fMLF-Induced Neutrophil Functions and Exhibits Anti-inflammatory Properties In Vivo. *Biochem. Pharmacol.* **2016**, *104*, 62–73.
6. Bulca, S. Çörek otunun bileşenleri ve bu yağın ve diğer bazi uçucu yağlarin antioksidan olarak gıda teknolojisinde kullanımı. *J. Adnan Menderes Univ. Agric. Faculty* **2014**, *11* (2), 29–36.
7. Burits, M.; Bucar, F. Antioxidant Activity of *Nigella sativa* Essential Oil. *Phytother. Res.* **2000**, *14* (5), 323–328.
8. Castro-Vargas, H. I.; Rodríguez-Varela, L. I.; Ferreira, S. R. S.; Parada-Alfonso, F. Extraction of Phenolic Fraction from Guava Seeds (*Psidium guajava* L.) Using Supercritical Carbon Dioxide and Co-Solvents. *J. Supercrit. Fluids* **2010**, *51* (3), 319–324.
9. Cheikh-Rouhou, S.; Besbes, S.; Hentati, B.; Blecker, C.; Deroanne, C.; Attia, H. *Nigella sativa* L.: Chemical Composition and Physicochemical Characteristics of Lipid Fraction. *Food Chem.* **2007**, *101* (2), 673–681.
10. Danh, L. T.; Mammucari, R.; Truong, P.; Foster, N. Response Surface Method Applied to Supercritical Carbon Dioxide Extraction of *Vetiveria zizanioides* Essential Oil. *Chem. Eng. J.* **2009**, *155* (3), 617–626.
11. Ebrahimi, S. S.; Oryan, S.; Izadpanah, E.; Hassanzadeh, K. Thymoquinone Exerts Neuroprotective Effect in Animal Model of Parkinson's Disease. *Toxicol. Lett.* **2017**, *276*, 108–114.
12. Erkan, N.; Ayranci, G.; Ayranci, E. Antioxidant Activities of Rosemary (*Rosmarinus officinalis* L.) Extract, Black Seed (*Nigella sativa* L.) Essential Oil, Carnosic Acid, Rosmarinic Acid and Sesamol. *Food Chem.* **2008**, *110* (1), 76–82.
13. Forouzanfar, F.; Bazzaz, B. S. F.; Hosseinzadeh, H. Black Cumin (*Nigella sativa*) and Its Constituent (Thymoquinone): A Review on Antimicrobial Effects. *Iran. J. Basic Med. Sci.* **2014**, *17* (12), 929–938.
14. Gharby, S.; Harhar, H.; Guillaume, D.; Roudani, A.; Boulbaroud, S.; Ibrahimi, M.; Ahmad, M.; Sultana, S.; Hadda, T. B.; Chafchaouni-Moussaoui, I.; Charrouf, Z. Chemical Investigation of *Nigella sativa* L. Seed Oil Produced in Morocco. *J. Saudi Soc. Agric. Sci.* **2015**, *14* (2), 172–177.
15. Ghazwani, M.; Zhang, Y.; Gao, X.; Fan, J.; Li, J.; Li, S. Anti-Fibrotic Effect of Thymoquinone on Hepatic Stellate Cells. *Phytomedicine* **2014**, *21* (3), 254–260.

16. Gholamnezhad, Z.; Havakhah, S.; Boskabady, M. H. Preclinical and Clinical Effects of *Nigella sativa* and Its Constituent, Thymoquinone: A Review. *J. Ethnopharmacol.* **2016**, *190*, 372–386.

17. Grosso, C.; Figueiredo, A. C.; Burillo, J.; Mainar, A. M.; Urieta, J. S.; Barroso, J. G.; Coelho, J. A.; Palavra, A. M. Enrichment of the Thymoquinone Content in Volatile Oil from *Satureja montana* Using Supercritical Fluid Extraction. *J. Sep. Sci.* **2009**, *32* (2), 328–334.

18. Güneş, A. The Production of Thymoquinone from Thymol and Carvacrol by Using Zeolite Catalysts. Master's Thesis, İzmir Institute of Technology, İzmir, 2005.

19. Huang, M.; Chen, Y.; Liu, J. Chromosomal Engineering of *Escherichia coli* for Efficient Production of Coenzyme Q10. *Chin. J. Chem. Eng.* **2014**, *22* (5), 559–569.

20. Hussain, S. Z.; Maqbool, K. GC–MS: Principle, Technique and Its Application in Food Science. *J. Curr. Sci.* **2014**, *13*, 116–126.

21. Ishtikhar, M.; Rahisuddin; Khan, M. V.; Khan, R. H. Anti-Aggregation Property of Thymoquinone Induced by Copper-Nanoparticles: A Biophysical Approach. *Int. J. Biol. Macromol.* **2016**, *93*, 1174–1182.

22. Jeevan Kumar, S. P.; Garlapati, V. K.; Dash, A.; Scholz, P.; Banerjee, R. Sustainable Green Solvents and Techniques for Lipid Extraction from Microalgae: A Review. *Algal Res.* **2017**, *21*, 138–147.

23. Jrah Harzallah, H.; Kouidhi, B.; Flamini, G.; Bakhrouf, A.; Mahjoub, T. Chemical Composition, Antimicrobial Potential against Cariogenic Bacteria and Cytotoxic Activity of Tunisian *Nigella sativa* Essential Oil and Thymoquinone. *Food Chem.* **2011**, *129* (4), 1469–1474.

24. Karacabey, E. Optimization of Microwave-Assisted Extraction of Thymoquinone from *Nigella sativa* L. Seeds. *Maced. J. Chem. Chem. Eng.* **2016**, *35*(2), 209–216

25. Khan, M. A.; Afzal, M. Chemical Composition of *Nigella sativa* Linn: Part 2 Recent Advances. *Inflammopharmacology* **2016**, *24* (2–3), 67–79.

26. Kiralan, M.; Özkan, G.; Bayrak, A.; Ramadan, M. F. Physicochemical Properties and Stability of Black Cumin (*Nigella sativa*) Seed Oil as Affected by Different Extraction Methods. *Ind. Crops Prod.* **2014**, *57*, 52–58.

27. Kislik, V. S. *Solvent Extraction: Classical and Novel Approaches*; Elsevier, Oxford, UK, 2012.

28. Lee, M. R.; Lin, C. Y.; Li, Z. G.; Tsai, T. F. Simultaneous Analysis of Antioxidants and Preservatives in Cosmetics by Supercritical Fluid Extraction Combined with Liquid Chromatography-Mass Spectrometry. *J. Chromatogr. A* **2006**, *1120* (1–2), 244–251.

29. Luengthanaphol, S.; Mongkholkhajornsilp, D.; Douglas, S.; Douglas, P. L.; Pengsopa, L.; Pongamphai, S. Extraction of Antioxidants from Sweet Thai Tamarind Seed Coat—Preliminary Experiments. *J. Food Eng.* **2004**, *63* (3), 247–252.

30. Lutterodt, H.; Luther, M.; Slavin, M.; Yin, J. J.; Parry, J.; Gao, J. M.; Yu, L. Fatty Acid Profile, Thymoquinone Content, Oxidative Stability, and Antioxidant Properties of Cold-Pressed Black Cumin Seed Oils. *Food Sci. Technol.* **2010**, *43* (9), 1409–1413.

31. Machmudah, S.; Shiramizu, Y.; Goto, M.; Sasaki, M.; Hirose, T. Extraction of *Nigella sativa* L. Using Supercritical CO_2: A Study of Antioxidant Activity of the Extract. *Sep. Sci. Technol.* **2005**, *40* (6), 1267–1275.

32. Malviya, R.; Bansal, V.; Pal, O. P.; Sharma, P. K. High Performance Liquid Chromatography: A Short Review. *J. Glob. Pharm. Technol.* **2010**, *2* (5), 22–26.

33. Milos, M.; Mastelic, J.; Jerkovic, I. Chemical Composition and Antioxidant Effect of Glycosidically Bound Volatile Compounds from Oregano (*Origanum vulgare* L. ssp. hirtum). *Food Chem.* **2000**, *71* (1), 79–83.

34. Mohammed, N. K.; Abd Manap, M. Y.; Tan, C. P.; Muhialdin, B. J.; Alhelli, A. M.; Meor Hussin, A. S. The Effects of Different Extraction Methods on Antioxidant Properties, Chemical Composition, and Thermal Behavior of Black Seed (*Nigella sativa* L.) Oil. *J. Evidence-Based Complementary Altern. Med.* **2016**, 5, 31–33.

35. Mujahid, Y. Synthesis Characterization and Anticancer Activity of Active Principle of *Nigella sativa* and Its Analogs. Ph.D. Dissertation, Abeda Inamdar Senior College of Arts Science & Commerce, 2014.

36. Ozer, E. K.; Goktas, M. T.; Toker, A.; Pehlivan, S.; Bariskaner, H.; Ugurluoglu, C.; Iskit, A. B. Thymoquinone Protects against the Sepsis Induced Mortality, Mesenteric Hypoperfusion, Aortic Dysfunction and Multiple Organ Damage in Rats. *Pharmacol. Rep.* **2017**, *69* (4), 683–690.

37. Pagola, S.; Benavente, A.; Raschi, A.; Romano, E.; Molina, M. A.; Stephens, P. W. Crystal Structure Determination of Thymoquinone by High-Resolution X-ray Powder Diffraction. *AAPS PharmSciTech* **2004**, *5* (2), 28.

38. Parry, J.; Su, L.; Luther, M.; Zhou, K.; Yurawecz, M. P.; Whittaker, P.; Yu, L. Fatty Acid Composition and Antioxidant Properties of Cold-Pressed Marionberry, Boysenberry, Red Raspberry, and Blueberry Seed Oils. *J. Agric. Food. Chem.* **2005**, *53* (3), 566–573.

39. Periasamy, V. S.; Athinarayanan, J.; Alshatwi, A. A. Anticancer Activity of an Ultrasonic Nanoemulsion Formulation of *Nigella sativa* L. Essential Oil on Human Breast Cancer Cells. *Ultrason. Sonochem.* **2016**, *31*, 449–455.

40. Piras, A.; Rosa, A.; Marongiu, B.; Porcedda, S.; Falconieri, D.; Dessì, M. A.; Ozcelik, B.; Koca, U. Chemical Composition and In Vitro Bioactivity of the Volatile and Fixed Oils of *Nigella sativa* L. Extracted by Supercritical Carbon Dioxide. *Ind. Crops Prod.* **2013**, *46*, 317–323.

41. Płotka-Wasylka, J.; Rutkowska, M.; Owczarek, K.; Tobiszewski, M.; Namieśnik, J. Extraction with Environmentally Friendly Solvents. *TrAC Trends Anal. Chem.* **2017**, *91*, 12–25.

42. Rafati, S.; Niakan, M.; Naseri, M. Anti-Microbial Effect of *Nigella sativa* seed Extract against Staphylococcal Skin Infection. *Med. J. Islamic Republic Iran.* **2014**, *28*, 42.

43. Roy, J.; Shakleya, D. M.; Callery, P. S.; Thomas, J. Chemical Constituents and Antimicrobial Activity of a Traditional Herbal Medicine Containing Garlic and Black Cumin. *Afr. J. Tradit., Complementary Altern. Med.* **2006**, *3* (2), 1–7.

44. Salem, A. A.; El Haty, I. A.; Abdou, I. M.; Mu, Y. Interaction of Human Telomeric G-Quadruplex DNA with Thymoquinone: A Possible Mechanism for Thymoquinone Anticancer Effect. *Biochim. Biophys. Acta* **2015**, *1850* (2), 329–342.

45. Salmani, J. M. M.; Asghar, S.; Lv, H.; Zhou, J. Aqueous Solubility and Degradation Kinetics of the Phytochemical Anticancer Thymoquinone; Probing the Effects of Solvents, pH and Light. *Molecules* **2014**, *19* (5), 5925–5939.

46. Saurabh Arora, D. B. *Introduction to High Performance Liquid Chromatography*; labtraining.com, 2014.

47. Schneider-Stock, R.; Fakhoury, I. H.; Zaki, A. M.; El-Baba, C. O.; Gali-Muhtasib, H. U. Thymoquinone: Fifty Years of Success in the Battle against Cancer Models. *Drug Discovery Today* **2014**, *19* (1), 18–30.

48. Siger, A.; Nogala-Kalucka, M.; Lampart-Szczapa, E. The Content and Antioxidant Activity of Phenolic Compounds in Cold-Pressed Plant Oils. *J. Food Lipids* **2008**, *15* (2), 137–149.

49. Soares, J. F.; Zabot, G. L.; Tres, M. V.; Lunelli, F. C.; Rodrigues, V. M.; Friedrich, M. T.; Pazinatto, C. A.; Bilibio, D.; Mazutti, M. A.; Carniel, N.; Priamo, W. L. Supercritical CO_2 Extraction of Black Poplar (*Populus nigra* L.) Extract: Experimental Data and Fitting of Kinetic Parameters. *J. Supercrit. Fluids* **2016**, *117*, 270–278.

50. Solati, Z.; Baharin, B. S.; Bagheri, H. Supercritical Carbon Dioxide (SC-CO_2) Extraction of *Nigella sativa* L. Oil Using Full Factorial Design. *Ind. Crops Prod.* **2012**, *36* (1), 519–523.

51. Solati, Z.; Baharin, B. S.; Bagheri, H. Antioxidant Property, Thymoquinone Content and Chemical Characteristics of Different Extracts from *Nigella sativa* L. Seeds. *J. Am. Oil Chem. Soc.* **2014**, *91* (2), 295–300.

52. Solati, Z.; Baharin, B. S. Antioxidant Effect of Supercritical CO_2 Extracted *Nigella sativa* L. Seed Extract on Deep Fried Oil Quality Parameters. *J. Food Sci. Technol.* **2015**, *52* (6), 3475–3484.

53. Sovova, H.; Sajfrtova, M.; Topiar, M. Supercritical CO_2 Extraction of Volatile Thymoquinone from *Monarda didyma* and *M. fistulosa* Herbs. *J. Supercrit. Fluids* **2015**, *105*, 29–34.

54. Stashenko, E.; Martínez, J. R. *Gas Chromatography-Mass Spectrometry*; In Tech, London, UK, 2014.

55. Su, X.; Ren, Y.; Yu, N.; Kong, L.; Kang, J. Thymoquinone Inhibits Inflammation, Neoangiogenesis and Vascular Remodeling in Asthma Mice. *Int. Immunopharmacol.* **2016**, *38*, 70–80.

56. Sultan, M. T.; Butt, M. S.; Anjum, F. M.; Jamil, A.; Akhtar, S.; Nasir, M. Nutritional Profile of Indigenous Cultivar of Black Cumin Seeds and Antioxidant Potential of Its Fixed and Essential Oil. *Pak. J. Bot.* **2009**, *41* (3), 1321–1330.

57. Şahin, S. Zeytin Ağaci Yapraklarindan Süperkritik-CO_2 İle Ekstrakt Eldesi ve Bileşimindeki Oleuropein Miktarinin İncelenmesi. Doktora Tezi, İstanbul Üniversitesi Fen Bilimleri Enstitüsü, 2011.

58. Şeleci, D. A.; Gümüş, Z. P.; Yavuz, M.; Şeleci, M.; Bongartz, R.; Stahl, F.; Coşkunol, H., Timur, S.; Scheper, T. A Case Study on In Vitro Investigations of the Potent Biological Activities of Wheat Germ and Black Cumin Seed Oil. *Turk. J. Chem.* **2015**, *39*, 801–812.

59. Taka, E.; Mazzio, E. A.; Goodman, C. B.; Redmon, N.; Flores-Rozas, H.; Reams, R.; Darling-Reed, S.; Soliman, K. F. Anti-inflammatory Effects of Thymoquinone in Activated BV-2 Microglial Cells. *J. Neuroimmunol.* **2015**, *286*, 5–12.

60. Tavakkoli, A.; Ahmadi, A.; Razavi, B. M.; Hosseinzadeh, H. Black Seed (*Nigella sativa*) and Its Constituent Thymoquinone as an Antidote or a Protective Agent against Natural or Chemical Toxicities. *Iran. J. Pharm. Res.* **2017**, *16*, 2–23.

61. Tenore, G. C.; Ciampaglia, R.; Arnold, N. A.; Piozzi, F.; Napolitano, F.; Rigano, D.; Senatore, F. Antimicrobial and Antioxidant Properties of the Essential Oil of *Salvia lanigera* from Cyprus. *Food Chem. Toxicol.* **2011**, *49* (1), 238–243.

62. Toma, C. C.; Olah, N. K.; Vlase, L.; Mogosan, C.; Mocan, A. Comparative Studies on Polyphenolic Composition, Antioxidant and Diuretic Effects of *Nigella sativa* L. (Black Cumin) and *Nigella damascena* L. (Lady-in-a-Mist) Seeds. *Molecules* **2015**, *20* (6), 9560–9574.

63. Venkatachallam, S. K. T.; Pattekhan, H.; Divakar, S.; Kadimi, U. S. Chemical Composition of *Nigella sativa* L. Seed Extracts Obtained by Supercritical Carbon Dioxide. *Int. J. Food Sci. Technol.* **2010**, *47* (6), 598–605.

64. Vladić, J.; Canli, O.; Pavlić, B.; Zeković, Z.; Vidović, S.; Kaplan, M. Optimization of *Satureja montana* Subcritical Water Extraction Process and Chemical Characterization of Volatile Fraction of Extracts. *J. Supercrit. Fluids* **2017**, *120*, 86–94.

65. Young, I.; Woodside, J. Antioxidants in Health and Disease. *J. Clin. Pathol.* **2001**, *54* (3), 176–186.

66. Yusufi, M.; Banerjee, S.; Mohammad, M.; Khatal, S.; Swamy, K. V.; Khan, E. M.; Aboukameel, A.; Sarkar, F. H.; Padhye, S. Synthesis, Characterization and Anti-Tumor Activity of Novel Thymoquinone Analogs against Pancreatic Cancer. *Bioorg. Med. Chem. Lett.* **2013**, *23* (10), 3101–3104.

67. Zzaman, W.; Silvia, D.; Abdullah, W. N. W.; Yang, T. A. Physicochemical and Quality Characteristics of Cold and Hot Press of *Nigella sativa* L. Seed Oil Using Screw Press. *J. Appl. Sci. Res.* **2014**, *10* (12), 36–45.

CHAPTER 8

SIMULTANEOUSLY EVAPORATED Al-DOPED Zn FILMS FOR OPTOELECTRONIC APPLICATIONS

YASIR BEERAN POTTATHARA[1,2*], OBEY KOSHY[1,2], and M. A. KHADAR[2]

[1]International and Inter University Centre for Nanoscience and Nanotechnology, Mahatma Gandhi University, Kottayam 686560, India

[2]Centre for Nanoscience and Nanotechnology, University of Kerala, Trivandrum 695037, India

*Corresponding author. E-mail: ptyasirbeeran@gmail.com

ABSTRACT

We report the effect of simultaneously evaporated and oxidized thin films of aluminum (Al)- and zinc (Zn)-containing different atomic weight percentages of Al (2, 4, and 6%) in an atmosphere of nitrogen. Grazing incidence angle X-ray diffraction technique observed that the crystallinity remains unaffected with increase in dopant concentration and the samples not showing special growth and have a typical nanocrystalline structures. Atomic force microscopy images revealed aggregation of the grains and Raman peaks claims hexagonal wurtzite structure with good crystal quality. Samples showed decreased optical transmittance with increase in Al concentration. Photoluminescence analysis indicated that Al-doped ZnO thin films did not introduce much defects and preserved good material quality with a high luminescence. The electrical resistance decreased with increase in temperature, showing semiconducting behavior.

8.1 INTRODUCTION

Nanostructured materials with unique physical and chemical properties have fascinated a great deal of consideration because of the numerous applications in the field of both nanoelectronics and optoelectronics.[1] The nanomaterials having morphological features within nanometer range and particles with nanometer sizes showing enormous technological interest. Zinc oxide (ZnO), an n-type II–VI group semiconducting material with direct band gap (3.37 eV at room temperature) and exciton binding energy of 60 meV with hexagonal wurtzite structure,[2,3] is a perfect material suitable for the production of optoelectronic and piezoelectric devices, UV light emitting diodes, and thin film transistors.[3] The improved efficiency of this material also finds applications in photocatalysis, photodetectors, and in solar cells and is being studied expansively.[4]

Dopants are playing important roles of ZnO on the optical, electronic, and structural properties. Generally, elements of group III of periodic table (such as B, Al, Ga, In, and N) have been used to improve the performance of ZnO.[5] Based on previous efforts, aluminum (Al) decorates promising dopant element for ZnO due to several benefits. Aluminum-doping concentration can be altered the optical band gap of ZnO.[6] Al-doped ZnO (AZO) thin films showed cost-effective and easy-production processes, plenty in nature, nonpoisonous, and prominent photovoltaics than typical indium tin oxide films.

Numerous techniques were reported for the thin film fabrications of ZnO, while their texture and morphology strongly hang on the deposition techniques and experimental set up. The techniques include modified chemical vapor deposition,[4] metal organic chemical vapor deposition,[7] pulsed laser deposition (PLD),[8] rf magnetron sputtering,[9] molecular beam epitaxy (MBE),[10] and thermal evaporation.[11] In rf sputtering, ion bombarding hits the substrates and change deposition conditioning such as sudden heating of the substrates etc. MBE gives excellent uniformity and very low defect concentration, and PLD meets the best deposition characteristics than all other methods. Compared to these vacuum deposition techniques, the vacuum thermal evaporation meets the benefit of cost-effective technique and has the advantage of depositing nanostructured thin films by manipulating the deposition conditions.

In the present study, simultaneously evaporated and oxidized thin films of aluminum- and zinc-containing different atomic weight percentages of Al (2, 4, and 6%) were fabricated, and their structure, morphology, optical, and electrical properties were investigated. Apart from the synthesis, characterization, and material quality of the AZO films, the present study highlights the response of transmittance, photoluminescent, and electrical properties to varying Al concentrations. By addressing such concentration dependent physical property variations within existing crystal structures, we can aim to improve the various technological applications of the same.

8.2   EXPERIMENTAL

In the present work, simultaneously evaporated and oxidized Al–Zn thin films were prepared containing different atomic weight percentage of Al at 2, 4, and 6 on glass substrates. Zn and Al were procured from Merck chemicals with 99.99% purity. The total weight of the source material was maintained as 0.3 g, whereas the atomic percentage of Al was kept at 2, 4, and 6. The substrates were washed well with running water and detergent, placed in dilute HNO_3 solution and methanol, and finally ultrasonicated using distilled water followed by acetone. AZO films were deposited at a moderate vacuum with high pressure of nitrogen gas (99.99%) of 1×10^{-4} mbar to ensure that the deposited films were in nanostructured range. After deposition, films were oxidized at 550°C for 60 min. The fabricated thin films with Al concentrations at 2, 4, and 6 at.% are given the sample codes 2AZO, 4AZO, and 6AZO, respectively.

The crystallinity and crystal structure of the films were examined by grazing incidence X-ray diffraction (GI-XRD). For recording GI-XRD patterns, a Brucker AXS D8 advance X-ray diffractometer–containing Kristalloflex 780, KF.4KE ($\lambda{\sim}1.54$ Å) X-ray source was used. An AMBIOS DT-4048M-VI XP-1 stylus profiler was used to estimate average thickness of the film samples. The EDX spectra of the 2AZO and 6AZO film samples were obtained using OXFORD SWIFT-7582 energy dispersive spectrometer. The surface topography of the samples was explored by atomic force microscope (AFM) (Digital Instruments Nanoscope-E) in contact mode, and the surface roughness and grain size were determined using WSxM software. Micro-Raman bands of AZO film samples were examined with a LaBRAM-HR 800 Raman spectrometer with Argon laser source of excitation wavelength 488 nm. The UV–visible spectra of the samples were analyzed using a JASCO V-650 double beam spectrophotometer. The photoluminescent spectra of 6AZO film were studied with various excitation wavelengths of 250, 270, and 290 nm using HITACHI F-7000 fluorescence spectrometer. The dc electrical resistivity of the film samples was calculated from 300 to 450 K with an indigenously established experimental system equipped with the help of an electrometer (Keithley 617) and a temperature controller (DRC 93c Lakeshore) in a warm up style.

8.3   RESULTS AND DISCUSSION

8.3.1   STRUCTURAL AND MORPHOLOGICAL STUDIES

Figure 8.1 shows the GI-XRD patterns of film samples of ZnO doped with 2, 4, and 6 at.% of Al, respectively (sample codes: 2AZO, 4AZO, and 6AZO, respectively) and oxidized at 550°C for 60 min. In all cases, the polycrystalline nature of samples was identified by the presence of several peaks in the GI-XRD pattern. The GI-XRD pattern of AZO film samples shows five prominent peaks at 2θ values of 31.8, 34.5, 36.3, 47.5, and 56.6°, which corresponds to the reflections of (1 0 0), (0 0 2), (1 0 1), (1 0 2), and (1 1 0) planes of hexagonal zinc oxide structure (36-1451-ICDD), respectively. The (1 0 1) peak has the highest intensity which is in good agreement with the highest intensity peak in the ICDD file (36-1415). No diffraction peaks that correspond to Al or Zn are observed which suggests that Al doping does not disturb the lattice structure of ZnO or possibly Al atom isolate to the amorphous regions in boundaries and formulae oxygen bond. No peak corresponding to Zn was observed indicating the complete transformation of Zn films to ZnO films while oxidizing at 550° for 60 min.[12] Diffraction peak positions are not clearly shifted which indicates that AZO films have no changes in structure with the dopant concentration.

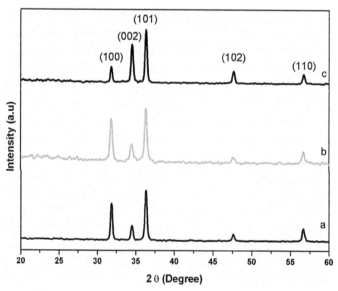

FIGURE 8.1   GI-XRD pattern of nanostructured (a) 2AZO, (b) 4AZO, and (c) 6AZO film samples oxidized at 550°C for 60 min.

There is no observed grain orientation of film samples by the relative intensity analysis of GI-XRD peaks (Fig. 8.1). The intensity of the (1 0 1) diffraction is the highest at 2AZO film sample, indicating an improvement in film crystallinity at this composition. However, at higher composition, the (1 0 1) peak intensity decreased, indicating the best crystallinity corresponding to the 2AZO film. Also, the relative intensity of (1 0 0) and (1 1 0) peaks decreased and that of (0 0 2) and (1 0 2) peaks increased. It is observed that there is no preferential deposition along any axis. The GI-XRD patterns indicate that the films actually do not have much preferential orientation. The preferred orientation, if any, of AZO films was determined by calculating the texture coefficient using the equation,[12]

$$TC_{(hkl)} = \frac{\left(I_{(hkl)} / I_{0(hkl)} \right)}{\left(1/n \sum \left(I_{(hkl)} / I_{0(hkl)} \right) \right)} \qquad (8.1)$$

where $I_{(hkl)}$ is the intensity of the diffraction peak from the $(h\ k\ l)$ plane of the present sample, $I_{0(hkl)}$ is the intensity of the equivalent diffraction peak in the reference XRD pattern (ICDD file) of randomly oriented grains of the material of the sample, and n is the total number of considered diffraction peaks. The calculated values of the texture coefficient for (1 0 1) plane, $TC_{(1\ 0\ 1)}$, for the AZO films oxidized at different temperatures showed that all films have texture coefficient greater than unity for (1 0 1) plane, which indicates that the films may have slight special orientation along (1 0 1) plane. The average crystallite size of the AZO films was estimated by the Debye–Scherrer equation.[13]

$$D = \frac{0.9\lambda}{(\beta \cos\theta)} \qquad (8.2)$$

where λ is the wavelength of X-ray (1.5406 Å), β is the full width at half maximum (FWHM) of the characteristic X-ray peak, and θ is the angle of diffraction. For calculating average crystallite size, the (1 0 1) peaks were used.

Table 8.1 shows average grain size, FWHM, and interplanar spacing of AZO film samples with different Al concentrations. The change in FWHM generally reflects the change in the crystallite size such as increased FWHM corresponds to the decreased crystallite size. The observed crystallite size of the 2AZO film sample oxidized at 550°C is 28 nm. The crystallite sizes ranged from about 28 nm for the 2AZO film sample to 31 nm for 6AZO film sample showing increment with an increase of Al content. The reason behind the observation is due to the fact that at higher temperatures than 327°C, the grain growth of zinc oxide is generally indorsed by the fast dispersal of zinc interstitials.[14] The GI-XRD patterns of AZO film show typical nanocrystalline hexagonal ZnO structure without obvious preferential growth.

TABLE 8.1   The XRD Data for AZO Film Samples Oxidized at 550°C for 60 min.

Sample name	2θ (°)	Obtained d value (Å)	ICDD d value (Å)	(h k l) plane	FWHM (°)	Crystallite size (nm)
2AZO	36.33	2.4708	2.4759	101	0.2971	28
4AZO	36.30	2.4728	2.4759	101	0.2902	28
6AZO	36.30	2.4728	2.4759	101	0.2502	31

Energy dispersive X-ray (EDX) analysis was used to determine the composition of AZO film samples. Figure 8.2(a) and (b) shows the EDX spectrum of 2AZO and 6AZO films deposited on glass substrates which indicate the presence of elements Zn and Al. The observed dopant concentrations in 2AZO and 6AZO film samples are about 1.318 and 5.003 atomic wt%, respectively. The comparatively lower value with the dopant concentrations in the source may be credited to the melting temperature differences of zinc and aluminum.[15] The peaks seen in the EDX spectra other than that of Zn and Al are from oxygen and from the glass substrate.

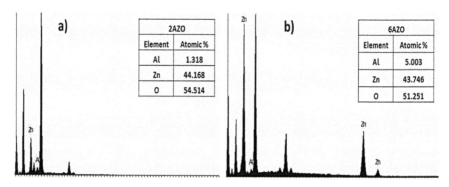

FIGURE 8.2   EDAX spectrum for nanostructured (a) 2AZO and (b) 6AZO film samples, respectively, oxidized at 550°C for 60 min.

To study the effect of Al doping on the vibrational characteristics of ZnO, micro-Raman spectroscopic analysis was done, as shown in Figure 8.3. Generally, the wurtzite ZnO has four primitive cell atoms foremost to twelve phonon branches including nine optical and three acoustic.[16] Group theory predicts the lattice optical phonons having the following equation:

$$\Gamma \text{opt} = 1A_1 + 2B_1 + 1E_1 + 2E_2,$$

where A_1 and E_1 are active modes in both IR and Raman spectra, E_2 is only active mode in Raman, whereas B_2 is inactive mode in both IR and Raman spectra. The bulk ZnO Raman modes for E_2 [E_2 (low) and E_2 (high)] is 101 and 437 cm^{-1}, respectively, and 574 cm^{-1} for A_1 (LO) mode.[17,18] The AZO film samples exhibit strong peaks at 99 and 437 cm^{-1} and peak like features around 104, 140, 334, and 580 cm^{-1}. Some of the Raman peaks are broad indicating considerable amorphous nature of the structure of the film samples and some weak bands with amorphous background confirming that grains are not fully crystalline nature.

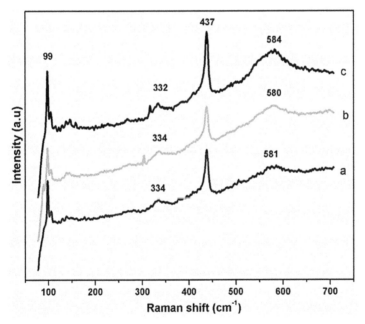

FIGURE 8.3 Micro-Raman spectra for nanostructured (a) 2AZO, (b) 4AZO, and (c) 6AZO film samples oxidized at 550°C for 60 min.

The detected intense Raman peaks at 99 and 437 cm^{-1} are allotted to the E_2 (low) and E_2 (high) optical phonon modes, respectively, reported for bulk ZnO.[19] The typical hexagonal wurtzite phase of ZnO was observed at

437 cm^{-1} for all samples, and its presence indicates the good crystal quality consistent with GI-XRD observations. Besides the weak and broad band observed at ~334 cm^{-1} in the AZO film, samples may be attributed to Raman second-order scattering usually originating from zone boundary phonon E_2 (M) in the Brillouin zone.[20] In nanostructured materials, a large fraction of atoms resides at grain boundaries, and as a result, surface optical phonons become observable with measurable intensity. One of the characteristic features of surface modes is that they appear in between the TO and LO modes. The surface phonon mode peak intensity should become weaker as the size of the grain increases. From the GI-XRD results, it is understood that the grain size of the AZO films declines with the increase in Al concentration. In addition to E_2 modes, weak Al (LO) mode is observed ~581 cm^{-1} with shifting to higher wavenumber can be accredited to the alteration of ZnO6 octahedra and this kind of shifting of the Raman bands to higher wavenumbers is due to the nanocrystalline nature of the film samples.

Figure 8.4 shows AFM images of AZO film samples, having almost uniformly distributed grains with well-defined boundaries and an observable change in surface morphology developed with higher Al content. The AFM images of 6AZO film show the distribution of smaller grains with well-defined boundaries compared to other samples. When the dopant concentration was increased, the grains are more densely packed with well-distinct boundaries.

The average grain size estimated from AFM data shows an enhancement from about 65 nm for the 2AZO film sample to about 105 nm for the 6AZO film. These values are found to be much larger to those obtained from the GI-XRD analysis. This discrepancy can be addressed by taking into account the fact that AFM measurements directly visualize the surface morphology of the agglomerated grains, while XRD gives the size of individual grains. Hence, it may be inferred that the grains of AZO film samples are aggregates of primary grain of smaller size. Increase in dopant concentration favors the formation of clusters and thereby causing an increase in the size of the aggregates in the samples.[21] The surface roughness, calculated using WSxM software, shows an increased root mean square (RMS) value of ~52 nm for 6AZO compared to ~42 nm for 2AZO films directing to the formation of clusters of various sizes and shapes. The aggregation of primary grains leads the increased roughness, confirming previous analysis that increased dopant concentration drops the smoothness of the film surface. The surface morphology of thin films has great applications in the field of photovoltaics such as dye sensitized solar cells.[22]

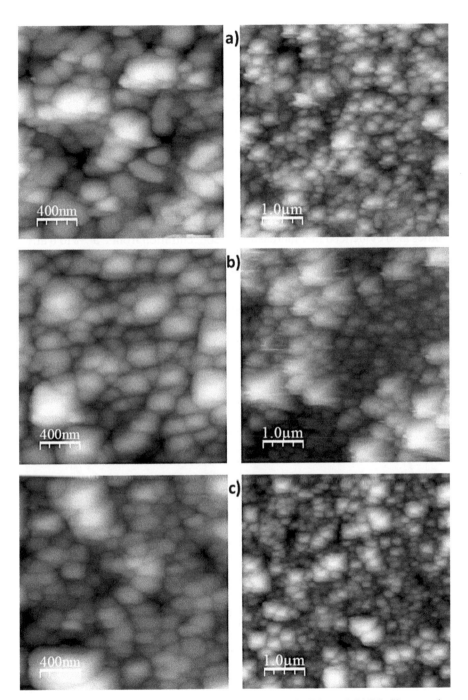

FIGURE 8.4 AFM images for nanostructured 2AZO, 4AZO, and 6AZO film samples oxidized at 550°C for 60 min.

8.3.2 OPTICAL STUDIES

To study the optical properties of AZO film samples, transmission spectra were recorded from 350 to 800 nm, shown in Figure 8.5(a). Generally, transmittance is associated to the surface roughness and crystallinity.[23] The optical transmittance of the AZO films decreased from 90% to 69% as the Al concentration was increased from 2 at.% to 6 at.%. The observed high transmittance in visible region indicates better crystallinity and fewer defects.[24] The maximum transmittance of ~90.17% was obtained for the film with the lowest Al concentration and transmittance decreasing with highest Al content could be because of the increased optical scattering generated by substantial Al doping as well as its rough surface morphology.[14,25] The AFM studies also evidenced that increasing Al concentration from 2 at.% to 6 at.% produced rough film surfaces. Crystalline nature and direct band gap of AZO films were identified by the sharp drop of transmittance around ~375 nm. The absorption edge, which is related to the optical energy gap, shows slight higher wavelength shift with increased dopant concentration.

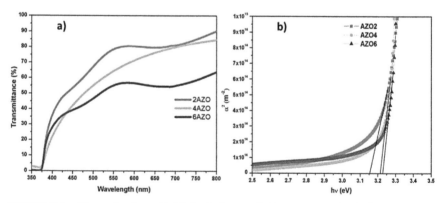

FIGURE 8.5 **(See color insert.)** (a) Optical transmission spectra and (b) α^2 as a function of the photon energy, $h\nu$, for nanostructured 2AZO, 4AZO, and 6AZO film samples oxidized at 550°C for 60 min.

The optical band gap (E_g) was determined from the extrapolation of the Tauc's plot (α^2 vs. $h\nu$), as shown in Figure 8.5(b) with the transmittance (T) and thickness (t) of the film samples using the relation

$$\alpha = \frac{\left(\ln\left(1/T\right)\right)}{t} \tag{8.3}$$

The average thickness was found to be 1000 Å, 1080 Å, and 1200 Å for the 2AZO, 4AZO, and 6AZO samples, respectively. The band gap of AZO films was increased from 3.19 eV for 2AZO film sample to 3.29 eV for 6AZO film sample. The slight widening of optical band gap of AZO film samples may be attributed to the Burstein–Moss shift, blocking of lowest states in the conduction band due to doping.[26] Band gap widening also specifies the increased amount of free charge carriers. In the case of ZnO, conduction mechanism is mainly subjected by free electrons which regulate the widening of optical band gap to relatively smaller values than bulk band gap of ~3.37 eV. The detected red shift in the case of AZO samples are taking into the account of smaller size of crystallites.[22,27]

8.3.3 PHOTOLUMINESCENT STUDIES

Luminescence of a material has a great connection with the crystallinity.[27] From the structural findings, 2AZO film sample has the highest improvement in film crystallinity. To verify the accuracy of photoluminescence (PL) spectra, the spectra recorded for three excitation wavelengths of (a) 290, (b) 270, and (c) 250 nm of 2AZO thin film oxidized at 550°C for 60 min are shown in Figure 8.6. There are no observable variations in the peak positions of the film sample, but the intensity of the peaks altered with the excitation wavelengths. The PL spectra of the 2AZO sample show high intensity near band edge emission (NBE) peak at 377 nm (3.29 eV) and a weak NBE peak at ~397 nm. The spectra also show weak PL emission in visible (violet and blue) range, usually related with deep level (DL) emission, in addition to the UV emission. Presence of strong NBE peak is the result of exciton recombination,[28] and emission in violet and blue region is due to Zn and O_2 vacancies, respectively.[29] For AZO films, the NBE emission peak is centered at 377 nm compared to 380 nm for bulk ZnO. The intensity of band edge luminescence in the present case is found to decrease with excitation wavelength. The strong NBE emission of the films is an indication of the good quality of the AZO films. Li et al. also reported strong intensity NBE emission without DL emission for AZO thin films.[27] Behra and Acharya have observed PL bands at 449 and 484 nm in the emission spectra for 5 at.% AZO thin films.[20] Besides strong NBE emission peak, some weak emission peaks at 450, 468, and 482 nm are observed in the AZO film sample. The PL analysis indicated that AZO thin films did not introduce much defects and preserved good material quality with a high luminescence. The strong NBE in AZO films make them suitable candidate for optoelectronic device fabrication.[18]

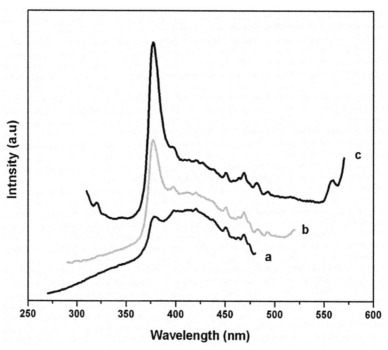

FIGURE 8.6   PL spectra for nanostructured 2AZO film sample oxidized at 550°C for 60 min for excitation wavelength of (a) 290, (b) 270, and (c) 250 nm.

8.3.4   ELECTRICAL STUDIES

The dc resistance of 2AZO and 6AZO films was carried out in the temperature range from 300 to 450 K, as shown in Figure 8.7(a) and (b). The average thickness was found to be 1000 Å and 1200 Å for the 2AZO and 6AZO samples, respectively. Generally, various factors including concentration of dopants, oxygen vacancies, material composition, and grain boundaries are influencing the resistivity of a material.[30] The decreased electrical resistivity with increased temperature indicates the semiconducting behavior of AZO films. It may be attributed that the increment of Al content may cause the substantial drop of carrier concentration and hall mobility.[31] The increased electrical conductivity of 2AZO film sample was due to the better crystallinity and dense structure which can trap more free electrons compared with 6AZO film sample.[32] Sometimes, the highest amount of doping can reduce the carrier mobility and thus increases the electrical resistivity.[33]

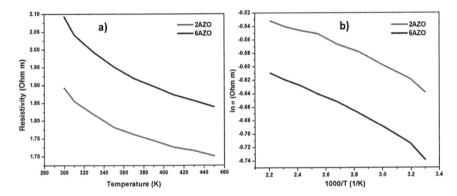

FIGURE 8.7 **(See color insert.)** (a) Electrical resistivity vs. temperature in the range of 300–450 K and (b) variation of electrical conductivity with reciprocal temperature of 2AZO and 6AZO film samples oxidized at 550°C for 60 min.

8.4 CONCLUSION

Simultaneously evaporated and oxidized thin films of aluminum (Al) and zinc (Zn) were fabricated with different atomic weight percentage of Al at 2, 4, and 6 on glass substrates at a pressure of 1×10^{-4}. The GI-XRD analysis shows that the AZO thin films have typical nanocrystalline hexagonal structure without obvious preferential growth. The observation of strong high frequency E_2 Raman mode supports the structural findings. The AFM images film surfaces show a uniformly distributed grain with well-defined boundaries. The transmittance of the film samples decreased and the optical band gap slightly widens with increased Al concentration. The PL analysis indicated that AZO thin films did not have much defects and preserved good material quality with high luminescence. The dc electrical resistivity of AZO film samples decreased with the increase in temperature, showing semiconducting behavior, and 2AZO films showing better electrical conductivity confirm the structural and morphological findings.

KEYWORDS

- physical vapor deposition
- thermal coevaporation
- ZnO
- photoluminescence
- optoelectronic applications

REFERENCES

1. Zhong, W.-W.; Liu, F.-M.; Cai, L.-G.; Peng-Ding; Zhou, C.-C.; Zeng, L.-G.; Liu, X.-Q.; Li, Y. *J. Alloys Compd.* **2011**, *509*, 3847.
2. Fang, G. J.; Li, D.; Yao, B.-L. *Thin Solid Films* **2002**, *418*, 156.
3. Ding, J. J.; Chen, H. X.; Ma, S. Y. *Physica E* **2010**, *42*, 1861.
4. Sun, T.; Qiu, J. *Fabrication of ZnO Microtube Arrays via Vapor Phase Growth*; 2008.
5. Díez-Betriu, X.; Jiménez-Rioboo, R.; Marcos, J. S.; Céspedes, E.; Espinosa, A.; de Andrés, A. *J. Alloys Compd.* **2012**, *536*, S445.
6. Kuhr, M.; Bauer, S.; Rothhaar, U.; Wolff, D. *Thin Solid Films* **2003**, *442*, 107.
7. Cui, Y.; Du, G.; Zhang, Y.; Zhu, H.; Zhang, B. *J. Cryst. Growth* **2005**, *282*, 389.
8. Zhu, B. L.; Sun, X. H.; Zhao, X. Z.; Su, F. H.; Li, G. H., Wu, X. G.; Wu, J.; Wu, R.; Liu, J. *Vacuum* **2008**, *82*, 495.
9. Kumar, S.; Gupta, V.; Sreenivas, K. *Nanotechnology* **2005**, *16*, 1167.
10. Chen, Y.; Bagnall, D. M.; Koh, H.; Park, K.; Hiraga, K.; Zhu, Z.; Yao, T. *J. Appl. Phys.* **1998**, *84*, 3912.
11. Girtan, M.; Rusu, G. G.; Dabos-Seignon, S.; Rusu, M. *Appl. Surf. Sci.* **2008**, *254*, 4179.
12. Koshy O.; Khadar, M. A. *J. Appl. Phys.* **2011**, *109*, 124315.
13. Cullity, B. D. *Phys. Today* **1957**, *10*, 50.
14. Wang, Z. L. *J. Phys. Condens. Matter* **2004**, *16*, R829.
15. Liu, Y.; Lian, J. *Appl. Surf. Sci.* **2007**, *253*, 3727.
16. Yadav, H. K.; Sreenivas, K.; Katiyar, R. S.; Gupta, V. *J. Phys. D. Appl. Phys.* **2007**, *40*, 6005.
17. Özgür, U.; Alivov, Y. I.; Liu, C.; Teke, A.; Reshchikov, M. A.; Doğan, S.; Avrutin, V.; Cho, S.-J.; Morkoç, H. *J. Appl. Phys.* **2005**, *98*, 041301.
18. Vinodkumar, R.; Navas, I.; Chalana, S. R.; Gopchandran, K. G.; Ganesan, V.; Philip, R.; Sudheer, S. K.; Mahadevan Pillai, V. P. *Appl. Surf. Sci.* **2010**, *257*, 708.
19. Alim, K. A.; Fonoberov, V. A.; Shamsa, M.; Balandin, A. A. *J. Appl. Phys.* **2005**, *97*.
20. Behera, D.; Acharya, B. S. *J. Lumin.* **2008**, *128*, 1577.
21. Grätzel, M. *J. Photochem. Photobiol. C: Photochem. Rev.* **2003**, *4*, 145.
22. Suchea, M.; Christoulakis, S.; Katharakis, M.; Kiriakidis, G.; Katsarakis, N.; Koudoumas, E. *Appl. Surf. Sci.* **2007**, *253*, 8141.

23. Zhu, B. L.; Wang, J.; Zhu, S. J.; Wu, J.; Zeng, D. W.; Xie, C. S. *Thin Solid Films* **2012**, *520*, 6963.

24. Zhong, W. W.; Liu, F. M.; Cai, L. G.; Zhou, C. C.; Ding, P.; Zhang, H. *J. Alloys Compd.* **2010**, *499*, 265.

25. Kim, D. K.; Kim, H. B. *J. Alloys Compd.* **2011**, *509*, 421.

26. Burstein, E. *Phys. Rev.* **1954**, *93*, 632.

27. Li, X.-Y.; Li, H.-J.; Wang, Z.-J.; Xia, H.; Xiong, Z.-Y.; Wang, J.-X.; Yang, B.-C. *Opt. Commun.* **2009**, *282*, 247.

28. Srikant, V.; Clarke, D. R. *J. Appl. Phys.* **1997**, *81*, 6357.

29. Bagnall, D. M.; Chen, Y. F.; Zhu, Z.; Yao, T.; Shen, M. Y.; Goto, T. *Appl. Phys. Lett.* **1998**, *73*, 1038.

30. Igasaki, Y.; Saito, H. *J. Appl. Phys.* **1991**, *70*, 3613.

31. Jeong, S. H.; Park, B. N.; Lee, S.-B., Boo, J.-H. *Surf. Coatings Technol.* **2007**, *201*, 5318.

32. Zhaochun, Z.; Baibiao, H.; Yongqin, Y.; Deliang, C. *Mater. Sci. Eng. B* **2001**, *86*, 109.

33. El Hichou, A.; Diliberto, S.; Stein, N. *Surf. Coatings Technol.* **2015**, *270*, 236.

CHAPTER 9

WATER VAPOR ADSORPTION, DESORPTION, AND PERMEATION THROUGH SODIUM CARBOXY METHYL CELLULOSE- AND HYDROXYPROPYL CELLULOSE-BASED FILMS

DIDEM BERKÜN, DEVRIM BALKÖSE*, FUNDA TIHMINLIOĞLU, and SACIDE ALSOY ALTINKAYA

Department of Chemical Engineering, Izmir Institute of Technology, Gülbahçe, 35437 Urla-Izmir, Turkey

Corresponding author. E-mail: devrimbalkose@gmail.com

ABSTRACT

Two types of films consisting of sodium salt of carboxy methyl cellulose (NaCMC) and hydroxypropyl cellulose (HPC) as film forming materials and glycerin as plasticizer were prepared and characterized, and their water vapor sorption, permeation, and mechanical properties were determined.

The water sorption isotherms of the cellulose-based films were measured by two different methods using the Environmental Chamber and Magnetic Suspension Balance, respectively. Water vapor permeability (WVP) of the films was measured using a permeation cell. The results show that diffusion of water vapor in NaCMC-based film is faster than that in HPC-based films, due to the heterogeneous structure and larger pore dimensions of the NaCMC films that was observed in the scanning electron micrographs. In addition, WVP of NaCMC films was higher than that of HPC-based films. Carrots coated with NaCMC had longer shelf life in terms of weight, dimensions, and color.

9.1   INTRODUCTION

The increased consumer demand for high quality, long shelf-life ready to eat foods has initiated the development of mildly preserved products that keep their natural and fresh appearance as long as possible. For this purpose; over the past 30 years, considerable research effort has been devoted to the uses of edible films and coatings. An edible coating or film has been defined as a thin, continuous layer of edible materials, which may be eaten together with the food, formed or placed, on or between foods or food components. Their function is to provide a barrier to mass transfer (water, gas, and lipids), to serve as a carrier of food ingredients and additives (pigments, flavors, etc.), or to provide mechanical and microbial protection. Edible film coatings represent a viable preservation technology; they often constitute a water vapor barrier that delays deterioration and maintain the product structural integrity. Hydroxy propylmethyl cellulose, a cellulose derivative used in pharmaceutical industries, was employed as the basic formulation component to coat fresh blueberries. Addition of carrageenan, plasticizer, and carnauba wax emulsion adds value to film structure and functionality.[14] The structural and barrier properties of edible films are affected by some parameters such as viscosity of film forming solution, film formation procedure, film thickness, water vapor sorption characteristics, etc. The water vapor permeability (WVP) is the most extensively studied property of edible films mainly because of the importance of the role of water in deteriorative reactions.[2,3,14,15] Water vapor adsorption data give the hydration properties of polymer; thus, sorption isotherms (adsorption and desorption) of these films have been extensively studied. The moisture sorption behavior of methyl cellulose films was investigated in order to evaluate some functional properties of films such as barrier property and stability of the films.[4,7,9,15,17] It was shown that diffusion of water vapor in sodium carboxy methyl cellulose-based film is faster than that in hydroxypropyl cellulose (HPC)-based films, due to the heterogeneous structure and larger pore dimensions of the NaCMC films.[6]

The objective of this work is to produce and characterize cellulose-based edible films that are water vapor permeable. For this purpose, sodium NaCMC and HPC were used as the cellulose-based film forming materials. Films were characterized by using scanning electron microscopy (SEM), differential scanning calorimetry (DSC), X-ray diffraction (XRD) analysis, and thermal gravimetric analysis (TGA) techniques. In addition, the water vapor sorption and permeability characteristics and mechanical properties of the films were determined. Increasing shelf life of carrots by coating with NaCMC was also investigated.

9.2 MATERIALS AND METHODS

9.2.1 MATERIALS

The film forming materials HPC (M_w = 370.000) and sodium salt of carboxy methyl cellulose (NaCMC) (having 250.000 g/mol molecular weight and with the degree of substitution of 1.2) used in this study were supplied by Sigma-Aldrich. Glycerol that was used as the plasticizer was purchased from Merck. Deionized water was used as a solvent throughout the experiment. Fresh carrots from the market were used for NaCMC-coating tests.

9.2.2 METHODS

9.2.2.1 FILM PREPARATION

To prepare films, 3–5 g of HPC or NaCMC was separately dissolved in 100 mL deionized water at room temperature by continuously stirring the solutions for 4 h. After complete dissolution of film forming materials, glycerol was added as a plasticizer at 10% of the dry weight, and the solutions were homogenized with a magnetic stirrer at room temperature for 1 h. Then, the solutions were kept in a vacuum oven at 25°C for about 48 h to remove air bubbles or dissolved air. HPC and NaCMC solutions were spread on glass plates by using an automatic film applicator (Sheen Automatic Film Applicator 1133N). The spread films were dried at room temperature for 48 h, then at 60°C in an oven for 25 min. Finally, they were detached from the glass plates, covered with paper, and stored at room temperature. The thickness of the films was measured with a digital micrometer at 10 different locations. Film thicknesses were approximately measured as 10, 20, and 30 μm for films prepared from solutions with 3, 4, and 5 wt% polymer concentration, respectively.

9.2.2.2 CHARACTERIZATION OF THE FILMS

Morphology of the films was examined by using SEM. Each film was mounted on aluminum stubs using aluminum sticky tape and coated with gold palladium film in a VG Microtech SC 7610 Sputter coater. Then, specimen was examined using a Philips XL 30S FEG electron microscope.

The crystal structure of the films was analyzed by using a Philips X'pert Pro. Diffractometer with CuK_α radiation. The scattering angle (2θ) was varied from 5 to 70°. DSC analysis was used to determine the glass transition temperature (T_g) of the films. For this purpose, film samples (3.4–5.2 mg) placed in aluminum crucibles were heated from room temperature to over 200°C at a heating rate of 10°C/min under nitrogen purge (flow rate = 40 mL/min) by using Shimadzu DSC-50 differential scanning calorimeter (DSC). Thermal properties of the NaCMC and HPC films were determined by the thermal gravimetric analyzer (Shimadzu TGA-51). ~10 mg and 11 mg of NaCMC and HPC film samples, respectively, were heated from room temperature up to 200°C at a heating rate of 10°C/min. Nitrogen gas was used as inert atmosphere in all thermal analysis. An Instron Universal Testing Instrument (Model 4210) was used to determine mechanical properties (TS and %E) of films in accordance with ASTM D882-83 (1984). The Environmental Chamber was used to condition each film specimens at 20°C and 50% relative humidity (RH) for 48 h. Testing film strips were 100 mm long and 10 mm wide. The initial grip separation was set at 100 mm and crosshead speed was 5 mm/min. Young's modulus (MPa) (E), stress at break (N/mm²) and strain at break (%), stress at yield (N/mm²), and strain at yield (%) parameters were collected and obtained directly from the computer. At least five replicates of each NaCMC and HPC films were tested.

9.2.3   MEASUREMENT OF SORPTION ISOTHERMS OF FILMS

9.2.3.1   ENVIRONMENTAL CHAMBER EXPERIMENTS

The water sorption isotherms of the cellulose-based films were measured by using the Environmental Chamber (Angelantoni, Cimacolle, Italy). This procedure is based on the determination of the moisture content of samples being at equilibrium in a closed chamber whose temperature (accuracy ±0.5°C) and RH (accuracy ±1%) are controlled with the vaporization of water at a given temperature.

Before measuring each sorption isotherm, films were dried at 80°C at 0% moisture in an oven (Nüve FN 500/TS 6073 model) and then weighed using an analytical balance (Sartorius BP 2215) with a precision of 0.1 mg in order to determine the mass of dried films. After each equilibrium condition, the mass of the films at a specific RH was measured again with an analytical balance which was placed in the chamber. Film samples were reached to equilibrium condition in nearly 15–20 min. The RH values between 20% and 90% at 25°C were increased at 10% steps after each equilibrium condition. During changing of RH values, all film samples were covered with an aluminum foil to protect the films from air ventilation in the chamber. The adsorption capacities of films were determined by the kinetics of water gain during equilibration at different levels of RH between 20% and 90% at 25°C as a function of time.

9.2.3.2 GRAVIMETRIC SORPTION EXPERIMENTS

The water vapor sorption isotherms were also determined by gravimetric sorption experiments conducted in a magnetic suspension balance (Rubotherm). The experimental set up was described by Altinkaya et al.[1] During these experiments, film samples were placed into the measuring cell with multitray sample holder. The column enclosing the sample was heated up to 60°C by using a water bath, and vacuum was applied to the column by rotary vane pump down to 0.0001 mbar to remove the water desorbed from the samples during the heating process for 4 h. Then, the column was cooled to 25°C, and the system was allowed to reach equilibrium in 24 h. Water was heated in a flask by using a constant temperature bath and water vapor was sent to the column until equilibrium is reached. When equilibrium was obtained, valves of the column were closed and the vapor pressure of the water was increased by increasing the temperature of water in the flask. Then, by opening the valves of the column, water vapor was sent to the column until new equilibrium is reached

9.2.4 WVP MEASUREMENTS OF EDIBLE FILMS

WVP of the films was determined by using the permeation cell shown in Figure 9.1. The permeation cell consists of lower and upper compartments. Water is present in an open container in the lower part and upper part contains a hygrometer. The film with an area of 16.6 cm² was placed between two compartments. The upper part of the permeation cell (volume = 56.53 cm³) was dried as much as possible by means of sweeping with dry air obtained by passing ambient air through a zeolite filled column. The valves of the permeation cell were closed after drying. The RH values in the upper cell were measured by a hygrometer (Datalogger SK—L200TH) and were recorded by a computer with respect to time. Three replicates of each film type were tested for WVP measurements.

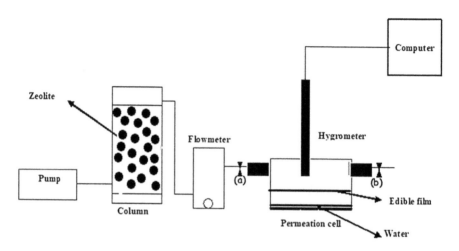

FIGURE 9.1 Schematic diagram of the experimental setup used for permeability measurements (a) and (b) are valves for isolating the upper part of the permeation cell.

9.2.5 COATING CARROTS WITH NaCMC AND TESTING SHELF LIFE

The effect of an edible film coating on preservation of mass, dimension, and color of fresh foods was shown by coating carrots with NaCMC solution. Carrots were washed and chosen by quality and uniformity. They were sliced in equal dimensions with an approximate thickness of 0.5 cm. Then, the slices were dipped to 3% or 4% NaCMC solution having 10% glycerol at 25°C for 5 s. The coating solutions were allowed to drain and afterward the applied coatings were dried by natural convection at room temperature. Control samples followed the same treatment but were dipped in distilled water. The coated and uncoated carrot samples were weighed daily for 3 days to measure the weight loss according to time. Also, their diameters were measured daily. The effect of the coating on carrot color was evaluated by measuring L, a, b values of the samples using an Avameter-Avamouse Colour Measurement device.

9.3   RESULT AND DISCUSSIONS

9.3.1   CHARACTERIZATION OF THE EDIBLE FILMS

9.3.1.1   MORPHOLOGY OF THE FILMS

SEM was used to determine the structure of the NaCMC- and HPC-based edible films. It can be seen from Figure 9.2 that both films have porous structures; on the other hand, HPC-based edible films have relatively more homogenous porous structure in comparison to NaCMC-based edible films. The cross-sectional micrographs of the film samples with magnification at 5000× indicate that HPC films contain many small pores with dimensions of ~0.5–1 μm.

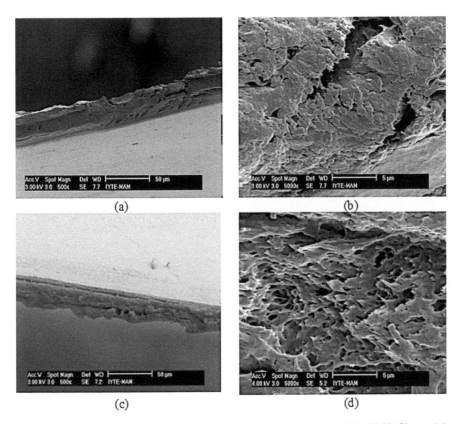

FIGURE 9.2 Scanning electron micrographs of the cross-section of NaCMC film at (a) 500× and (b) 5000× magnification and HPC film at (c) 500× and (d) 5000× magnification.

9.3.1.2   CRYSTALLINITY OF THE FILMS

XRD analyses were applied to investigate the crystallinity of the NaCMC and HPC-based edible films. Cellulose can be present in two different crystal forms. Cellulose I had peaks at 2θ values of 9.0°, 14.7°, 16.4°, and 22.6° and Cellulose II had peaks at 9.5°, 12.0°, 20°, and 21.5°.[1] XRD diagram of HPC film (Fig. 9.3a) has mainly two broad peaks at 2θ values of 9° and 16° indicating it has Cellulose I structure. The crystallinity of NaCMC was lowered compared to starting cellulose as the degree of substitution was increased.[11] However, XRD analysis of NaCMC film in the present study showed sharp crystalline peaks at 2θ values of 14°, 16.9°, and 25.2° (Fig. 9.3b). These sharp diffraction peaks could be due to orientation of molecules during solvent casting. NaCMC also appear to have Cellulose I structure.

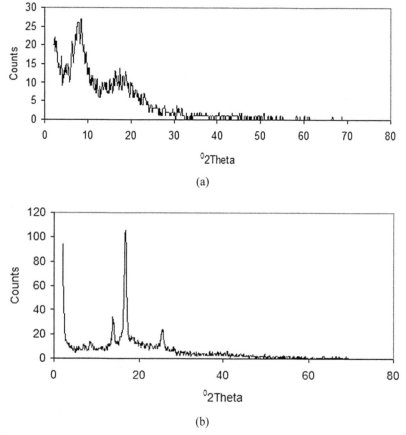

FIGURE 9.3   X-ray diffraction diagrams of the (a) HPC- and (b) NaCMC-based films.

9.3.1.3   DSC ANALYSIS

Figure 9.4a shows DSC curves for NaCMC-based edible films prepared from three different concentrations. One endotherm was observed due to dehydration and evaporation of water from the films that are above their glass transition temperature. Indeed, all films were prepared with 10 wt% glycerin to lower their T_g value below room temperature. The DSC analysis of all NaCMC edible films shows the endotherm maximum at 83–98°C. On the other hand, the DSC curves of HPC films also show a broad endotherm being maximum in the range of 70–85°C, corresponding to the loss of residual water as illustrated in Figure 9.4b.

DSC curves also indicated that NaCMC- and HPC-based films had T_g values that were lower than room temperature. Thus, rubbery films were used in the sorption and desorption experiments.

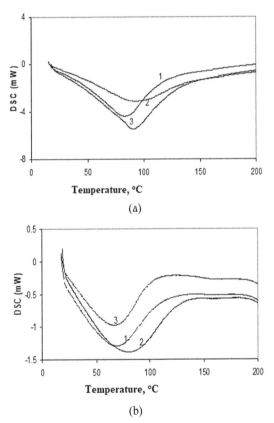

FIGURE 9.4   DSC curves of (a) NaCMC- and (b) HPC-based edible films from (1) 3%, (2) 4%, (3) 5% polymer solution.

9.3.1.4 THERMAL GRAVIMETRIC ANALYSIS

The weight of the films decreases on heating due to evaporation of water present in them. The TGA curves of the films obtained from 3 wt% polymer solutions shown in Figure 9.5 indicate that weight losses of the HPC film and NaCMC film up to 200°C were 3% and 15%, respectively. Drying of HPC film was completed up to 100°C, but NaCMC film continued to dry even at 200°C.

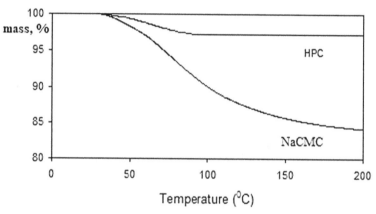

FIGURE 9.5   TGA curves of NACMC- and HPC-based films prepared from 3% polymer solution.

9.3.1.5 MECHANICAL PROPERTIES OF EDIBLE FILMS

Table 9.1 clearly shows the differences in the mechanical properties of NaCMC- and HPC-based edible films. The results in Table 9.1 indicate that NaCMC-based films had higher yield stress (33–68 MPa) and tensile stress (35–68 MPa) values than those (8.7–15.5 MPa; 8.5–15.5 MPa, respectively) of HPC films, so they are stronger than HPC-based films. On the other hand, strain at break values of HPC-based edible films are higher than NaCMC films, so these films are more elastic than NaCMC-based films. As seen from Table 9.1, as the concentration of polymers in film forming solutions increased, tensile strength of the films decreased, and their elongation values increased due to an increase in overall porosity of the films. NaCMC films prepared with the lowest NaCMC concentration had high tensile strength, high Young modulus, and low elongation values indicating that these films are strong; however, they have stiff, brittle structures. Increase in elongation values with the increasing NaCMC concentration indicated that the films were more ductile. The mechanical properties of HPC-based edible films showed similar behavior with increased HPC concentration in the film forming solution.

Tensile strength of NaCMC-based edible films is higher than HPC-based films due to the presence of crystals in these films.

Elongation at break values of (58–91%) of the HPC-based films are higher than those (4.4–10%) of NaCMC-based films, so it can be concluded that HPC polymers produce more ductile films. Higher elongation of the HPC-based films is due to the absence of crystalline phase in these films. Young modulus values give information about the stiffness of the films. According to the results in Table 9.1, NaCMC-based edible films were stiffer than HPC films. Lower Young modulus of HPC film indicated their more elastic behavior.

TABLE 9.1 Mechanical Properties of NaCMC and HPC Films.

Polymer	Solution conc. (wt%)	Yield stress (MPa)	Yield strain (%)	Young modulus (MPa)	Stress at break (MPa)	Strain at break (%)
NaCMC	3	68 ± 12	4.4 ± 1.8	3329 ± 334	68 ± 12	4.4 ± 1.8
NaCMC	4	38.9 ± 4.5	5.5 ± 1.7	2282 ± 228	39 ± 4	7 ± 1.6
NaCMC	5	33 ± 4.4	7.8 ± 2.6	1827 ± 260	35 ± 3.6	10 ± 2.5
HPC	3	15.5 ± 1.8	3.8 ± 0.9	729 ± 88	15.5 ± 0.7	58 ± 21
HPC	4	10.4 ± 4.6	3.5 ± 1.8	487 ± 213	13 ± 15.45	82 ± 20
HPC	5	8.7 ± 2.8	4.5 ± 0.5	399 ± 117	8.5 ± 13	91 ± 20

NaCMC, sodium salt of carboxy methyl cellulose; HPC, hydroxypropyl cellulose.

9.3.2 SORPTION ISOTHERMS OF EDIBLE FILMS

The relationship between water activity (a_w) and the moisture content of the films (at constant temperature) was described by moisture sorption isotherms. Figure 9.6 shows the equilibrium water content of the NaCMC- and HPC-based edible films as a function of water activity at 25°C and the effect of polymer concentration on the moisture-sorption isotherms of the films. NaCMC-based edible films have higher sorption capacity than HPC-based edible films. As seen from Figure 9.6, water vapor sorption capacity of the NaCMC- and HPC-based films was 70 wt%, and 25 wt%, respectively, at 90% RH As evident in Figure 9.6, the sorption curves of HPC- and

NaCMC-based edible films were typical of cellulose films. The curves showed a relatively small slope at a low water activity, with an exponential increase at high water activity (above 0.7) solely due to the higher sorption of water molecules by NaCMC and HPC.

Moisture sorption isotherms were also determined with the microbalance for NaCMC and HPC films prepared with 3% polymer concentration in the solution. As seen in Figure 9.6, the moisture content of the films determined from gravimetric sorption experiments was found to be lower than that obtained from the experiments conducted in the humidity chamber. This could be the result of lower drying temperature (60°C) in microbalance than that (80°C) in humidity chamber experiments; thus, higher residual water content in the films used in the microbalance. The differences in moisture content of the films dried at 60°C and 80°C for a period of 4 h were 1% and 3.5% for HPC and NaCMC films, respectively. The sorption isotherms of NaCMC and HPC films showed Type II isotherm that describes the sorption of water on macroporous solids. The macroporous structures of both films were confirmed with the scanning electron micrographs.

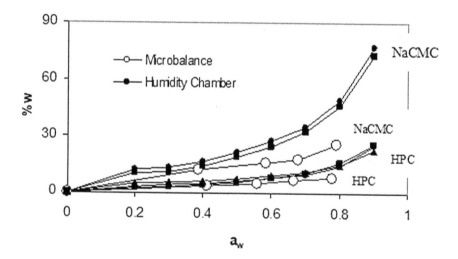

FIGURE 9.6 Moisture sorption isotherms of NaCMC- and HPC-based edible films prepared from solution concentrations of (•) 3%; (■) 4%; (▲) 5%. Filled symbols correspond to the data obtained from humidity chamber, while open circle (o) represents the data collected in microbalance using the films prepared with 3% polymer concentration.

Water sorption isotherms obtained in humidity chamber were fitted with four different models. These are Bruanuer–Emmet–Teller (BET),[1]

Guggenheim–Anderson–de Boer (GAB),[16] Smith,[1] and Halsey[1] models which are represented in eqs 9.1–9.4, respectively.

BET model: $1/\left[(1-a_w)\times M\right]=1/M_m+\left[1/(C\times M_m)\right]\times\left[(1-a_w)/a_w\right]$ (9.1)

where M_m is the monolayer moisture content and C is the constant related to the net heat of sorption. M is the equilibrium moisture content on a dry basis and a_w is the water activity.

GAB model: $M=\dfrac{m_0 Cka_w}{(1-ka_w)(1-ka_w+Cka_w)}$ (9.2)

where M is the equilibrium moisture content (g/g polymer), at a water activity a_w, m_0 is the monolayer value, and C and k are the constants. k is assumed to be less than 1.[8] According to the van der Wel and Adan,[18] GAB equation is effective for fitting data of nonideal water sorption in polymers (Type II or III) over activity ranging from 0 to 0.95, which is not the case for the classic BET model.

Smith model: $M=M_b-M_a\times[\ln(1-a_w)]$ (9.3)

where M_b and M_a are constants. From a linear regression of M versus $\ln(1-a_w)$, the Smith constants can be computed.

Halsey Model: $\ln(M)=a+b\times\{\ln[-\ln(a_w)]\}$ (9.4)

where a and b are Halsey constants, which can be estimated from a linear plot of $\ln(M)$ versus $\ln[-\ln(a_w)]$. Halsey model is another sorption isotherm model that expresses condensation of a multilayer at a relatively large distance from the surface.

In the present study, together with r^2 values, standard error (SE) and mean relative deviation (MRD) values were also calculated to measure the accuracy of the r^2 values. High r^2 value gives information about the suitability of the model, but besides high r^2 values, low SE and MRD (<0.5) values should be obtained to determine best fit model. The calculated parameters of the models are listed in Tables 9.2–9.5. As a result of the calculations, Halsey model was chosen as the best model for representing the sorption isotherms of HPC films, while for NaCMC films, GAB model was determined as the best fit model. The comparison of the experimental data and correlation of the sorption isotherms of NaCMC and HPC films prepared from 3 wt% polymer solution is shown in Figures 9.7 and 9.8, respectively.

TABLE 9.2   Optimized Values of the Parameters for GAB Model.

Polymer	Concentration (wt%)	Constants			R^2	SE	MRD
		M_o	C	k			
NaCMC	3	12.96	10.03	0.93	0.98	0.61	−0.0006
NaCMC	4	12.22	6.06	0.93	0.99	0.83	0.0015
NaCMC	5	17.83	1.36	0.88	1.00	1.02	−0.0197
HPC	3	3.64	77.85	0.95	0.95	0.38	0.0000
HPC	4	416	4.11	0.94	0.99	0.30	0.0031
HPC	5	4.33	60.36	0.88	0.97	0.21	0.0004

SE, standard error; MRD, mean relative deviation; NaCMC, sodium salt of carboxy methyl cellulose; HPC, hydroxypropyl cellulose.

TABLE 9.3   Optimized Values of the Parameters for BET Model.

Polymer	Concentration (wt%)	Constants		R^2	SE	MRD
		m_o	C			
NaCMC	3	11.43	15.65	0.98	0.42	−0.0010
NaCMC	4	10.19	11.15	0.99	0.57	−0.0016
NaCMC	5	14.68	1.28	1.00	0.78	−0.0021
HPC	3	11.43	15.65	0.99	10.85	−1.2340
HPC	4	10.19	11.15	1.00	9.49	−1.3800
HPC	5	14.68	1.28	1.00	6.78	−0.4630

SE, standard error; MRD, mean relative deviation; NaCMC, sodium salt of carboxy methyl cellulose; HPC, hydroxypropyl cellulose.

TABLE 9.4   Optimum Parameters for Smith Model.

Polymer	Concentration (wt%)	Constants		R^2	SE	MRD
		M_a	M_b			
NaCMC	3	31.28	1.08	0.93	2.96	0.0230
NaCMC	4	30.35	−0.84	0.98	2.93	0.0290
NaCMC	5	32.03	−4.81	0.99	1.88	0.0430

SE, standard error; MRD, mean relative deviation; NaCMC, sodium salt of carboxy methyl cellulose.

TABLE 9.5 Optimum Parameters for Halsey Model.

Polymer	Concentration (wt%)	Constants		R^2	SE	MRD
		A	B			
HPC	3	1.71	−0.65	0.99	0.04	−0.0002
HPC	4	1.46	−0.84	0.99	0.07	−0.0040
HPC	5	1.84	−0.55	1.00	0.02	−0.0002

SE, standard error; MRD, mean relative deviation; HPC, hydroxypropyl cellulose.

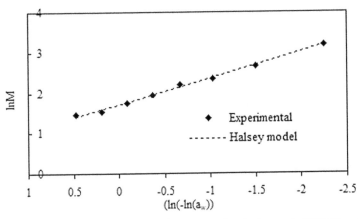

FIGURE 9.7 Experimental and predicted water sorption isotherms in HPC films prepared with 3 wt% HPC in the solution.

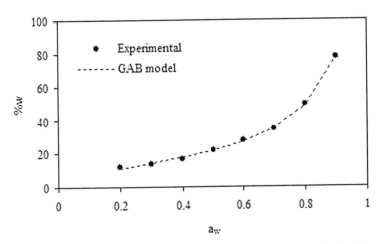

FIGURE 9.8 Experimental and predicted water sorption isotherms in NaCMC films prepared with 3 wt% NaCMC in the solution.

9.3.3 *WATER VAPOR DIFFUSION IN FILMS*

Figures 9.9 and 9.10 show the mass uptake (M_t/M_∞) versus $t^{1/2}$ graphs of NaCMC-based films at different RH levels. The first uptake curve in the microbalance was obtained at RH of 40%. Large water sorption into the NaCMC film at 40% RH level caused an S-shaped uptake curve as shown in Figure 9.10. When RH in the column was increased from 60% to 80%, the rate of sorption into the NaCMC films decreased significantly due to increased mass transfer resistance in the swollen film. Figure 9.10 indicates that mass uptake curves of HPC film are all concave with respect to $t^{1/2}$ axis even at 40% RH level due to lower water sorption levels in this film. The initial parts of all uptake curves shown in Figures 9.9 and 9.10 are linear.

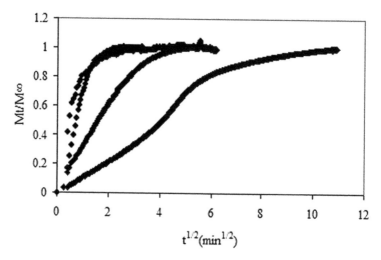

FIGURE 9.9 Mass uptake curves of NaCMC film at different relative humidity values in microbalance. The films were prepared with 3% NaCMC in the solution.

The effective diffusivities of water vapor, D, through the films with an initial thickness of h have been determined from the slope of the linear region of the M_t/M_∞ against $t^{1/2}$ curve by using by the following equation:

$$\frac{M_t}{M_\infty} = \left(\frac{16D}{h^2 \pi}\right)^{1/2} \cdot t^{1/2} \qquad (9.5)$$

where M_t and M_∞ are amount of water adsorbed at time t and at equilibrium, respectively.[10]

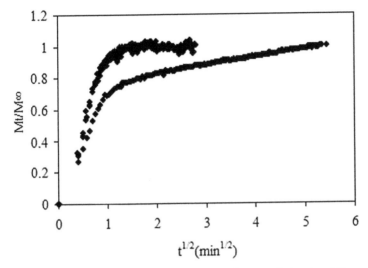

FIGURE 9.10 Mass uptake curves of HPC at different relative humidity values in microbalance. The films were prepared with 3% HPC in the solution.

The calculated diffusion coefficients of water in NaCMC- and HPC-based films prepared from 3 wt% polymer solution are listed in Table 9.6.

TABLE 9.6 Diffusion Coefficients of Water Vapor in NaCMC- and HPC-Based Edible Films Prepared from 3% Polymer Concentration.

RH (%)	$D \times 10^{14}$ (m²/s)	
	NaCMC	**HPC**
40	7	165
50	275	283
60	818	241
80	88	235

RH, relative humidity; NaCMC, sodium salt of carboxy methyl cellulose; HPC, hydroxypropyl cellulose.

HPC-based edible films have a narrower range of diffusion coefficient values (165×10^{-14}–283×10^{-14} m²/s) than that of (7×10^{-14}–818×10^{-14} m²/s) NaCMC-based films. This is due to presence of crystalline phase in NaCMC films which act as a barrier for diffusion of water. The diffusion coefficient of water in NaCMC film decreased significantly when RH was increased from 60% to 80%. This may be due to clustering of water in the film structure or gel formation which could also hinder diffusion. It should also be noted that eq 9.5 gives approximate values for the diffusivities in NaCMC films corresponding to 40% and 80% RH due to strong swelling of the films.

9.3.4 WVP ANALYSIS OF THE FILMS

RH versus time data was obtained using the set up shown in Figure 9.1. The permeability (P_e) of water through the films was calculated from the slope (s) of linear portion of $\ln[(P_L - P_{Ui})/(P_L - P_{U(t)})]$ versus time graph (Fig. 9.11) using the following equation and the values are listed in Table 9.7:

$$P_e = \frac{L \times V \times s}{R \times A \times T} \qquad (9.6)$$

where L is the film thickness, A film area, V is the volume of the upper compartment, T is the absolute temperature, P is the pressure at temperature T, subscripts L and U represent upper and lower sides of the film, subscripts i and t represent initial time and time t, respectively.

Figure 9.12 shows that WVP values of NaCMC- and HPC-based films increased with increasing polymer concentration in the film forming solution as a result of the increasing hydrophilicity of the films. Similar trend was observed for polysaccharide-based edible films in the study of Aydınlı and Tutaş.[3] On the other hand, Park et al.[15] have found no concentration dependence for WVP through HPC films.

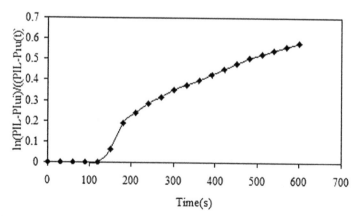

FIGURE 9.11   The change of $\ln(P1_L - P1_{ui})/(P1_L - P1_{u(t)})$ vs. time for the HPC film prepared with 3% HPC in the solution.

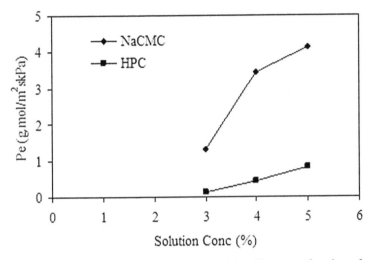

FIGURE 9.12   Water vapor permeability values of the films as a function of polymer concentration in the film forming solution.

TABLE 9.7   Water Vapor Permeability Values of NaCMC- and HPC-Based Edible Films Prepared from Different Concentration Solutions.

Polymer	Concentration of polymer solution (%)	$P_e \times 10^{10}$ (g mol H_2O/s cm kPa)
NaCMC	3	0.058 ± 0.0200
NaCMC	4	0.098 ± 0.0120
NaCMC	5	0.12 ± 0.0003
HPC	3	0.033 ± 0.0011
HPC	4	0.049 ± 0.0056
HPC	5	0.056 ± 0.0087

NaCMC, sodium salt of carboxy methyl cellulose; HPC, hydroxypropyl cellulose.

As seen from Table 9.7, permeability of NaCMC-based films is higher than that of HPC-based edible films. Factors such as accessibility of polar groups, the relative strength of water–water versus the water–polymer bonds, crystallite size, shape, and degree of crystallinity of the polymer matrix affect the WVP of the polymers. The mass transfer of water vapor in a semicrystalline polymer is primarily a function of the amorphous phase, because the crystalline phase is usually assumed to be impermeable.[13] High crystalline structure may provide more dense and compact structure which contains less-free volume for water molecule migrations, so WVP decrease.

Even though NaCMC-based films have crystalline structure, their higher WVP values are due to more hydrophilic nature of this polymer, hence, higher water sorption capacities.

9.3.5   EFFECT OF COATING OF CARROTS ON THE WEIGHT LOSS, DIMENSION CHANGE, AND COLOR CHANGE

All carrot samples have weight loss with time, but the carrot samples which were not coated have the highest weight loss compared to coated ones as seen in Figure 9.13. 4% NaCMC-coated carrot samples showed lower weight loss than 3% NaCMC-coated carrot samples. This result indicates that the 3% NaCMC coating is more permeable than the 4% NaCMC coating. The diameter of the carrot slices decrease with time as seen in Figure 9.14. Coated samples have lower diameter changes than uncoated ones. Table 9.8 shows the L, a, and b values of the uncoated and coated carrot samples. a Value which is the measure of the redness of the sample was taken as a reference for the color change of carrot slices with time. The red intensity of the samples decreases when they were coated with NaCMC. However, a value of the uncoated carrot sample was decreasing with time, but for the coated samples, this value was increasing with time as seen in Figure 9.15. Thus, coating carrots with NaCMC decreased the rates of weight loss and diameter with time in storing them in ambient air. Even if the coating reduces the redness of the carrots initially, it increases with time. Thus, coating carrots with NaCMC increased the shelf life of the carrots.

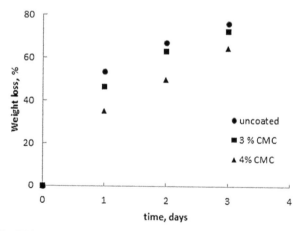

FIGURE 9.13   Weight change of uncoated and coated carrots with time.

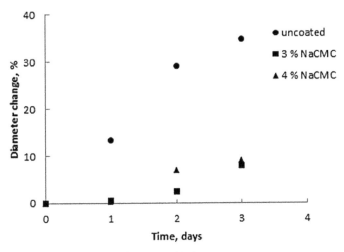

FIGURE 9.14   Diameter change of carrots with time.

TABLE 9.8 L, a, b Values of Uncoated and Coated Carrot Samples.

	1st day			2nd day			3rd day		
	Uncoated	3% NaCMC	4% NaCMC	Uncoated	3% NaCMC	4% NaCMC	Uncoated	3% NaCMC	4% NaCMC
L	60.22	42.63	44.66	58.79	43.05	45.81	63.15	44	45.11
a	28.36	17.44	20.18	16.17	23.97	24.34	12.53	30.6	30.61
b	43	31.35	35.53	19.11	30.19	38.82	20.16	41.97	42.32

NaCMC, sodium salt of carboxy methyl cellulose.

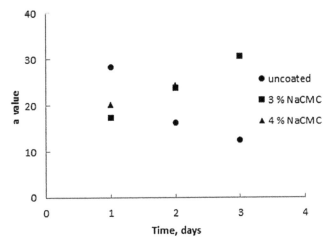

FIGURE 9.15   Color change of carrots with time.

9.4 CONCLUSIONS

In this study, NaCMC- and HPC-based edible films were prepared and characterized to obtain information about the packaging properties of these films. Scanning electron microscope pictures have shown that HPC-based edible films had higher porosity with more homogenous distribution of the pores than those of NaCMC-based films. Based on XRD analysis, it was observed that HPC films have mainly an amorphous structure, while NaCMC films have a crystalline order. DSC indicated the presence of one endotherm being maximum between 83 and 98°C for NaCMC films and 70–85°C for HPC films due to dehydration and evaporation of water from the films. In addition, glass transition temperatures of both films were found to be below room temperature. Measurement of mechanical property of the films indicated that NaCMC-based films have higher tensile strength and Young modulus than those of HPC films. HPC films showed more elastic structure compared to NaCMC films which was brittle with low elongation at break values.

Water vapor sorption characteristics of NaCMC and HPC-based films were determined at 25°C with sorption and desorption analysis in a microbalance and humidity chamber. Water sorption capacity of the films was found to be lower from the experiments conducted in the microbalance which may be due to insufficient drying of the films before the experiments. According to sorption experiments conducted in the humidity chamber, water vapor sorption capacity of the NaCMC-based films (70 wt%) was determined to be higher than that of HPC-based films (25 wt%). GAB and Halsey models were found to give the best models to correlate the sorption data of NaCMC and HPC films, respectively.

NaCMC-based edible films showed lower diffusion coefficient values than HPC-based films especially at low RH levels due to the presence of crystalline phase in these films. On the other hand, WVP of NaCMC-based films were measured to be higher than that of HPC films since NaCMC-based films have more hydrophilic groups; hence, their water sorption capacities are higher. Both films showed an increased WVP with increasing polymer concentration in the film forming solution. Based on these results, it can be concluded that HPC-based films show better barrier property for decreasing the respiration rate from the food.

Carrots coated with NaCMC had longer shelf life in terms of weight, dimensions, and color.

ACKNOWLEDGMENT

The authors acknowledge the contributions of Asst. Prof. Yilmaz Yürekli for gravimetric adsorption experiments and Beyza Keklik and Elif Karaduman for carrot coating experiments.

KEYWORDS

- cellulose films
- surface characterization
- water vapor permeation
- sorption
- diffusion

REFERENCES

1. Altinkaya, S. A.; Topcuoglu, O.; Yurekli, Y.; Balkose, D. The Influence of Binder Content on the Water Transport Properties of Waterborne Acrylic Paints. *Prog. Org. Coat.* **2010**, *69* (4), 417–425.
2. Anker, M.; Berntsen, J.; Hermansson, A. M.; Stading, M. Improved Water Barrier of Whey Protein Films by Addition of an Acetylated Monoglyceride. *Innovative Food Sci. Emerg. Technol.* **2002**, *3*, 81–92.
3. Aydinli, M. ; Tutas, M. Water Sorption and Water Vapor Permeability of Polysaccharide (Locust Bean Gum) Based Edible Films. *Lebensm. Wiss. Technol.* **2000**, *33*, 63–67.
4. Ayranci, E.; Büyüktaş, B. S.; Cetin, E. E. The Effect of Molecular Weight of Constituents on Properties of Cellulose-Based Edible Films. *Lebensm. Wiss. Technol.* **1997**, *30*, 101–104.
5. Baldev, R.; Eugene, A.; Kumar, K. R.; Siddarramaiah. Moisture-Sorption Characteristics of Starch/Low-Density Polyethylene Films. *J. Appl. Polym. Sci.* **2001**, *84*, 1193–1202.
6. Berkun, D., Balkose, D., Tıhmınlıoğlu, F.; Altınkaya, S. A. Sorption and Diffusion of Water Vapor on Edible Films. *J. Therm. Anal. Calorim.* **2008**, *94* (3), 683–686.
7. Buonocore, G. G.; Del Nobile, M. A.; Di Martino, C.; Gambacorta, G.; La Notte, E.; Nicolais, L. Modeling the Water Transport Properties of Casein-Based Edible Coating. *J. Food Eng.* **2003**, *60*, 99–106.
8. Cho, S. Y.; Rhee, C. Sorption Characteristics of Soy Protein Films and their Relation to Mechanical Properties. *Lebensm. Wiss. Technol.*, **2002**, *35* (2), 151–157.
9. Coupland, J. N.; Shaw, N. B.; Monahan, F. J.; O'Riordan, E. D. ; O'Sullivan, M. Modelling the Effect of Glycerol on the Moisture Sorption Behaviour of Whey Protein Edible Films. *J. Food Eng.* **2000**, *43*, 25–30.

10. Crank, J. *The Mathematics of Diffusion;* Oxford University Press: Oxford, 1976.
11. Kamide, K. *Cellulose and Cellulose Derivatives Molecular Characterization and Its Applications;* Elsevier: Amsterdam, 2005.
12. McHugh, T. H.; Avena-Bustillos, R.; Krochta, J. M. Hydrophylic Edible Films: Modified Procedure for Water Permeability and Explanation of Thickness Effects. *J. Food Sci.* **1993**, *58* (4), 899–903.
13. Miller, K. S.; Krochta, J. M. Oxygen and Aroma Barrier Properties of Edible Films; A Review. *Trends Food Sci. Technol.* **1997**, *8*, 228–237.
14. Osorio, F. A.; Molina, P.; Matiacevich, S.; Enrione J.; Skurty O. Characteristics of Hydroxy Propyl Methyl Cellulose (HPMC) Based Edible Film Developed for Blueberry Coatings. *Procedia Food Sci.* **2011**, *1*, 287–293.
15. Park, H. J.; Weller, C. L.; Vergano, P. J.; Testin, R. F. Permeability and Mechanical Properties of Cellulose Based Edible Films. *J. Food Sci.* **1993**, *58*, 1361–1365.
16. Roman-Gutierrez, A. D.; Guilbert, S.; Cuq, B. Distribution of Water between Wheat Flour Components: A Dynamic Water Vapor Adsorption Study. *J. Cereal Sci.* **2002**, *36*, 347–355.
17. Turhan, K. N.; Şahbaz, F. Water Permeability, Tensile Properties and Solubility of Methylcellulose-Based Edible Films. *J. Food Eng.* **2003**, *61*, 459–466.
18. van der Wel, G. K.; Adan, O. C. G. Moisture in Organics—A Review. *Prog. Org. Coat.* **1999**, *37*, 1–14.

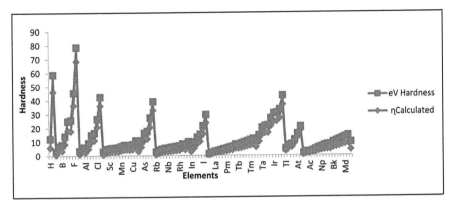

FIGURE 2.1 Comparative analysis between our computed data (mdyne) with atomic hardness of Ghosh et al.

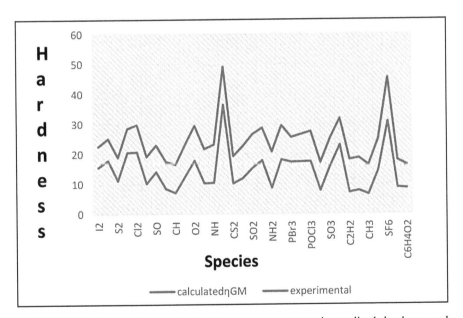

FIGURE 2.2 Comparative analysis between our computed equalized hardness and experimental counterparts.

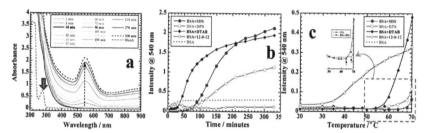

FIGURE 6.1 (a) UV–visible scans of BSA+SDS+HAuCl$_4$ mixture with (BSA/residue)/(SDS) mole ratio = 88 at 70°C. The block arrow shows the absorbance due to tryptophan residues, while the dotted indicates the increasing absorbance of AuNPs due to SPR with time. Blank means when no HAuCl$_4$ is added. (b) and (c) Plots of intensity at 540 nm versus reaction time and temperature, respectively, for different mixtures in the presence and absence of surfactants. (Reproduced with permission from Reference [12]. © 2012 American Chemical Society.)

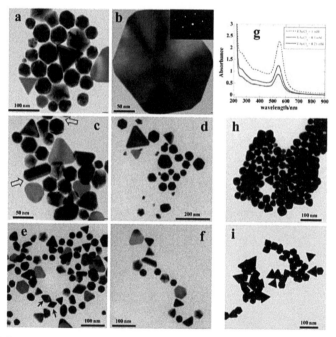

FIGURE 6.2 (a) TEM images of purified AuNPs prepared with BSA+SDS+HAuCl$_4$ mixture at 70°C. Part (b) shows a single hexagonal NP with diffraction image in inset. (c) and (d) TEM images of purified AuNPs prepared with BSA+DPS+HAuCl$_4$ and BSA+DTAB+HAuCl$_4$ mixtures, respectively, at 70°C. Empty block arrows in (c) indicate the presence of thin BSA coating around each NP. (e) and (f) TEM images of purified AuNPs prepared with BSA+TPS+HAuCl$_4$ at 70°C. A pearl necklace arrangement of BSA conjugated NPs in (f) is due to the fibrillation of BSA. (g) UV–visible scans of different samples of as-prepared 16-2-16+HAuCl$_4$ mixture at 70C with different concentrations of gold salt. Parts (h) and (i) show their respective TEM images with HAuCl$_4$ = 0.25 and 1 mM, respectively. (Reproduced with permission from Reference [12]. © 2012 American Chemical Society.)

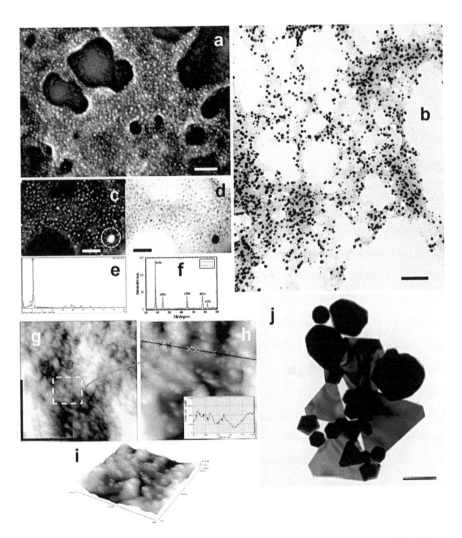

FIGURE 6.3 (a) FESEM (scale bar = 200 nm) and (b) TEM micrographs of AuNPs as bright spots on unfolded BSA as a soft template of a sample with Au/BSA–mole ratio 167 synthesized at 80°C (scale bar = 100 nm). (c) and (d) Bright field (scale bar = 250 nm) and dark field images (scale bar = 250nm), respectively, differentiating between AuNPs and unfolded BSA. (e) and (f) EDS spectrum and XRD patterns of AuNPs, respectively. (g) AFM height image of unfolded BSA as a soft template bearing AuNPs (scale bar = 870 nm). (h) Close-up image showing the line analysis of various AuNPs (scale bar = 100 nm). (i) Topography of the BSA film bearing AuNPs showing peaks and valleys. (j) TEM micrograph of large-plate-like NPs of a sample with Au/BSA–mole ratio 17 prepared at 80°C (scale bar = 100 nm). (Reproduced with permission from Reference [13]. © 2011 American Chemical Society.)

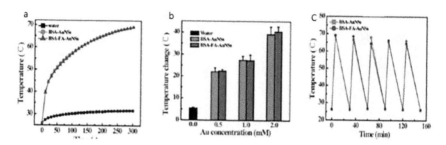

FIGURE 6.13 (a) Temperature increase in water, an aqueous solution of BSA-AuNSs or BSA-FA-AuNSa; (b) temperature change in an aqueous solution of BSA-AuNSs or BSA-FA-AuNSs at different Au concentrations; (c) temperature change in an aqueous solution of BSA-AuNSs or BSA-FA-AuNSs over five laser on–off cycles. (Reproduced with permission from Reference [42]. © 2015 Royal Society of Chemistry.)

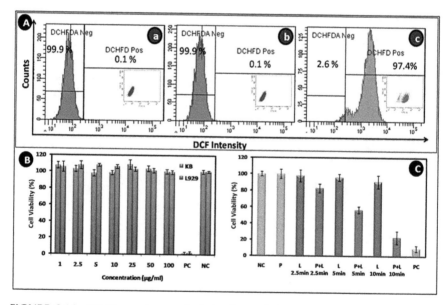

FIGURE 6.14 (A) Flow cytogram depicting intracellular ROS generation by DCFH-DA assay in KB cells treated with (a) media alone (negative control), (b) Alb-GNS (100 mg mL⁻¹), and (c) 30 mM H_2O_2 (positive control); (B) graphical representation of cell viability tests on KB and L929 cells treated with Alb-GNS at varying concentration for 24 h; (C) graphical representation of cell viability test on KB cells treated with 100 mg mL⁻¹ of Alb-GNS (P) and laser (L) for different time period. NC and PC represent negative control (untreated cells) and positive control (cells treated with 1% Triton X-100), respectively. (Reproduced with permission from Reference [25]. © 2016 Royal Society of Chemistry.)

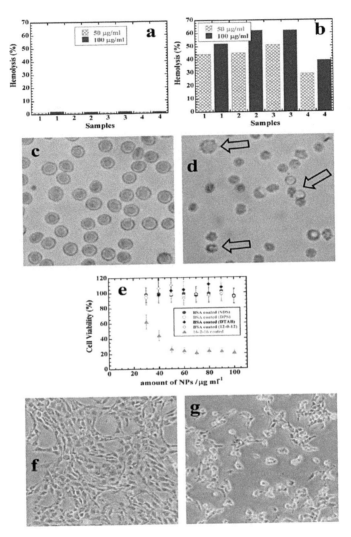

FIGURE 6.15 (a) Percentage hemolysis of purified BSA-coated AuNPs prepared with different mixtures, BSA+SDS+HAuCl$_4$ (1); BSA+DPS+HAuCl$_4$ (2); BSA+DTAB+HAuCl$_4$ (3); BSA+12−0−12+HAuCl$_4$ (4), with doses of 50 and 100 µg mL^{-1}. (b) Percentage hemolysis of purified 16-2-16-coated AuNPs prepared with 16-2-16+HAuCl$_4$ mixtures at 70°C at fixed HAuCl$_4$ = 0.25 mM and with different concentrations of 16-2-16 = 2 mM (1); 4 mM (2); 8 mM (3); and at HAuCl$_4$ = 1 mM and 16-2-16 = 4 mM (4), with doses of 50 and 100 µg mL^{-1}. (c) and (d) Bright field optical microscopic images of RBCs without and with 16-2-16-coated AuNPs, respectively, of "sample 2" of (b). Empty block arrows in (d) indicate broken cells with released contents. (e) Percentage cell viability of glioma cell lines with different amounts of BSA and 16-2-16-coated NPs. (f) and (g) Bright field optical microscopic images of glioma cell lines without and with 16-2-16-coated AuNPs, respectively, of "sample 2" of (b). (Reproduced with permission from Reference [12]. © 2012 American Chemical Society.)

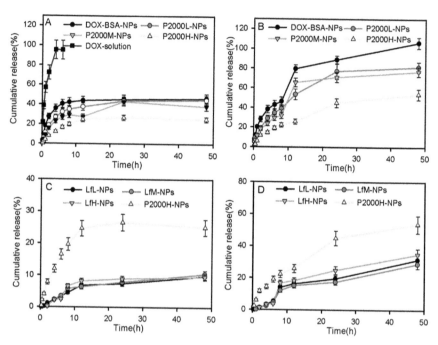

FIGURE 6.16 In vitro release curve of DOX from DOX solution and DOX-loaded NPs: (A) and (C) in saline and (B) and (D) in saline with trypsin. Data were presented as the mean $\pm$ SD ($n = 3$). (Reproduced with permission from Reference [27]. © 2014 American Chemical Society.)

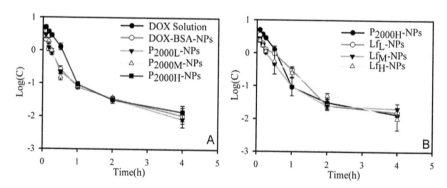

FIGURE 6.17 Plasma concentration–time curves of DOX in rats after intravenous administration of (A) DOX solution, DOX-BSA-NPs, P_{2000}L-NPs, P_{2000}M-NPs, and P_{2000}H-NPs and (B) P_{2000}H-NPs, LfL-NPs, LfM-NPs, and LfH-NPs. Data were presented as the mean $\pm$ SD ($n = 6$). (Reproduced with permission from Reference [27]. © 2014 American Chemical Society.)

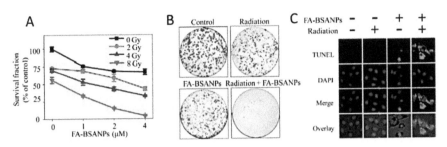

FIGURE 6.18 FA-BSA NPs in combination with X-ray radiation increased G2/M phase arrest and apoptosis in HeLa cells. (A) Combined treatment of FA-BSA NPs and X-ray radiation decreased HeLa cells survival detected by clonogenic assay. HeLa cells were pretreated FA-BSA NPs at different concentrations for 2 h and then were irradiated by X-ray radiation at different dosages. Values expressed were means ± SD of triplicate. (B) Colony formation of HeLa cells under the cotreatment of FA-BSA NPs and radiation (8 Gy). (C) Fluorescence images of DNA fragmentation and nuclear condensation after exposure to FA-BSA NPs or/and X-ray radiation. Cells pretreated with FA-BSA NPs (4 µM) and X-ray radiation (8 Gy) were stained with TUNEL working buffer and DAPI for DNA fragmentation and nucleus visualization, respectively. (Reproduced with permission from Reference [28]. © 2014 American Chemical Society.)

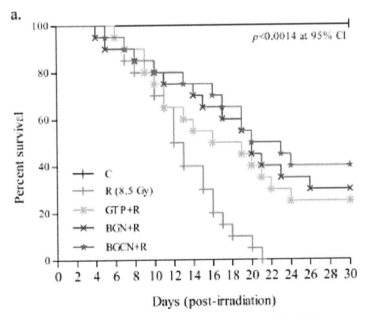

FIGURE 6.20 Effect of GTP, BGN, and BGCN on radiation-induced damage to hematopoietic system and mortality. (Reproduced with permission from Reference [29]. © 2016 American Chemical Society.)

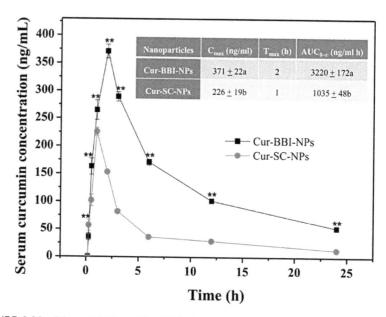

Nanoparticles	C_{max} (ng/ml)	T_{max} (h)	$AUC_{0-\infty}$ (ng/ml h)
Cur-BBI-NPs	$371 \pm 22a$	2	$3220 \pm 172a$
Cur-SC-NPs	$226 \pm 19b$	1	$1035 \pm 48b$

FIGURE 6.30 Bioavailability of Cur-BBI-NPs and Cur-SC-NPs. All values are presented as the mean ± SD, $n = 6$.** Indicates an extremely significant difference between two groups ($p < 0.01$). (Inset) Pharmacokinetics parameters of two curcumin nanoparticles. AUC, area under the plasma concentration–time curve from 0 h to ∞; C_{max}, peak concentration; T_{max}, time to reach peak concentration. Means of symbols marked with different letters indicate an extremely significant difference between two groups ($p < 0.01$). (Reproduced with permission from Reference [49]. © 2017 American Chemical Society.)

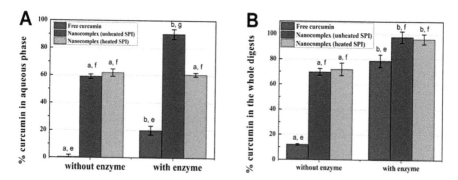

FIGURE 6.32 (A) Percentage of curcumin remaining in the aqueous phase for free curcumin and curcumin nanocomplexes with unheated and heated (at 95°C) SPI after the whole simulated digestion of 180 min in the absence or presence of proteases. (B) Percentage of curcumin in the whole digests of free curcumin and curcumin nanocomplexes with unheated and heated SPI after the digestion of 180 min in the absence or presence of proteases. (Reproduced with permission from Reference [50]. © 2015 American Chemical Society.)

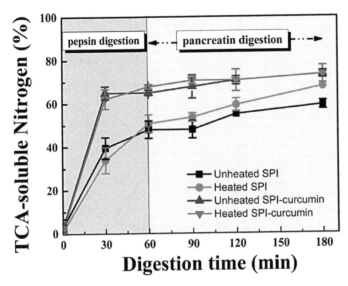

FIGURE 6.33 Release kinetics of TCA-soluble nitrogen during the sequential SGF and SIF digestion of unheated and heated SPI and its nanocomplexes with curcumin. Each data point is the mean and standard deviation of triplicate measurements on separate samples. (Reproduced with permission from Reference [50]. © 2015 American Chemical Society.)

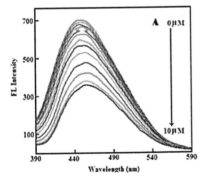
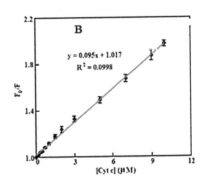

FIGURE 6.35 (A) Fluorescence emission spectra upon addition of different concentrations of Cyt C (0, 0.16, 0.32, 0.48, 0.70, 1.0, 1.5, 2.0, 3.0, 5.0, 7.0, 9.0, and 10 μM). The spectra were recorded at time intervals of 5 min. (B) Stern–Volmer plot of fluorescence quenching of the Hb/AuNCs by Cyt C. All experiments were carried out in 10 mM PBS of pH 7.4 with Hb/AuNCs concentration of 2.2 μM. Error bars represented as ±3σ. (Reproduced with permission from Reference [96]. © 2016 American Chemical Society.)

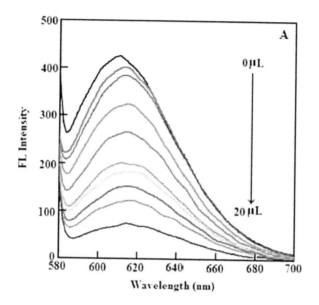

FIGURE 6.36 Fluorescence emission spectra of the Cyt *C*-2/AgNCs in the presence of variable concentrations of cancer cell extracts treated with (a) hypotonic buffer and (b) with CCLR buffer (both in the presence of 0, 3.0, 5.0, 7.0, 9.0, 11.0, 13.0, 15.0, 17.0, and 20 μL aliquots taken from initial 10,000 cells mL^{-1} solution). The spectra were recorded at time intervals of 20 min. (Reproduced with permission from Reference [96]. © 2016 American Chemical Society.)

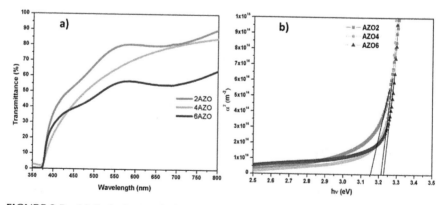

FIGURE 8.5 (a) Optical transmission spectra and (b) α^2 as a function of the photon energy, *hv*, for nanostructured 2AZO, 4AZO, and 6AZO film samples oxidized at 550°C for 60 min.

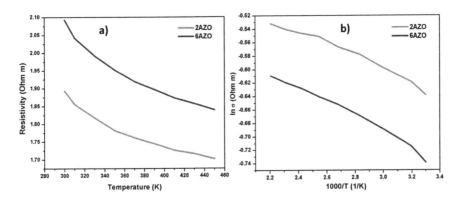

FIGURE 8.7 (a) Electrical resistivity vs. temperature in the range of 300–450 K and (b) variation of electrical conductivity with reciprocal temperature of 2AZO and 6AZO film samples oxidized at 550°C for 60 min.

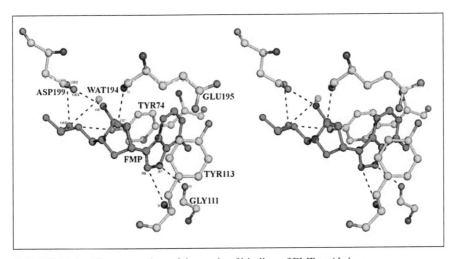

FIGURE 10.1 The stereo-view of the mode of binding of FMP to Abrin-a.

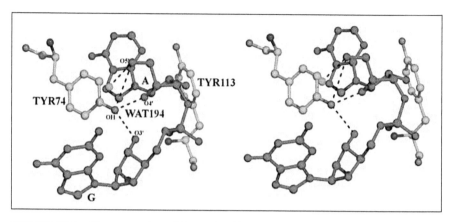

FIGURE 10.2 The stereo-view of the mode of binding of APG to Abrin-a.

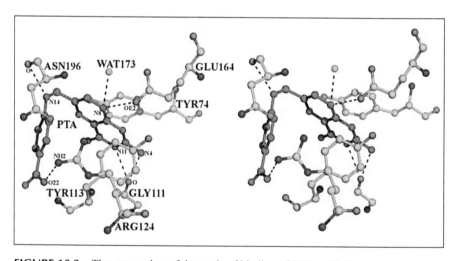

FIGURE 10.3 The stereo-view of the mode of binding of PTA to Abrin-a.

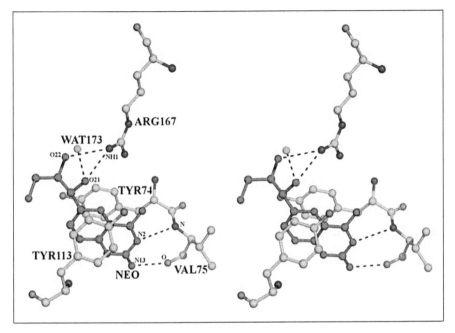

FIGURE 10.4   The stereo-view of the mode of binding of NEO to Abrin-a.

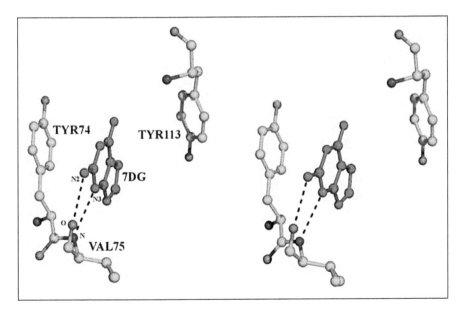

FIGURE 10.5   The stereo-view of the mode of binding of 7DG to Abrin-a.

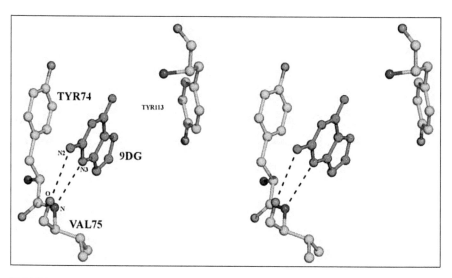

FIGURE 10.6   The stereo-view of the mode of binding of 9DG to Abrin-a.

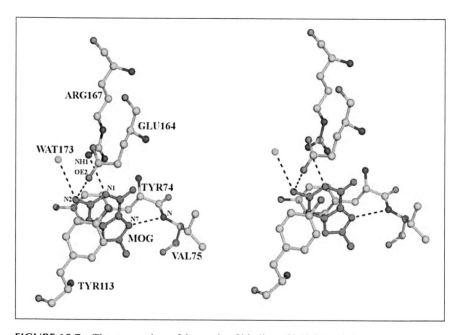

FIGURE 10.7   The stereo-view of the mode of binding of MOG to Abrin-a.

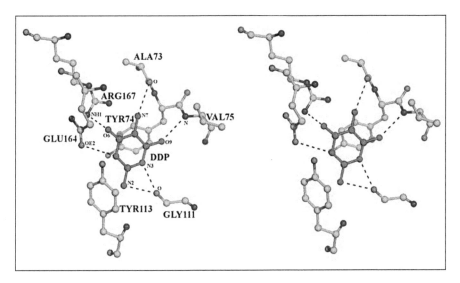

FIGURE 10.8   The stereo-view of the mode of binding of DDP to Abrin-a.

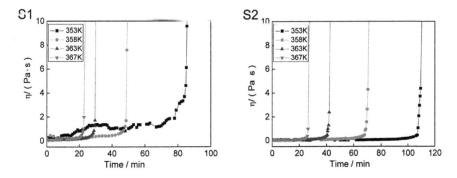

FIGURE 12.4   Viscosity of the silica sol system with a specific composition (samples S1 and S2) at different temperatures. (Reproduced from Reference [66] with permission.)

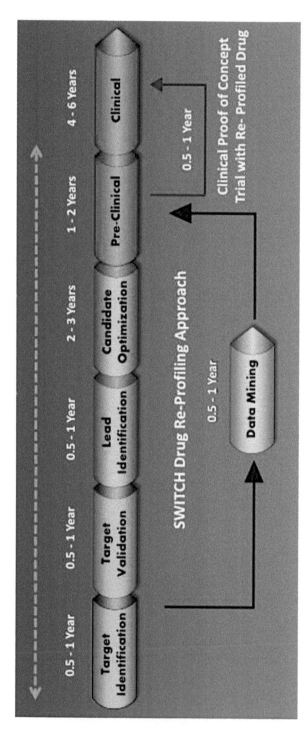

FIGURE 15.1 The traditional drug discovery and development process.

CHAPTER 10

RECOGNITION OF ADENINE-LIKE RINGS BY THE ABRIN—A BINDING SITE: A FLEXIBLE DOCKING APPROACH

ASHIMA BAGARIA*, MUKESH SARAN, and JAGDISH PARIHAR

Department of Physics, Manipal University Jaipur, Jaipur, India

Corresponding author. E-mail: ashima.bagaria@jaipur.manipal.edu

ABSTRACT

Plant toxins and the bacterial toxins are the best studied cytotoxic proteins. The plant toxins Abrin and Ricin are exceptionally toxic and form a part of the ribosome inactivating proteins (RIPs) that are a large family of N-glycosidases, exhibiting exquisite specificity in their catalytic hydrolysis of a single adenine base from among nearly 7000 nucleotides found in mammalian ribosomes. Owing to the remarkable cytotoxicity of Abrin-a, like Ricin, there is a considerable interest in the search for strong competitive inhibitors. Abrin has been widely used in the design of antitumor agents but has also been used as a potent bio-warfare agent. Effective inhibitors might be useful to help control nonspecific toxicity in treatments with immunotoxins and could also serve as antidotes in poisonings. This report deals with the in silico modeling of Abrin-a-inhibitors complexes using the docking method. Various ringed compounds were analyzed to obtain at least a qualitative understanding of the nature of interactions between the protein and the inhibitors. The details are discussed in the various sections of this chapter.

10.1   RECENT ADVANCES IN THE SEARCH FOR INHIBITOR OF CYTOTOXIC PLANT PROTEINS

Plant toxins and the bacterial toxins are the best studied cytotoxic proteins. The plant toxins Abrin and Ricin are exceptionally toxic and form a part of the ribosome inactivating proteins (RIPs) that are a large family of N-glycosidases, exhibiting exquisite specificity in their catalytic hydrolysis of a single adenine base from among nearly 7000 nucleotides found in mammalian ribosomes.

In recent years, these plant RIPs which inhibit protein synthesis have become a subject of intense investigation not only because of the possible role played by them in synthesizing immunotoxins that are used in cancer therapy but also because they serve as model system for studying the molecular mechanism of transmembrane translocation of proteins.[1,2] Abrin-a is one of the four isoabrins isolated from the seeds of *Abrus precatorius* and is anticarcinogenic.[3,4] The anticarcinogenic activity includes the inhibition of protein biosynthesis[5] by the cleavage of N-glycosidase bond of adenosine residue at position A43234 residues from 28 S rRNA.[6] Depurination occurs at the first adenine, in the loop sequence GAGA, located in a highly conserved single stranded rRNA hairpin thereby arresting protein synthesis and causing cytotoxicity.[6,7] Abrin-a is a heterodimer consisting of a cytotoxic A-chain linked by a disulfide bond to a B-chain, which binds to galactose residues present on various cell surface glycoproteins and glycolipids. The Abrin-a molecule shares major structural similarities with Ricin despite differences in their primary structures. The three-dimensional crystal structure of Abrin-a at 2.14 Å resolution has already been solved.[8] The active site consists of the residues Tyr74, Tyr113, Glu164, Arg167, Trp198 and Asn72, Arg124, Gln160, Gln195 which are well conserved among all RIPs. Recent analysis of crystal structures of Ricin,[9] Trichosanthin,[10] pokeweed antiviral protein[11] (PAP), Abrin-a,[8] *A. precatorius* agglutinin[12] indicates that the overall architecture of the active site cleft remains constant in all these proteins. In Abrin-a, the aromatic rings of Tyr74 and Tyr113 are almost parallel. This orientation of both invariant tyrosine side-chains is most appropriate for sandwiching the planar adenine. In Ricin, PAP, and Trichosanthin, it is necessary to rotate the tyrosine side-chains to accommodate the adenine between them.[13–15]

According to the existing concept of the protein inhibitory activity of RIPs, the key residues that are directly involved in catalytic processes are invariant arginine and glutamic acid residues.[13] The role of tyrosine rings is probably that of forming a route for the substrate by keeping adenine strictly parallel to the plane of their rings. This makes a favorable orienta-tion for adenine, so that its N3 can approach arginine and make a hydrogen bond with it. In addition, such an orientation of adenine will allow O-3' of ribose to approach invariant glutamic acid residue and form a hydrogen bond. Site directed mutagenesis of these tyrosine rings to phenylalanine in Ricin have shown reduction in enzymatic activity.[16] Due to similarity of active site structures specially the positions of Glu164 and Arg167 in Abrin-a and Ricin, the mechanism of action proposed for Ricin[13] is also applicable to Abrin-a.

Owing to the remarkable cytotoxicity of Abrin-a, like Ricin, there is a considerable interest in the search for strong competitive inhibitors. Abrin has been widely used in the design of antitumor agents[17] but has also been used as a potent bio-warfare agent. Effective inhibitors might be useful to help control nonspecific toxicity in treatments with immunotoxins and could also serve as antidotes in poisonings. In case of Ricin, which is also a cytotoxic protein like Abrin, only modest success has been achieved to date for small molecules,[18] being used as inhibitors. There is more promise for a transition state mimic containing a modified 14-mer ribonucleotide hairpin[19] though its molecular size limits its practical use as a therapeutic inhibitor. The failure of smaller similar ribonucleotide loops to inhibit Ricin[20] indicates that many issues remain to be clarified regarding ligand size requirements for inhibition.

Our study aims to identify novel inhibitors for Abrin-a molecule and facilitates detailed analysis of protein ligand complex in various ways to ascertain them as potential ligands. The in silico work is important since the experimentally determined three dimensional structures of Abrin-a in complex with ligands are currently not available. Like for Ricin, we exam-ined small ringed compounds for Abrin-a as well.

In this chapter, the binding behavior and binding energies of different compounds such as adenyl (3'–5') guanosine (APG), formycin-5'-monophosphate (FMP), 7-deazaguanine (7DG), 9-deazaguanine (9DG), 2,5-diamino-4,6-dihydroxypyrimidine (DDP), 8-methyl-9-oxoguanine (MOG), neopterin (NEO), pteroic acid (PTA) are discussed.

10.2 COMPUTATIONAL PROCEDURES AND METHODS

Auto Dock[21] is an automated docking approach developed to provide an automated procedure for predicting the interaction of ligands with the bio-macromolecular targets. It starts a ligand molecule in an arbitrary orientation and position and finds favorable docked configurations in a protein binding site. Simulated annealing is used to explore the conformational space. The ligand performs a random search around the static protein. At each step in the simulation, a small random displacement is applied to each of the degrees of freedom of the substrate: translation, orientation, and rotation around each of its flexible internal dihedral angles. The energy of the new configuration is compared to the energy of the preceding step. If the energy is lower, the new configuration is accepted. If the new energy is higher, then the configuration to be accepted or rejected is based upon a probability expression dependent on a user-defined temperature. After each simulation step, the interaction energy of ligand and the protein is calculated using atomic affinity potentials computed on a grid similar to that described by Goodford.[22].

To model Abrin-a inhibitor complexes, the coordinates of Abrin-a at 2.14 Å resolution were taken from the Protein Data Bank (PDB code 1ABR). *The docking procedure was carried out with and without active site water molecules. The protein was not allowed to move during the docking, while ligand was allowed to rotate in the conformational space around the active site.* Some of the synthetic compounds suggested here for inhibitor complex studies of Abrin-a have already been tested for Ricin, another toxin belonging to the class of RIPs. Table 10.1 shows the molecules that were complexed with Abrin-a. Hydrogen atoms were fixed to the protein atoms at pH 7 using the Biopolymer module in INSIGHT II molecular modeling package (INSIGHTII version 2000, Molecular Modeling System, User Guide, Accelrys 2000, San Diego, CA) and the KOLLUA partial charges were assigned to the protein using the same package. Docking was performed using AUTODOCK 3.0. The docked complexes were analyzed for hydrogen bonds, hydrophobic contacts, and other ligand–protein interactions using LIGPLOT,[23] HBPLUS,[24] and LPC-CSU[25] server. The LPC-CSU server was used to calculate the surface complementary index. The Lipinski rule for various ligands was calculated using MOE (Molecular Operating Environment; http://www.chemcomp.com) software.

Schematic diagrams of all the synthetic ligands are shown in Table 10.1 and their related PDBs are shown in Table 10.2. The coordinates of the ligands were taken from the corresponding PDB files. All these ligands were docked with Abrin-a using docking software AUTODOCK 3.0.[21] Default parameters were used except for the grid center, which was set to "auto" in most of the cases and "specified" in other cases. The complexed structures of various ligands described above were sorted on the basis of number of conformations present in the cluster and lowest docked energy. All the synthetic molecules, except APG, were docked efficiently with Abrin-a.

These computationally predicted lowest energy complexes of Abrin-a are stabilized by intermolecular hydrogen bonding and van der Waals and stacking interactions. *In all these computed complexes, the active site residues participate extensively, in intermolecular hydrogen bonding, and the substrate is sandwiched between the specificity pocket residues, namely, Tyr74 and Tyr113.*

TABLE 10.1 Various Ligands Along with Their Chemical Structures.

Ligand	Chemical structure
FMP	
APG	
PTA	
NEO	

TABLE 10.1   *(Continued)*

Ligand	Chemical structure
7DG	
9DG	
MOG	
DDP	

FMP, formycin-5′-monophosphate; APG, adenyl (3′–5) guanosine; PTA, pteroic acid; NEO, neopterin; 7DG, 7-deazaguanine; 9DG, 9-deazaguanine; MOG, 8-methyl-9-oxoguanine; DDP, 2,5-diamino-4,6-dihydroxypyrimidine.

The Lipinski rule[26] of five was calculated for these ligands and the values are tabulated in Table 10.3. The Lipinski values were calculated in order to check whether the particular ligand has the potential to be used as inhibitory drugs for future designs. They are classified as the Lipinski drug like ligand and Lipinski violating ligands on the basis of these values. *The two active site water molecules were also considered and the effect of these solvent molecules in inhibitor binding was studied.*

TABLE 10.2 Ligand Molecules Complexed with Abrin-a A-Chain.

Molecule	PDB ID[a]	Identifier
Formycin-5'-monophosphate	1FMP	FMP
Adenyl (3'–5') guanosine	1APG	APG
Pteroic acid	1BR6	PTA
Neopterin	1BR5	NEO
7-Dezaguanine	1IL3	7DG
9-Dezaguanine	1IL4	9DG
8-Methyl-9-oxaguanine	1IL9	MOG
2,5-Diamino-4,6-dihydroxypyrimidine	1IL5	DDP

[a]The coordinates of the ligand molecule were taken from these PDB files with the indicated codes.

TABLE 10.3 The Lipinski Rule of Five for the Ligands and Their Classification as Lipinski Drug like or Lipinski Violating Ligands.

Ligand	Molecular weight (<500)	No. of hydrogen bond donors (<10)	No. hydrogen bond acceptors (<5)	Log P (O/W) (<5)	Lipinski-drug like	Lipinski-violation
FMP	347.22	11	8	−2.70	0	2
APG	606.41	15	4	−4.87	0	2
PTA	312.29	7	5	0.37	1	0
NEO	253.218	7	5	−3.04	1	0
7DG	150.14	3	2	−0.76	1	0
9DG	150.14	2	3	−0.53	1	0
MOG	166.14	3	2	−1.12	1	0
DDP	142.12	4	4	−1.26	1	0

Table 10.4 shows the lowest docked energies of interaction of different inhibitors complexed with Abrin-a with and without the explicit solvent molecule.

10.3 BINDING MODES OF DIFFERENT LIGANDS TO ABRIN-A

10.3.1 FORMYCIN-5'-MONOPHOSPHATE

Figure 10.1 shows the stereo view of the most energetically favored binding mode of FMP to Abrin-a in the presence of solvent molecules. It is clearly seen that the nucleotide (adenine here) binds into the specificity pocket.

TABLE 10.4 Lowest Docked Energies of Abrin-a + Inhibitor Complexes With and Without Solvent Molecule.

Ligand name	Explicit solvent				Without solvent			
	No. of distinct cluster	Lowest docked energy (kcal/mol)	No. in best docked cluster	No. of distinct cluster	Lowest docked energy (kcal/mol)	No. in best docked cluster		
FMP	5	−8.60	76	5	−8.10	59		
APG (1)	3	−8.89	67	3	−8.53	84		
APG (2)	3	−8.18	32	3	−6.99	5		
PTA	5	−9.62	89	10	−7.41	57		
NEO	2	−9.53	64	3	−9.89	93		
7DG	4	−7.75	83	3	−7.71	85		
9DG	3	−7.66	62	4	−7.62	83		
MOG	2	−7.93	69	4	−7.82	62		
DDP	6	−6.69	44	3	−6.57	58		

APG (1) and (2) correspond to the two best docked conformers of APG.

The adenine ring stacks itself parallel to the Tyr74 and Tyr113 rings thereby paving way for the inhibitory activity on Abrin-a. The intermolecular hydrogen bonds were calculated using the HBPLUS[24] software and are shown in Table 10.5. N-6 and N-7 of formycin donate hydrogen bonds to the specific receptor oxygen atoms of Gly111. The ribose O-2' and O-3' make hydrogen bonds with the backbone carbonyl oxygen of Glu195 and Asp199, respectively. Difference in lowest docked energy is clearly visible in the presence and absence of active site solvent molecule (Table 10.4). The presence of active site water molecule further stabilizes the binding of FMP in the specificity pocket with strong hydrogen bonds as shown in Figure 10.1 and Table 10.5. But the binding of FMP is facilitated both in presence and absence (figure not shown) of active site water. Table 10.6 contains the surface complementarity (SC) indices of various Abrin-inhibitor complexes calculated using the LPC-CSU server. *The SC index values for FMP is greater than 0.5, as seen in Table 10.6, thereby suggesting a high affinity of Abrin-a toward FMP.* On the contrary to this, there is Lipinski rule violation as is clear from Table 10.3. *Even though FMP interacts with Abrin-a, the Lipinski parameter values fall out of range, thereby making it less likely that FMP is a pharmaceutically useful inhibitor of Abrin-a.*

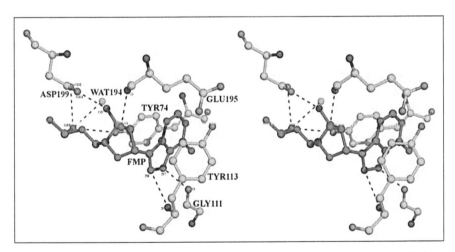

FIGURE 10.1 **(See color insert.)** The stereo-view of the mode of binding of FMP to Abrin-a.

10.3.2 ADENYL (3'–5') GUANOSINE

APG, a dinucleotide, could not be firmly docked with Abrin-a. It is less occupied than FMP in the specificity pocket. The docking results were not much promising as seen for other ligands. APG was also docked with Ricin,[2] and the APG-RTA complex results reveal the presence of the guanine moiety at the binding site region which has been confirmed experimentally. *In case of APG-Abrin-a complex, the best docking result showed the presence of guanine ring at the substrate binding region and is evident from the SC values of adenine and guanine (Table 10.6). The second best docking run instead exhibited the presence of adenine at the binding site stacked in between Tyr74 and Tyr113 leaving the guanine ring outside.* The lowest docked energies for the two runs are comparable as seen from Table 10.4. The surface complementary indices for the two rings are in accordance with the docking results with SC values being less for guanine and more for adenine (Table 10.6).

The docking of the APG molecule was carried out by specifying the grid center manually. The adenine and the guanine rings were thus seen making strong hydrogen bonds like in FMP. The O-2' of the adenine is hydrogen bonded with the OE2 and the carbonyl oxygen atom of Glu164 and Glu195, respectively. The N-3 of adenine is hydrogen bonded with the NH1 of the Arg167. The ribose oxygen O-4' and O-5' of the adenine ring make hydrogen bonds with the hydroxyl group of Tyr74. Other intermolecular hydrogen bonds of Abrin-a and APG complex are shown in Table 10.5. The stereo view (Fig. 10.2) shows the binding of APG with Abrin-a in the presence of solvent molecules. Table 10.3 shows the values of Lipinski Rule of five. APG violates all the rules strongly and therefore fails to exhibit drug-like properties according to the rule.

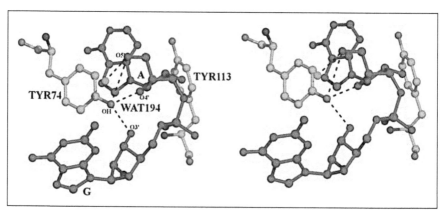

FIGURE 10.2   (See color insert.) The stereo-view of the mode of binding of APG to Abrin-a.

10.3.3   PTEROIC ACID

Among the compounds predicted to have hydrogen bond donors and acceptors in an orientation favorable for interacting with Abrin-a are pterin derivatives[18,28] PTA and NEO were considered to be good choices. Figure 10.3 illustrates stereo view of the interactions between PTA and the active site of Abrin-a. *The pterin ring roughly binds in the position specific to substrate adenine and is sandwiched between the two aromatics Tyr74 and Tyr113.* The distance between N-4 of the PTA and the carbonyl oxygen atom of Gly111 is 2.75 Å, suggesting that the PTA is stabilized in a tautomeric form with hydrogen atom on N-4 as shown in Table 10.5. The 2-amino group of pterin, N-11, also donates a hydrogen bond to the carbonyl oxygen atom of Gly111. Thus, it roughly mimics the role of 6-amino group on adenine substrate. *The benzoate moiety of PTA binds around the side chain of Tyr113 making a few nonpolar contacts with it. It appears to bind at the surface of the pocket hypothesized as the second recognition site that might accommodate guanine base of natural rRNA substrate.* A strong hydrogen bond is made between the carboxylate group of benzoate moiety and Arg124. Active site water further stabilizes the binding of PTA in the specificity pocket via strong hydrogen bond between N8 of PTA and WAT173.

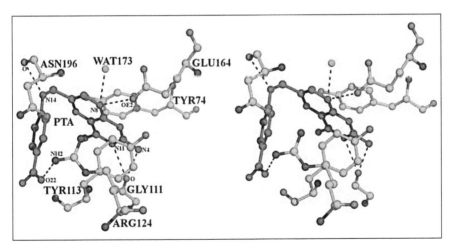

FIGURE 10.3   (See color insert.) The stereo-view of the mode of binding of PTA to Abrin-a.

10.3.4 NEOPTERIN

NEO is a pterin derivatized at position 6 with propane triol. A stereo view of the binding mode of NEO to Abrin-a is shown in Figure 10.4. The orientation of pterin ring here is similar to that of PTA. Strong hydrogen bonds are made with Arg167 at the 4-oxo group of NEO and the proximal hydroxyl group of propane triol.

The propane triol moiety of NEO is smaller than the benzoate moiety of PTA and does not interact with either Tyr74 or Tyr113 and lacks the van der Waals contribution to binding energy as compared to PTA. Strong hydrogen bonds that stabilize the ligand binding in the pocket are seen between O21 of NEO and WAT173 (Table 10.5). Pterin derivatives are consistent with the Lipinski rule of five thereby fulfilling the criteria of being potent ligands (Table 10.3).

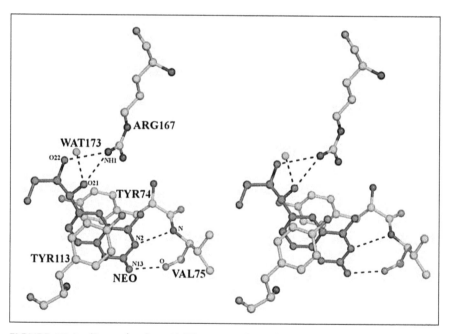

FIGURE 10.4 (See color insert.) The stereo-view of the mode of binding of NEO to Abrin-a.

10.3.5 7DG, 9DG, AND MOG

The pterin-like compounds are poorly soluble. So like in Ricin, search was done for guanine-like compounds[29] *for Abrin-a.* The chemical name of the above two compounds are 2-amino-3,4-dihydroxy-4-oxo-7*H*-pyrrole [2,3-*d*]-pyrimidine (7DG) and 2-amino-3,4-dihydroxy-4-oxo-9*H*-pyrrole [2,3-*d*]-pyrimidine (9DG) and MOG, respectively. These three compounds were found to be stacked between Tyr74 and Tyr113, and the stability is attributed to strong hydrogen bonds and hydrophobic contacts. The SC index has a high value thereby showing affinity of 7DG and 9DG for binding at the active site of Abrin-a (Table 10.6). The SC index for MOG is less compared to 7DG and 9DG (Table 10.6). For these compounds, very few interactions are seen with the protein molecule. The guanine ring makes hydrogen bonds with Arg167, Glu164, and Val75. Figures 10.5–10.7 show the stereo-view of the binding modes of 7DG, 9DG, and MOG. From Table 10.3, it is clear that these ligands fulfill the criterion for Lipinski rule of five.

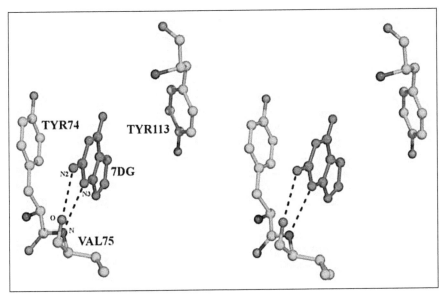

FIGURE 10.5 **(See color insert.)** The stereo-view of the mode of binding of 7DG to Abrin-a.

10.3.6 *2,5-DIAMINO-4,6-DIHYDROXYPYRIMIDINE (DDP)*

The stereo view of the binding modes of the inhibitor DDP to the active site of Abrin-a is shown in Figure 10.8. *Like the other ligands, DDP is also sandwiched between the two aromatics, Tyr74 and Tyr113, and interacts with the side chain of Tyr74.* Table 10.5 shows various hydrogen bonds of the ligand with the protein molecule. The SC index of DDP with Abrin-a is

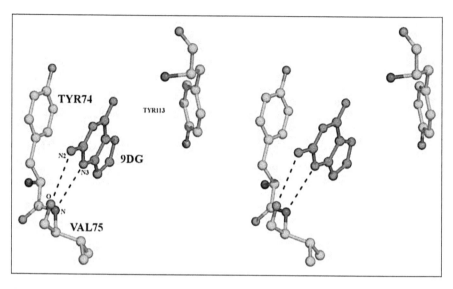

FIGURE 10.6 (See color insert.) The stereo-view of the mode of binding of 9DG to Abrin-a.

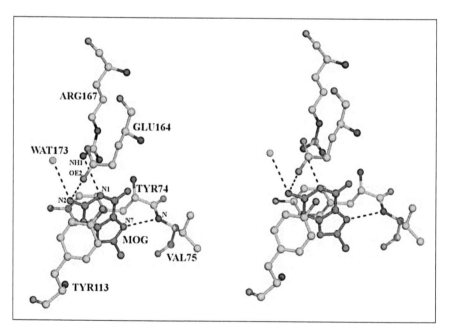

FIGURE 10.7 (See color insert.) The stereo-view of the mode of binding of MOG to Abrin-a.

high (Table 10.6) and the binding mode can be seen in Figure 10.8. In case of Ricin, DDP does not displace the side chain of Tyr80 from its apoenzyme conformation and thus fails to open the specificity pocket[29] thereby failing to be a strong inhibitor of Ricin. Whether this single-ring compound can act as an inhibitor for Abrin-a needs to be investigated experimentally.

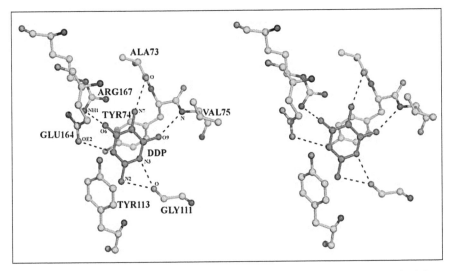

FIGURE 10.8   (See color insert.) The stereo-view of the mode of binding of DDP to Abrin-a.

TABLE 10.5   Hydrogen Bonds Formed Between the Ligand and the Binding Site Residues.

Ligand	Donor				Acceptor				Distance
	Name	Res	Atom	Chain	Name	Res	Atom	Chain	
FMP	FMP	301	O3P	B	HOH	303	O	B	2.98
	FMP	301	O3P	B	ASP	199	OD2	A	3.24
	FMP	301	O3'	B	ASP	199	OD1	A	2.55
	FMP	301	O2'	B	GLU	195	O	A	2.62
	TYR	113	N	A	FMP	301	N8	B	3.13
	FMP	301	N7	B	GLY	111	O	A	2.52
	FMP	301	O3P	B	TYR	74	OH	A	3.22
APG-1st	TYR	110	OH	A	A	701	O5'	B	2.5
	G	702	O3'	B	TYR	74	OH	A	2.94
	ASN	72	ND2	A	A	701	O5'	B	2.97

TABLE 10.5   *(Continued)*

Ligand	Donor				Acceptor				Distance
	Name	Res	Atom	Chain	Name	Res	Atom	Chain	
APG-2nd	A	701	O5'	B	HOH	704	O	B	2.43
	G	702	O3'	B	TYR	74	OH	A	3
	TYR	74	OH	A	A	701	O4'	B	2.39
	TYR	74	OH	A	A	701	O5'	B	3.35
PTA	PT1	301	N8	B	HOH	302	O	B	2.81
	PT1	301	N14	B	ASN	196	O	A	3.26
	PT1	301	N8	B	GLU	164	OE2	A	3.19
	ARG	124	NH2	A	PT1	301	O22	B	2.7
	PT1	301	N4	B	GLY	111	O	A	2.75
	PT1	301	N11	B	GLY	111	O	A	2.57
NEO	NEO	500	O21	B	HOH	501	O	B	2.42
	ARG	167	NH1	A	NEO	500	O22	B	3.09
	ARG	167	NH1	A	NEO	500	O21	B	2.87
	NEO	500	N13	B	VAL	75	O	A	2.7
	VAL	75	N	A	NEO	500	N2	B	3.25
7DG	7DG	301	N2	B	VAL	75	O	A	3.05
	VAL	75	N	A	7DG	301	N3	B	3.02
9DG	9DG	301	N2	B	VAL	75	O	A	3.20
	VAL	75	N	A	9DG	301	N3	B	3.06
MOG	MOG	301	N2	B	HOH	302	O	B	3.21
	ARG	167	NH1	A	MOG	301	N1	B	3.13
	MOG	301	N2	B	GLU	164	OE2	A	2.90
	VAL	75	N	A	MOG	301	N7	B	3.09
DDP	ARG	167	NH1	A	DDP	301	O6	B	2.60
	DDP	301	N1	B	GLU	164	OE2	A	3.32
	DDP	301	N2	B	GLY	111	O	A	2.81
	DDP	301	N3	B	GLY	111	O	A	2.67
	VAL	75	N	A	DDP	301	O9	B	2.77
	DDP	301	N7	B	ALA	73	O	A	2.95

TABLE 10.6 Surface Complementarity Index Values in Abrin–Ligand Complex.

Ligand	Explicit solvent			Without solvent		
	Theoretical maxm.	Actual value	Normalized complementarity	Theoretical maxm.	Actual value	Normalized complementarity
FMP	493	325	0.66	495	348	0.70
APG(1)-A	426	149	0.35	422	141	0.34
APG(1)-G	480	303	0.63	480	298	0.62
APG(2)-A	423	344	0.81	426	354	0.83
APG(2)-G	480	190	0.41	479	199	0.42
PTA	497	281	0.56	501	313	0.63
NEO	422	316	0.75	423	259	0.61
7DG	299	223	0.74	495	348	0.70
9DG	299	256	0.86	299	265	0.89
MOG	318	143	0.45	319	138	0.43
DDP	283	203	0.72	283	217	0.77

10.4 ANALYSIS OF THE BINDING MODES OF VARIOUS INHIBITORS OF ABRIN-A

Based on the docking studies, we can identify the possible ligands, which could inhibit Abrin-a. Tables 10.2 and 10.3 show various ligands and their corresponding Lipinski parameter values. From these tables (Tables 10.2 and 10.3), it is clear that FMP fails to pass the Lipinski criteria of potential ligands. Earlier studies suggest that FMP could not compete for strong binding of the ribosomes. FMP complexed with Ricin does not show any inhibitory activity. But the possibility of FMP not being an inhibitor of Abrin-a can be ruled out only if suggested by biochemical assays.

For APG, the guanine ring is found to stabilize at the substrate recognition site in Abrin-a for the best docked conformer, thereby putting a question on the depurination mechanism that involves an adenine ring. Table 10.6 also suggests high SC index value of guanine moiety for the conformer with minimum docked energy. Though the second best docked conformer reveals the presence of adenine moiety at the substrate binding region and is evident from the SC index values (Table 10.6). The Lipinski rule parameters (Table 10.3) show less probability of APG being a ligand as all the parameters are outside the range of specified cut off values.

Compounds like pterins and their derivatives were also tested for acting as the inhibitors of Abrin-a. High SC index values and strong interactions suggest them to be potent inhibitors of Abrin-a. The pterin rings of PTA and NEO roughly take the position of adenine ring at the substrate binding site and are seen sandwiched between Tyr74 and Tyr113 in a similar fashion to the adenine ring. PTA seems to have high binding energy with contributions from van der Waals interactions by the benzoate moiety and strong hydrogen bonds. The propane triol group of NEO on the other hand lacks hydrophobic interaction, thereby showing less binding energy. Thus, from the binding modes with the static structure of Abrin-a and the interaction energies though not fully definitive, it is clear that PTA is a better inhibitor exhibiting tight binding. Thus, the design of a molecule with groups covalently linked to the Pterin ring and which have strong and specific interactions with Abrin-a will be an attractive approach toward creating an efficient inhibitor. The major drawback of using Pterin rings as potential ligands is their weak solubility.

Owing to this certain guanine like compounds like 7DG, 9DG, MOG also formed the part of our in silico studies. Table 10.4 shows that the lowest docked energies for guanine like compounds are higher as compared to the adenine like compounds. The probable reason to this could be that the Abrin-a binding site is specific to adenine though guanine residues are present. The reason for this specificity may arise mainly due to the steric hindrance caused by the guanine like compounds.[29]

10.5 CONCLUSIONS

Abrin-a is being tested extensively for the design of therapeutic immunotoxins. In these constructs Abrin-a is chemically or genetically linked to an antibody to form a "magic bullet" which can preferentially target the cell lines carrying antigenic markers recognized by the antibody. Thus, the design of a potent inhibitor of Abrin-a could facilitate immunotoxin treatment and can also be used as antidotes in poison attacks.

Our work highlights the analysis of the binding modes of various adenine like rings structures to the active site of Abrin-a. A few synthetic compounds like APG, MOG depart from the category of inhibitors for Abrin-a on the basis of low SC index values and lack of strong interactions with the active site residues of Abrin-a. FMP is not an inhibitor of Ricin[27] as seen from earlier studies, and this is perhaps true for Abrin-a, can be confirmed by three-dimensional crystal structure determination and the biochemical analysis. APG also shows weaker binding of the adenine ring but shows strong affinity of guanine ring toward the binding site. Pterin rings show high affinity toward binding when compared with adenine. Additional contacts and high interaction energies (computational) pave the way for these rings acting as strong inhibitors of Abrin-a.

It is not necessary that the ligands which interact with every point at the binding site will be strong inhibitors, the additional groups commonly known as pendant groups, outside the binding pocket have to be further exploited to develop potent inhibitors of Abrin-a. These pendant groups can further enhance the properties like solubility and activity of the ligand. Thus, the in silico studies provide useful pointers to design of efficient inhibitors of Abrin-a.

10.6   FUTURE PROSPECTS

The above findings suggest that Abrin includes a fairly rigid receptor that binds specific complementary ring structures, which necessarily are very similar to adenine. But like Ricin, we see that it is possible for compounds such as pterins to bind to Abrin. The pendant groups that lie outside the binding pocket have to be further exploited to develop potent inhibitors of Abrin-a. The in silico analysis reported here points to experimental work such as X-ray crystal structure analysis of potential inhibitors in complex with Abrin-a and biochemical characterization of inhibitory activity.

KEYWORDS

- plant toxins
- ribosome inactivating proteins
- *N*-glycosidases
- cytotoxicity
- inhibitors
- ringed compounds

REFERENCES

1. Day, P. J.; Owens, S. R.; Wesche, J.; Olsnes, S.; Roberts, L. M.; Lord, J. M. An Interaction between Ricin and Calreticulin that May Have Implications for Toxin Trafficking. *J. Biol. Chem.* **2001**, *276*, 7202–7208. DOI: 10.1074/jbc.M009499200. pmid:11113144.
2. Bolognesi, A.; Tazzari, P. L.; Olivieri, F.; Polito, L.; Falini, B.; Stripe, F. Induction of Apoptosis by Ribosome-Inactivating Proteins and Related Immunotoxins. *Int. J. Cancer* **1996**, *68* (3), 349–355. DOI: 10.1002/(SICI)1097-0215(19961104)68:3<349::AID-IJC13>3.0.CO;2-3. pmid:8903477.
3. Lin, J. Y.; Lee, T. C.; Hu, S.; Tung, T. C. Isolation of Four Isotoxic Proteins and One Agglutinin from Jequiriti Bean (*Abrus precatorius*). *Toxicon* **1981**, *19* (1): 41–51. pmid:7222088.
4. Lin, J. Y.; Lee, T. C.; Tung, T. C. Inhibitory Effects of Four Isoabrins on the Growth of Sarcoma 180 Cells. *Cancer Res.* **1982**, *42* (1), 276–279. pmid:7053854.
5. Olsnes, S.; Eiklid, K. Isolation and Characterization of Shigella Shigae Cytotoxin. *J. Biol. Chem.* **1980**, *255* (1), 284–289. pmid:7350160.
6. Endo, Y.; Tsurugi, K. RNA *N*-Glycosidase Activity of Ricin A-Chain. Mechanism of Action of the Toxic Lectin Ricin on Eukaryotic Ribosomes. *J. Biol. Chem.* **1987**, *262* (17), 8128–8130. pmid:3036799.
7. Sperti, S.; Montanaro, L.; Mattioli, A.; Stirpe, F. Inhibition by Ricin of Protein Synthesis In Vitro: 60 S Ribosomal Subunit as the Target of the Toxin. *Biochem. J.* **1973**, *136* (3): 813–815. pmid:4360718.
8. Tahirov, T. H.; Lu, T. H.; Liaw, Y. C.; Chen, Y. L; Lin, J. Y. Crystal Structure of Abrin-A at 2.14 Å. *J. Mol. Biol.* **1995**, *250* (1), 354–367. DOI: 10.1006/jmbi.1995.0382. pmid:7608980.
9. Rutenber, E.; Katzin, B. J.; Ernst, S.; Collins, E. J.; Mlsna, D.; Ready, M. P.; Robertus, J. D. Crystallographic Refinement of Ricin to 2.5 A. *Proteins 1991*, *10* (3): 240–250. DOI: 10.1002/prot.340100308. pmid:1881880.
10. Zhou, K.; Fu, Z.; Chen, M.; Lin, Y.; Pan, K. Structure of Trichosanthin at 1.88 A Resolution. *Proteins* **1994**, *19* (1), 4–13. DOI: 10.1002/prot340190103. pmid:8066085.
11. Zeng, Z. H.; He, X. L.; Li, H. M.; Hu, Z.; Wang, D. C. Crystal Structure of Pokeweed Antiviral Protein with Well-Defined Sugars from Seeds at 1.8 Å Resolution. *J. Struct. Biol.* **2003**, *141* (2), 171–178. pmid:12615543.
12. Bagaria, A.; Surendranath, K.; Ramagopal, U. A.; Ramakumar, S.; Karande, A. A. Structure-Function Analysis and Insights into the Reduced Toxicity of *Abrus precatorius* Agglutinin I in Relation to Abrin. *J. Biol. Chem.* **2006**, *281* (45), 34465–34474. DOI: 10.1074/jbc.M601777200. pmid:16772301.
13. Monzingo, A. F.; Robertus, J. D. X-Ray Analysis of Substrate Analogs in the Ricin A-Chain Active Site. *J. Mol. Biol.* **1992**, *227* (4), 1136–1145. pmid:1433290.
14. Monzingo, A. F.; Collins, E. J.; Ernst, S. R.; Irvin, J. D.; Robertus, J. D. The 2.5 A Structure of Pokeweed Antiviral Protein. **1993**, *233* (4), 705–715. DOI: 10.1006/jmbi.1993.1547. pmid:8411176.
15. Xiong, J. P.; Xia, Z. X.; Wang, Y. Crystal Structure of Trichosanthin–NADPH Complex at 1.7 A Resolution Reveals Active–Site Architecture. *Nat. Struct. Mol. Biol.* **1994**, *1* (10), 695–700. pmid:7634073.
16. Ready, M.; Wilson, K.; Piatak, M.; Robertus, J. D. Ricin-Like Plant Toxins are Evolutionarily Related to Single-Chain Ribosome-Inhibiting Proteins from Phytolacca. *J. Biol. Chem.* **1984**, *259* (24): 15252–15256. pmid:6511792.

17. Reddy, V. V.; Sirsi, M.; Effect of *Abrus precatorius* L. on Experimental Tumors. *Cancer Res.* **1969**, *29* (7), 1447–1451. pmid:5799161.
18. Yan, X.; Hollis, T.; Svinth, M.; Day, P.; Monzingo, A. F.; Milne, G. W.; Robertus, J. D. Structure-Based Identification of a Ricin Inhibitor. *J. Mol. Biol.* **1997**, *266* (5), 1043–1049. DOI: 10.1006/jmbi.1996.0865. pmid:9086280.
19. Chen, X. Y.; Link, T. M.; Schramm, V. L. Ricin A–Chain: Kinetics, Mechanism, and RNA Stem–Loop Inhibitors. *Biochemistry* **1998**, *37* (33), 11605–11613. DOI: 10.1021/bi980990p. pmid:9708998.
20. Link, T.; Chen, X. Y.; Niu, L. H.; Schramm, V. L. Ricin-A Chain Activity and Inhibitor Studies with Oxycarbenium Analogues. *FASEB J.* **1995**, *9* (6), A1423.
21. Morris, G. M.; Goodsell, D. S.; Halliday, R. S.; Huey, R.; Hart, W. E.; Belew, R. K.; Olson, A. J. Automated Docking using a Lamarckian Genetic Algorithm and an Empirical Binding Free Energy Function. *J. Comput. Chem.* **1998**, *19* (14), 1639–1662.
22. Goodford, P. J. A Computational Procedure for Determining Energetically Favorable Binding Sites on Biologically Important Macromolecules. *J. Med. Chem.* **1985**, *28* (7), 849–857. DOI: 10.1021/jm00145a002. pmid:3892003.
23. Wallace, A. C.; Laskowski, R. A.; Thornton, J. M. LIGPLOT: A Person to Generate Schematic Diagrams of Protein-Ligand Interactions. *Prot. Eng.* **1995**, *8* (2), 127–134. pmid:7630882.
24. McDonald, I. K.; Thornton, J. M. Satisfying Hydrogen Bonding Potential in Proteins. *J. Mol. Biol.* **1994**, *238* (5), 777–793. DOI: 10.1006/jmbi.1994.1334. pmid:8182748.
25. Sobolev, V.; Sorokine, A.; Prilusky, J.; Abola, E. E, Edelman, M. Automated Analysis of Interatomic Contacts in Proteins. *Bioinformatics* **1999**, *15* (4), 327–332. pmid:10320101.
26. Lipinski, C. A.; Beryl, F. L.; Paul, W. D.; Feeney, J. Experimental and computational approaches to estimate solubility and permeability in drug discovery and development settings. Advanced Drug Delivery Reviews 1997, 23 (1-3), 3-25. pmid: 11259830.
27. Olson, M. A. Ricin A-Chain Structural Determinant for Binding Substrate Analogues: A Molecular Dynamics Simulation Analysis. *Proteins* **1997**, *27* (1), 80–95. pmid:9037714.
28. Yan, X.; Day, P.; Hollis, T.; Monzingo, A. F.; Schelp, E.; Robertus, J. D.; Milne, G. W.; Wang, S. Recognition and Interaction of Small Rings with the Ricin A-Chain Binding Site. *Proteins* **1998**, *31* (1), 33–41. pmid:9552157.
29. Miller, D. J.; Ravikumar, K.; Shen, H.; Suh, J. K.; Kerwin, S. M., Robertus, J. D. Structurebased Design and Characterization of Novel Platforms for Ricin and Shiga Toxin Inhibition. *J. Med. Chem.* **2002**, *45* (1), 90–98. pmid:11754581.

CHAPTER 11

BIOCARBON STORAGE POTENTIAL OF TEA PLANTATION OF NILGIRIS, INDIA, IN RELATION TO LEAF CHLOROPHYLL AND SOIL PARAMETERS

L. ARUL PRAGASAN[*] and R. SARATH

Department of Environmental Sciences, Bharathiar University, Coimbatore 641046, India

[*]*Corresponding author. E-mail: arulpragasan@yahoo.co.in*

ABSTRACT

Biocarbon storage is the sinking of atmospheric carbon in plants. It helps to reduce the alarming level of increase in greenhouse gases particularly carbon dioxide in the atmosphere. The concentration of atmospheric CO_2 was 270 ppm before the industrial revolution period, and now, it has crossed 500 ppm. This is one of the main reasons for global warming and climate change. Currently, there are lots of measures taken for mitigation of this global threat, and one such measure is biocarbon storage. This chapter aims to provide the biocarbon storage potential of tea plantation in Gudalur located in Nilgiris district, Tamil Nadu, India.

11.1 INTRODUCTION

Biocarbon storage or sequestration refers to the storage of carbon in a stable system that occurs through direct fixation of atmospheric carbon dioxide (CO_2) by means of plant photosynthesis. In this process, plants absorb CO_2 from atmosphere and store as biomass, and thereby reduce the air pollution caused by the greenhouse gas. It is well known that the increasing concentration of greenhouse gases particularly CO_2 is the reason behind global warming and climate change. Reduction of such harmful gas is vital for mitigation of the drastic change in climate. Thus, biocarbon storage is a handy technique which helps in the reduction of atmospheric carbon concentration.

Although biocarbon storage is important, so far only a few number of studies particularly related to the effect of crop plantations (such as coffee, cocoa, and rubber) on carbon fixation in tropical regions were carried out.[14,26] Besides the growing awareness, information to support the sustainable management of crop plantations, conditions of biocarbon storage, and dynamics in the systems are still poorly understood.[15]

Tea (*Camellia sinensis* L.) is a perennial evergreen broad-leaved cash crop cultivated for continuous growth of young shoots. Tea leaves contain more than 700 chemicals; among those, useful compounds for human health include flavonoids, amino acids, vitamins (C, E, and K), caffeine, and polysaccharides. Recently, it is proven that tea drinking has been associated with cell-mediated immune function in human, tea improves beneficial intestinal microflora, provides immunity against intestinal disorders, and protects cell membrane from oxidative damage.[3] In India, tea is best known as a medical plant and not as a drink for pleasure until the 19th century during when tea plantations were established by the British. Tea is one of the three common beverages (coffee, tea, and cocoa) in the world, consumed by a large number of people. Due to high growing demand, it is considered as one of the major components of world beverage market. Nevertheless, tea cultivation is restricted to certain regions of the world due to specific climate and soil conditions.[15]

Worldwide, about 36.9 lakhs hectares of land area is used for tea cultivation. With increasing global tea trade, area of tea cultivation as well as tea production have increased significantly. As of 2013, the world annual tea production has crossed 5.34 million tonnes which is 66% higher than the value for the year 2003. Globally, China (1.94 million tonnes) and India (1.21 million tonnes) are the top tea-producing nations, and they account for 59% of the total tea production of the world.[8] India is the world's largest tea-drinking nation; however, the per capita of tea consumption is average 750 g/person/year. In India, tea plantations are concentrated in Assam, Himachal Pradesh, West Bengal, and parts of Karnataka, Kerala, and Tamil Nadu. The total tea harvest area as of 2003 was 516,000 ha, and it had increased to 563,980 by 2013, and the tea production for 2013 was 1,208,780 t which was 846,000 t for the year 2003.[8] Tea industries in India provide job opportunity for about one million people.[11]

Although, today, tea production for beverage is the main reason for the establishment of tea plantations, its long rotational lifecycle (from 40 to 90 years) may represent a high potential for biocarbon storage. Unlike the intensively studied biocarbon storage dynamics of forest systems,[9,10,16–22,24,25,27] tea plantations are poorly understood in terms of its biocarbon storage. To understand the biocarbon storage in tea crop, present work was carried out in a tea plantation in India. The main objectives of the present study were (1) to determine biocarbon storage of tea plantation and (2) to find the relationship between biocarbon storage of tea plant with leaf chlorophyll content and soil parameters such as pH and conductivity, nitrogen (N), phosphorus (P), and potassium (K).

11.2 MATERIALS AND METHODS

11.2.1 STUDY AREA

The present study was carried out in Gudalur (11.50° N, 76.50° E) located at Nilgiris district, Tamil Nadu, India (Fig. 11.1). The human population of Gudalur (survey 2011) was 49,535 individuals with a sex ratio of 1032 females for every 1000 males, and the density of population was 200/km². Gudalur is located at an altitude ca. 1100 m asl. This region has hills with varied altitudinal range, rich in diverse flora and fauna.

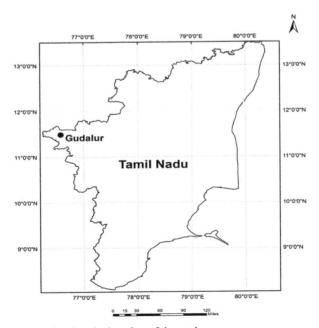

FIGURE 11.1   Map showing the location of the study area.

Besides tea (Fig. 11.2) and coffee plantation, vegetables such as potato, cabbage, and cauliflower, spices such as cardamom, pepper, and ginger are cultivated in this area. Tea cultivation was introduced by British during the preindependence period. Many small and large private tea estates exist in the region. In recent years, potato and other root growers in the region have switched to tea cultivation.

FIGURE 11.2   Tea plantation of the study area.

11.2.2   CLIMATE

Gudalur falls in tropical climatic zone. This region experiences four seasons, namely, southwest monsoon (SWM), northeast monsoon (NEM), winter (WS), and summer (SS) seasons (Fig. 11.3). The mean annual rainfall for a 10-year period (1999–2008) available for the district was 1652 ± 353 mm (±SD). Rainfall was maximum during SWM (836 ± 214 mm) followed by NEM (494 ± 129 mm), SS (277 ± 97 mm), and WS (46 ± 53 mm). The mean monthly maximum temperature (for 2008–2009) was 21.4 ± 5.6°C and the mean monthly minimum temperature for the same period was 15.0 ± 3.8°C.

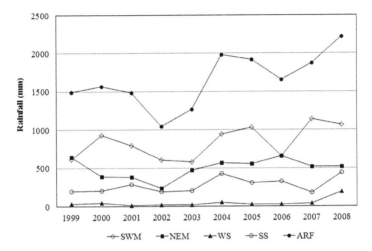

FIGURE 11.3   Rainfall pattern for a 10-year period available for the study area.

11.2.3   SOIL

The soil type of this region can be categorized into five major types, namely, alluvial and colluvial soil, black soil, lateritic soil, red loam, and red sandy soil. The predominant type being lateritic soil, in contrast the red loam and red sandy soils are seen in small patches. In the valleys, where the water-logging is common during the monsoons are dominated by the black soil. While, alluvial and colluvial soils are common along the valleys and river courses.

11.2.4   METHODS

Biocarbon storage of tea plantation was calculated based on field survey and laboratory analysis. To determine the total biocarbon storage (tC/ha) for tea plantation, density (individuals/ha), biomass (kg/individual), and carbon content (%) data were used. Density of tea crop was determined by quadrat method. Ten quadrats of size 10 m × 10 m were randomly placed and all the tea crops were counted from each quadrat to determine the density. Biomass of a single tea crop (13.6 kg/individual; leaf—5%, stem—84%, root—11%) was adopted from a 43-year-old tea plantation study[12] as destruction sampling was not possible. From each quadrat, leaf and stem samples were collected to determine the carbon content (%) of leaf, stem, and root components for calculation of biocarbon storage. Chlorophyll content of tea crop was detected using Chlorophyll meter (SPAD-502 Plus). Soil samples were collected to determine the soil pH, conductivity, and soil nutrients such as nitrogen (N, g/kg), phosphorus (P, %), potassium (K, %).

In the laboratory, the collected leaf and stem samples were air dried for 2 days. Then, the samples were dried in hot air oven for 24 h at 105°C to get dry weights. Then, 1 g of oven dried grind samples was taken separately in preweighted crucibles. The crucibles were placed in the furnace at 550 ± 5°C for 2 h. Then, the crucibles were cooled slowly inside the furnace. After cooling, the crucibles with ash were weighted for determination of carbon content using the following equations adopted earlier:[16]

$$C_c = (100 - A_{sh}) \times 0.58$$

$$A_{sh} = \frac{(W_3 - W_1)}{(W_2 - W_1)} \times 100$$

where C_c is the carbon content (%), A_{sh} is the ash percentage (%), W_1 is the weight of crucible (g), W_2 is the sum of the weight of oven dried grind sample and crucible (g), and W_3 is the sum of the weight of ash and crucible (g).

Soil pH and conductivity were estimated using the following procedure. One gram of soil sample was mixed with 50 mL of water and equilibrated for 1 h using a mechanical shaking incubator (Wisecube, Daihan Scientific, Korea) at 200 rpm at 35°C. The supernatant was analyzed for pH using digital pH meter (Elico—LI 127) and conductivity using digital conductivity meter (Elico—CM 180). Estimation of soil N, P, and K was carried out using standard methods such as Kjeldahl method, Bray's method, and Flame photometry method, respectively.

Further, carbon dioxide sequestration was determined from biocarbon storage data using conversion ratio of 1 kg of carbon = 3.68 kg of carbon dioxide, as this ratio was arrived by dividing the atomic weight of CO_2 by the atomic weight of carbon.

11.2.5 STATISTICAL ANALYSIS

Statistical tool, analysis of variance (ANOVA) was carried out to check for significance in variation of carbon content of tea plant components such as leaf, root, and stem samples. Linear regression was adopted to find the relationship between biocarbon storage of tea plantation with tea leaf chlorophyll content and soil parameters such as pH, conductivity, N, P, and K.

11.3 RESULTS

11.3.1 DENSITY

Density of plants calculated for the tea plantation was $18{,}540 \pm 250$ individuals/ha. Tea crop in the plantation had approximately 60–75 cm interdistance between two individuals. The abundance of tea plant for the ten 10 m × 10 m sample plots varied from 179 individuals to 187 individuals. The mean abundance per sample was 185.4 ± 2.5 (±SD).

11.3.2 CARBON CONTENT

The carbon content for both the plant components of tea crop was determined through loss on ignition method. The biocarbon storage assessed for the leaf samples ($n = 30$) ranged from 53.94% to 54.52%, and the average value was $54.15 \pm 0.28\%$. The carbon content determined for the stem samples ($n = 20$) ranged from 56.26% to 57.42%, the average carbon storage for the stem samples was $56.78 \pm 0.26\%$. The carbon content calculated for root ($n = 20$) varied from 55.1% to 55.68%, and the average value was $55.49 \pm 0.19\%$. One-way ANOVA revealed that there existed no significant variation in carbon content values among the leaf ($F(9, 20) = 1.095$, $p > 0.05$), stem ($F(9, 10) = 1.000$, $p > 0.05$) as well as for root ($F(9, 10) = 1.603$, $p > 0.05$) samples.

11.3.3 CHLOROPHYLL CONTENT

The chlorophyll content observed for the 10 sample plots ranged from 61.71 ± 23.3 to 83.55 ± 28.9. The average chlorophyll content for the total leaf samples ($n = 100$) of tea plantation was $69.31 \pm 20.54\%$. One-way ANOVA performed to check for variation in chlorophyll content showed no significant variation in chlorophyll content among the leaf samples ($F(9,90) = 1.188$, $p > 0.05$).

11.3.4 SOIL PARAMETERS

The soil parameters (pH, conductivity, N, P, and K) were determined using standard methods. Soil pH is considered a master variable in soils as it controls many chemical processes that take place. It specifically affects plant nutrient availability by controlling the chemical forms of the nutrient. The optimum pH range for most plants is between 5.5 and 7.0. The soil pH value for the tea plantation ($n = 5$) varied from 4.9 to 5.73, and the mean pH value was 5.4 ± 0.4. The soil conductivity determined for the tea plantation ($n = 5$) varied from 13.7 to 17.6 μs, and the average value was 16.31 ± 1.84 μs. Soil N, P, and K values ranged from 1.68 to 3.43 g/kg, 0.05 to 0.29%, and 0.02 to 0.04%, respectively, for the total samples ($n = 5$ for each parameters) collected from the tea plantation. The average value for the soil N, P, and K was 2.42 ± 0.65 g/kg, $0.13 \pm 0.09\%$, and $5.38 \pm 0.37\%$, respectively.

11.3.5 BIOCARBON STORAGE

The total biocarbon storage (tC/ha) determined for the tea plantation of Gudalur was 142.47 ± 1.92 tC/ha, and the biocarbon value varied from 137.56 tC/ha to 143.71 tC/ha among the 10 sample plots. The total biocarbon storage of tea plantation was maximum shared by stem (84%, 120.26 ± 1.62 tC/ha), followed by root (11%, 15.39 ± 0.21) and leaf (5%, 6.83 ± 0.09) (Fig. 11.4). One-way ANOVA revealed a strong significant variation in biocarbon stock among the three plant components of tea plant ($F(2,27) = 44{,}521.42$, $p < 0.00001$).

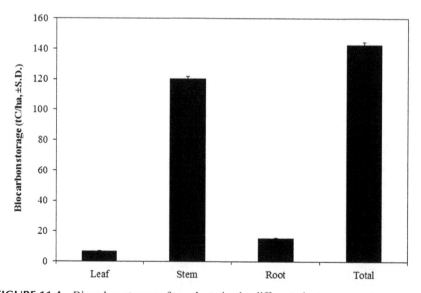

FIGURE 11.4   Biocarbon storage of tea plantation by different plant components.

11.3.6 RELATIONSHIP OF BIOCARBON STORAGE WITH LEAF CHLOROPHYLL AND SOIL PARAMETERS

Regression analysis done between the biocarbon storage (tC/ha) values of tea plantations with leaf chlorophyll and soil parameters such as pH, conductivity, N, P and K of the tea plantation under the study revealed that biocarbon storage (tC/ha) of tea plantation had positive relation with only soil N (Table 11.1). This reveals that N is a necessary component for sequestering more atmospheric carbon as biocarbon in tea plantations.

TABLE 11.1 Relationship of Biocarbon Storage (tC/ha) with Other Parameters.

Parameters	Leaf carbon	
	Equation	R^2
Leaf chlorophyll	$y = 0.0327x + 140.57$	0.0300
Soil pH	$y = -0.7892x + 147.02$	0.0867
Soil conductivity	$y = 0.1499x + 140.34$	0.0754
Soil N	$y = 1.2947x + 139.64$	0.7091
Soil P	$y = 3.7166x + 142.31$	0.1228
Soil K	$y = 18.592x + 142.33$	0.0213

11.3.7 CARBON DIOXIDE SEQUESTRATION

The estimated value of CO_2 sequestration by the tea plantation varied between 506.22 t/ha and 528.84 t/ha, and the average value was 524.32 ± 7.1 t/ha. For the different components of tea plant, the CO_2 sequestration value was 25.12 ± 0.3, 442.56 ± 6.0, and 56.64 ± 0.8 t/ha for leaf, stem and root part, respectively.

11.4 DISCUSSIONS

Quantifying biocarbon storage is a necessary task when assessing the particular carbon budget of an ecosystem. Tea is an important cash crop, and tea plantations take large amounts of cultivable land worldwide. Plants trap atmospheric carbon dioxide through photosynthesis and store in leaves, stems, and roots. The tea plantations absorb more carbon and serve as a good source for mitigation for global warming. The present study revealed the biocarbon storage potential of tea plantation in Gudalur, Nilgiris district, Tamil Nadu, India was high 142.47 ± 1.92 tC/ha when compared to a few other studies available in literature (Table 11.2). The biocarbon storage (tC/ha) of the tea plantation of Gudalur falls within the range reported for tropical forest of West Papua, Indonesia;[10] higher than the value reported for tea plantation in China (44.99–60.64 tC/ha[15]), Kenya (43.0–72.0 tC/ha[12]), tropical forest in Costa Rica (13.2–65.4 tC/ha[9]), natural forest and plantations in Puducherry (19.5–131.8 tC/ha[27]), noncommercial plantation in China (8.72–8.80 tC/ha[6]), agroforestry in India (31.95–83.07 tC/ha[23]), Scots pine plantation in Finland (13.87–22.58 tC/ha[5]), subtropical Pinus forest in Himachal Pradesh, India (47.3 tC/ha[24]); and lesser than the value reported for subtropical forest in Garhwal Himalayas, India (203.02–230.84 tC/ha[25]). The difference in biocarbon values (tC/ha) may be influenced by different factors including ecosystem type, plant density, rainfall, temperature, and soil parameters.

From a few available literature (Table 11.2), linear regression analysis was performed to check for any relation between biocarbon storage (tC/ha) and annual rainfall (mm) and temperature (°C). The analysis revealed that there existed no significant relationship of biocarbon storage (tC/ha) with annual rainfall (Fig. 11.5) as well as temperature (Fig. 11.6). In the present study, the plant density was $18,540 \pm 250$ individuals/ha which is almost three-fold higher than a 43-year-old tea plantation at China (6730 individuals/ha[12]), and the biocarbon storage (tC/ha) of Gudalur (142.47 tC/a) is also three-fold higher than the value (43.0 tC/ha) reported for the 43-year-old tea plantation at China,[12] and this supports a positive relation between tea plant density (individuals/ha) and biocarbon storage (tC/ha). However, more in-depth studies are warranted to draw a clear conclusion on the relationship between biocarbon storage and other influencing factors.

Among the three plant components, the total biocarbon storage of tea plantation in Gudalur was maximum shared by stem (120.26 ± 1.62 tC/ha), followed by root (15.39 ± 0.21) and leaf (6.83 ± 0.09), and a strong significant ($p < 0.00001$) variation in the biocarbon value was observed among them. The amount of biocarbon stock is directly proportional to the total biomass stock. It is widely applied that 50% of the total biomass is considered as the total carbon stock of tropical tree species.[2,4,7,17,28,29] In the present study, the carbon content (%) of the three plant components was 54.15%, 56.78%, and 55.49% for leaf, stem, and root, respectively. While the value recorded respectively for the three components was 51.97%, 49.73%, and 46.62% for the tea plantation at China.[15]

The results of regression analysis carried out between the biocarbon storage (tC/ha) and leaf chlorophyll content (%) and soil parameters, such as pH, conductivity, N, P, and K, for the present study revealed that biocarbon storage (tC/ha) of tea plantation at Gudalur had positive relation with only

TABLE 11.2 Comparison of Biocarbon Storage of Tea Plantation of Gudalur with Other Available Studies.

Location	Ecosystem	tC/ha	Annual rainfall (mm)	Temperature (°C)	References
Gudalur, India	Tea plantation	142.47	1652	15.0–21.4	Present study
Mali, West Africa	Agroforestry	0.70–54.00	300–700	29.0	[28]
Guanacaste, Costa Rica	Silvopastoral	3.50–12.50	1725	28.0	[1]
West Papua, Indonesia	Tropical forest	54.90–153.40	2394	26.9	[10]
Xiamen, China	Forest	4.56–13.52	1100	21.0	[22]
Costa Rica	Tropical forest	13.20–65.40	3420–6840	25.0–27.0	[9]
Bodamalai Hills, India	Tropical forest	10.94	1058	28.3	[18]
Kalrayan Hills, India	Tropical forest	38.88	1058	28.3	[19]
Shervarayan Hills, India	Tropical forest	56.55	1058	28.3	[20]
Chitteri Hills, India	Tropical forest	58.55	1058	28.3	[17]
Kolli Hills, India	Tropical forest	73.70	1058	28.3	[21]
Coimbatore, India	Eucalyptus plantation	27.72	646	26.3–38.8	[16]
Coimbatore, India	Mixed species plantation	22.25	646	26.3–38.8	[16]
Puthupet, Puducherry	Anacardiumoccidentale plantation	19.50	1311	28.5	[27]
Puthupet, Puducherry	Casuarina equisetifolia plantation	23.90	1311	28.5	[27]
Puthupet, Puducherry	Cocosnucifera plantation	83.30	1311	28.5	[27]
Puthupet, Puducherry	Mangifera indica plantation	70.50	1311	28.5	[27]
Puthupet, Puducherry	Natural forest	131.80	1311	28.5	[27]

TABLE 11.2 (Continued)

Location	Ecosystem	tC/ha	Annual rainfall (mm)	Temperature (°C)	References
Saharanpur (UP), India	Agroforestry	31.95	–	23.3	[23]
Haryana, India	Agroforestry	83.07	970	–	[23]
Garhwal Himalayas, India	Subtropical Pinus forest	203.02–230.84	960	10.0–23.0	[25]
Himachal Pradesh, India	Subtropical Pinus forest	47.30	1000	15.0–36.0	[24]
Kericho, Kenya	Tea plantation	43–72	2150	15.5–18.0	[12]
China	Tea plantation	50.90	700–2600	12.0–22.0	[15]
Yunnan Province, China	Noncommercial plantation	8.72–8.80	–	–	[6]
Finland	Scots pine plantation	13.87–22.58	–	–	[5]

soil N (R^2 = 0.71, Table 11.1), and this suggests that soil N is an influencing factor for biocarbon storage in tea plantation. Studies have proved earlier that nutrients, especially nitrogen, are known to be a key determinant of biocarbon storage in forest ecosystems.[12,13]

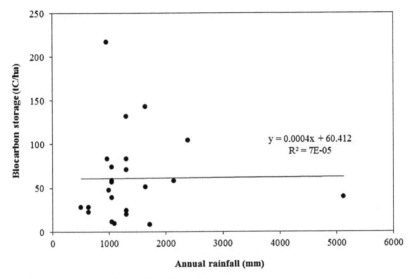

FIGURE 11.5   Impact of rainfall on biocarbon storage.

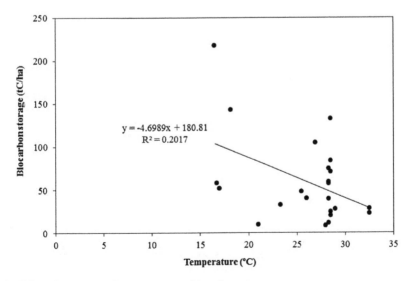

FIGURE 11.6   Impact of temperature on biocarbon storage.

In terms of CO_2 sequestration, the value determined for the tea plantation at Gudalur was 524.32 ± 7.1 tCO_2/ha. This suggests that the tea plantation in Gudalur traps 524 t of atmospheric carbon dioxide as biocarbon storage.

11.5   CONCLUSIONS

The present study signifies the potential role of tea plantation in biocarbon storage in the present scenario of global warming and climate change. The results of the present study conclude that the biocarbon storage in the tea plantation of Gudalur was 142.47 tC/ha, which is high when compared to other available studies. The biocarbon storage (tC/ha) potential of tea plantation in Gudalur was positively influenced by soil N which is in accordance with Kimmins[13] and Kamau et al.[12]. The global review of tea in terms of land cover, production, yield, exports, and imports indicated overall increase in the quantity of tea in the world market in recent years. If the existing tea plantations are managed sustainably, definitely tea plantations can work wonders in reduction of the alarming increase in the concentration of the atmospheric carbon dioxide gas that leads to global warming one of the grave environmental problems at present. Hence, tea plantations are not only an economic resource but also provide a long-term cost-effective means of biocarbon storage and thereby play a pivotal role in mitigation of global warming and climate change. Hence, creating awareness on biocarbon storage and encouragement of tea plantation and sustained harvest of tea leaf is required for improved biocarbon storage in order to mitigate global warming and climate change.

KEYWORDS

- biocarbon storage
- biomass
- chlorophyll
- density
- tea plantation
- Nilgiris

REFERENCES

1. Andrade, H. J.; Brook, R.; Ibrahim, M. Growth, Production and Carbon Sequestration of Silvopastoral Systems with Native Timber Species in the Dry Lowlands of Costa Rica. *Plant Soil* **2008**, DOI 10.1007/s11104-008-9600-x.
2. Atjay, G. L.; Ketner, P.; Duvignead, P. Terrestrial Primary Production and Phytomass. In *The global Carbon Cycle*; Bolin, B., Degens, E. T., Kempe, S., Eds.; Wiley and Sons: New York, USA, 1979; pp 129–182.
3. Bhagat, R. M.; Baruah, R. D.; Safique, S. Climate and Tea (*Camellia sinensis* (L.) O. Kuntze) Production with Special Reference to North Eastern India: A Review. *J. Environ. Res. Develop.* **2010**, *4*, 1017–1028.
4. Brown, S.; Lugo, A. E. The Storage and Production of Organic Matter in Tropical Forests and Their Role in the Global Carbon Cycle. *Biotropica* **1982**, *14*, 161–187.
5. Cao, T.; Valsta, L.; Makela, A. A Comparison of Carbon Assessment Methods for Optimizing Timber Production and Carbon Sequestration in Scots Pine Stands. *For. Ecol. Manage.* **2010**, *260*, 1726–1734.
6. Chen, X.; Zhang, X.; Zhang, Y.; Wan, C. Carbon Sequestration Potential of the Stands Under the Grain for Green Program in Yunnan Province, China. *For. Ecol. Manage.* **2009**, *258*, 199–206.
7. Dixon, R. K.; Brown, S.; Solomon, R. A.; Trexler, M. C.; Wisniewski, J. Carbon Pools and Flux of Global Forest Ecosystems. *Science* **1994**, *263*, 185–190.
8. FAO. *Food and Agriculture Association of the United Nations (FAO)*; Rome, 2016. (www.faostat-fao.org).
9. Fonseca, W.; Benayas, J. M. R.; Alice, F. E. Carbon Accumulation in the Biomass and Soil of Different Aged Secondary Forests in the Humid Tropics of Costa Rica. *For. Ecol. Manage.* **2011**, *262*, 1400–1408.
10. Hendri; Yamashita, T.; Kuntoro, A. A.; Lee, H. S. Carbon Stock Measurements of a Degraded Tropical Logged-Over Secondary Forest in Manokwari Regency, West Papua, Indonesia. *For. Stud. China* **2012**, *14* (1), 8–19.
11. Kalita, R. M.; Das, A. K.; Nath, A. J. Allometric Equations for Estimating above- and Belowground Biomass in Tea (*Camellia sinensis* (L.)O. Kuntze) Agroforestry System of Barak Valley, Assam, Northeast India. *Biomass Bioenerg.* **2015**, *83*, 42–49.
12. Kamau, D. M.; Spiertz, J. H. J.; Oenema, O. Carbon and Nutrient Stocks of Tea Plantations Differing in Age, Genotype and Plant Population Density. *Plant Soil* **2008**, *307*, 29–39.
13. Kimmins, J. P. Importance of Soil and Role of Ecosystem Disturbance for Sustainable Productivity of Cool Temperate and Boreal Forest. *Soil Sci. Soc. Am. J.* **1996**, *60*, 1643–1654.
14. Li, H. M.; Ma, Y. X.; Aide, T. M.; Liu, W. J. Past, Present and Future Land-Use in Xishuangbanna China and the Implications for Carbon Dynamics. *For. Ecol. Manage.* **2008**, *255*, 16–24.
15. Li, L.; Wu, X.; Xue, H.; et al. Quantifying Carbon Storage for Tea Plantations in China. *Agri. Ecosyst. Environ.* **2011**, *141*, 390–398.
16. Pragasan, L. A.; Karthick, A. Carbon Stock Sequestration by Tree Plantations in University Campus at Coimbatore, India. *Int. J. Environ. Sci.* **2013**, *3*, 1700–1710.
17. Pragasan, L. A. Carbon Stock Assessment in the Vegetation of the Chitteri Reserve Forest of the Eastern Ghats in India based on Non-Destructive Method Using Tree Inventory Data. *J. Earth Sci. Clim. Change* **2014**, *S11*, 001.

18. Pragasan, L. A. Tree Carbon Stock Assessment from the Tropical Forests of Bodamalai Hills Located in India. *J. Earth Sci. Clim. Change* **2015**, *6*, 314.

19. Pragasan, L. A. Assessment of Tree Carbon Stock in the Kalrayan Hills of the Eastern Ghats, India. *Walailak J. Sci. Technol.* **2015**, *12* (8), 659–670.

20. Pragasan, L. A. Total Carbon Stock of Tree Vegetation and its Relationship with Altitudinal Gradient from the Shervarayan Hills Located in India. *J. Earth Sci. Clim. Change* **2015**, *6*, 273.

21. Pragasan, L. A. 2016. Assessment of Carbon Stock of Tree Vegetation in the Kolli Hill Forest Located in India. *Appl. Ecol. Environ. Res.* **2016**, *14* (2), 169–183.

22. Ren, Y.; Wei, X.; Wei, X.; et al. Relationship between Vegetation Carbon Storage and Urbanization: A Case Study of Xiamen, China. *For. Ecol. Manage.* **2011**, *261*, 1214–1223.

23. Rizvi, R. H.; Dhyani, S. K.; Yadav, R. S.; Singh, R. Biomass Production and Carbon Stock of Popular Agroforestry Systems in Yamunanagar and Saharanpur Districts of Northwestern India. *Curr. Sci.* **2011**, *100* (5), 736–742.

24. Shah, S.; Sharma, D. P.; Pala, N. A.; et al. Temporal Variations in Carbon Stock of *Pinusroxburghii* Sargent Forests of Himachal Pradesh, India. *J. Mt. Sci.* **2014**, *11* (4), 959–966.

25. Sheikh, M. A.; Kumar, S.; Kumar, M. Above and Below Ground Organic Carbon Stocks in a Sub-Tropical *Pinusroxburghii* Sargent Forest of the Garhwal Himalayas. *For. Stud. China.* **2012**, *14* (3), 205–209.

26. Steffan-Dewenter, I.; Kessler, M.; Barkmann, J.; et al. Tradeoffs between Income, Bio-Diversity, and Ecosystem Functioning during Tropical Rainforest Conversion and Agroforestry Intensification. *PNAS* **2007**, *104*, 4973–4978.

27. Sundarapandian, S. M.; Amritha, S.; Gowsalya, L.; et al. Estimation of Biomass and Carbon Stock of Woody Plants in Different Land-Uses. *For. Res.* **2013**, *3*, 115.

28. Takimoto, A.; Nair, P. K. R.; Nair, V. D. Carbon Stock and Sequestration Potential of Traditional and Improved Agroforestry Systems in the West African Sahel. *Agric. Ecosyst. Environ.* **2008**, *125*, 159–166.

29. Timilsina, N.; Escobedo, F. J.; Staudhammer, C. L.; Brandeis, T. Analyzing the Causal Factors of Carbon Stores in a Subtropical Urban Forest. *Ecol. Complex.* **2014**, *20*, 23–32.

CHAPTER 12

SYNTHESIS, CHARACTERIZATION, AND APPLICATIONS OF SILICA SPHERES

LAVANYA TANDON and POONAM KHULLAR*

Department of Chemistry, B.B.K. D.A.V. College for Women, Amritsar 143001, Punjab, India

Corresponding author. E-mail: virgo16sep2005@gmail.com

ABSTRACT

Silica spheres provide large specific area and uniformity, and hence, these are used as a catalyst, adsorbent, and host materials. These spheres have large applications in chromatography and cosmetics. Hollow mesoporous nanospheres are used in drug delivery. Nanocoated nanoparticles are used in optics, optoelectronics, chemical engineering, pharmaceutics, etc. and have high catalytic activities for hydrogenation, hydroformylation, carbonylation, etc. Surfactants and electrolytes have profound effect on the catalytic properties of nanoparticles on the silica sphere. Synthetic approach has been devised to convert Stöber silica spheres into mesoporous silica spheres. Silica spheres can be synthesized under both acidic and basic conditions. Various templating approaches have been adopted to synthesize the porous materials.

12.1 INTRODUCTION

In the past few years, great excitement and expectations have been created by the science and technology of nanomateials.[1] This subject is of immense academic interest. The word "nanotechnology" became popularized in 1980 by K. Eric Drexler.[2] Nanotechnology deals with the building of things using bottom-up approach with atomic precision and is referred as a general purpose technology.[3] Its advanced form has significant impact on all industries and on all areas of society. It offers better built, longer lasting, cleaner, and safer and smarter products for home, communications, medicines, and industries. It is offering improved efficiency in every facet of life and is of dual use as it has commercial as well as military use, that is, for making weapons and tools of surveillance. High-quality products are made at low cost using nanotechnology. This technology deals with the substances less than approximately 100 nm, that is, 100,000 times smaller than diameter of a human hair.

These possess unique optical,[4] magnetic,[5] electrical,[6] and other properties. There are four generations that deal with nanotechnology. The current era deals with the passive nanostructures, that is, materials designed to perform one task, example products incorporating nanostructures like coatings, nanoparticles reinforced composites, ceramics, polymers, etc. The second phase deals with nanostructures for multitasking, for example, drug delivery devices, sensors. The third generation deals with the nanosystems with thousands of interacting components like 3D networking, new hierarchical architecture, and robotics. The fourth deals with hierarchical system within the system like the mammalian cell. Nanotechnology makes the use of small artifacts which are the important product of this technology.

In the past few years, the science and technology of nanomaterials have created great excitement and expectations in the past few years. There has been a great progress in the synthesis and assembly and fabrication of nanophase materials like nanocrystals, nanowires, nanotubes, nanorods, nanoribbon, nanoflasks, nanospheres, nanocoatings, and nanodots. Nanostructure materials have a great potential to be used in the wide variety of applications.[7] Thus, a new field is being developed which requires the cooperation of chemists, physicists, and engineers. Chemists are mainly concerned with synthesis, structures, and enhanced properties.[8] At present, different metallic nanomaterials are being produced using copper,[9] zinc,[10] titanium,[11] magnesium,[12] gold,[13] and silver[14] and are used for various purposes like medical treatments,[15] used in various branches of industry production like solar[16] and oxide fuel batteries for energy storage so as to widen the incorporation into diverse material of everyday use like cosmetics or clothes.[17]

Silver nanoparticles have good antimicrobial efficacy against bacteria and viruses and are also used in textile industries for water treatment, sunscreen lotions, etc. Gold nanoparticles (AuNPs) have many applications in the field of identification of protein interaction, used in lab tracers in DNA fingerprinting to detect the presence of DNA in sample. Nanorods of gold are used for diagnosis of different classes of bacteria and also used in diagnosis of cancer.[18,19]

Magnetic nanoparticles like magnetite (Fe_3O_4) is biocompatible and are used in magnetic resonance imaging[20] drug delivery, DNA analysis, and gene therapy. These selectively target the desired organs or tissues inside the body and accumulate the certain concentration of nanoparticles by means of application of an external magnetic field.[21] Materials designed for purpose of drug targeting are based on core–shell nanoparticles, whereas magnetic nanocrystals are coated by diverse species so as to avoid the contact between magnetic core and the tissue and maintaining the colloidal suspension stability within the biological environment. Mesoporous materials exhibit bulk morphologies.[22] Aerosol-assisted route is used to synthesize silica mesoporous spheres containing magnetic nanoparticles. The magnetic nanoparticles neither undergo crystallinity changes nor superparamagnetic loss in the encapsulation process. This chapter deals with synthesis, characterization, and applications of silica spheres.

Silica exists in various allotropic forms of silicon dioxide and makes up 12% of earth's crust and is found in human as well as in unicellular marine algae and in plant resources. Synthetically, various types of silica can be produced. It is the most versatile colloidal material[23] and gives a myriad of diverse application from formation of synthetic opals and their inverse[24] to colloidal templating[25] and silica coating for stabilization and protection of metal and semiconductor nanoparticles.

12.1.1 STÖBER SYNTHESIS

Porous silica has broad applications in the area of catalysis, controlled release, and separation science. Nonporous monodispersed spheres of silica were first synthesized by Stöber et al. For the preparation of silica particles, a base catalyst like ammonia in a system, including water, alcohol, and tetra alkoxysilane, is required. Micro/mesoporous silica sphere is the modified method of Stöber method, and the cationic surfactant was introduced as the directing agent. Tetraorthosilicate (TEOS) was chosen as silica source.[26] Ethanol was used as the solvent. Cetyltrimethyl ammonium bromide (CTAB) and n-hexadecylpyridinium chloride were used as the cationic surfactants. Aqueous ammonia was used as the base catalyst. Ordered mesoporous MCM-4z[27,28] were produced when the surfactant was removed by calcinations at 823 K. Same method is used to produce spherical silica with heteroatoms incorporated in the framework.[29,30] Different mesophases on the lamellar phase spherical particles were obtained at room temperature by varying the amount of ethanol in the system.[31,32] Lower concentration of ethanol has limited effect on the external morphology, on increasing the concentration of ethanol, it acted as cosolvent producing spherical particles.

Large-sized monodispersed silica spheres were also obtained by the seed growth method. In this method, most of the particles grow up to 1–1.6 μm,[33] where the seed particles have an average diameter of 300 nm. Monodispersed fluorescent core–shell silica spheres particles with diameter of 1.5 μm were successfully synthesized[34] by attaching the fluorescent dye to silane coupling agent to prepare seed particles. These were nonporous silica particles.

TEOS/octadecyltrimethoxysilane that is progen mixture was added to the suspension of prepared nonporous silica spheres, and submicrometer-sized solid core/mesoporous shell silica spheres were produced after the removal of progen by calcinations. By increasing the amount of progen up to a maximum value of 350 m^2/g,[35] the specific area increased. C$_n$TAB (n = 12, 14, 16, 18) were used as the structure directing agent for the mesoporous shell on the solid silica sphere with mesoporous channel perpendicular to the surface.[36] Core silica mesoporous microspheres were made using an aerosol assisted process in the presence of nonionic surfactant pluoronic P123. Silica spheres are also used as the packaging materials for chromatographic applications. By using small silica microspheres, <2 μm, the column efficiency could be improved and is enhanced by the factor of 5–10 for capillary electrochromatography.[37]

Ethanol was added to highly porous silica sphere so as to produce spherical particles.[38] In the modified Stöber process,[39] the cationic surfactants, CTAB, and cetyltrimethylammonium chloride (CTAC) were used, and the average size of particles was found to be around 2 nm. These particles are too small for separation sciences but are good for catalysis, sensors, and controlled delivery. For high-performance liquid chromatography, the size of the silica particle should be in the range of 6–50 nm.

Mesopores can be expanded by introducing the swelling agent in the structure directing template either in the preparation step[40] or in the hydrothermal post treatment.[41] Mizutani et al. reported the expansion of mesospheres from 2.5 to 5.5 nm and the spherical morphology of hexagonal mesopores were retained.[42] Nanometer-sized particles (50–300 nm) with pore size of 5–30 nm[43] range were synthesized by using organic swelling agent 1,3,5-trimethylbenzene along with fluorocarbon surfactant which were used as templates. Through oil-in-water macroemulsions, spherical siliceous mesocellular foam particles were synthesized in which triblock copolymer pluronoic 123 and swelling agent TMB were used as templates.

For high speed exclusion chromatography silica framework with ultra-large pore size (20–50 nm) and high pore volume (up to 2.6 cm³/g) were produced. Columbic interactions between silicate anions and positively charged nitrogen atoms on the micelle surface are observed during the synthesis of porous silica spheres, in which CTAB was used as the template. Hydrophilic polymers are used in the condensation of TEOS at acidic pH to form the macroscopic sol–gel materials.[44]

For the preparation of silica/polymer hybrid particles, emulsion polymerization was used in which polymers were miscible with the silica source and then dispersed in the continuous phase. Water-in-oil or oil-in-water emulsions were used to obtain silica/poly(dimethylsiloxane), silica polystyrene, and silica poly(ethylene glycol) particles with particle size in range of 200–500 nm.[45] When the polymer was removed, silica particles with low surface area with pore size distribution of 37 and 70 nm was observed.[46] Modified Stöber process is being used to prepare highly porous monodispersed silica spheres.

Stöber protocol has provided the information for the size controlled preparation of SiO₂. From this protocol, silica can be synthesized routinely into nondispersed spheres at a desired rate because the starting chemicals are available at low cost and no organic additives are required. Templating methods are being used for the fabrication of core–shell structures with multiple functionalities where silica spheres is synthesized into shells for molecular screening or metal catalyst protection.[47] Stöber method is used in the case of amorphous silica and mesoporosity is introduced into the Stöber's sphere, there is a development[48] of surfactant induced pseudomorphic transformation. These transformations are the reaction in which the shape of a solid material is preserved but the chemical composition changes. Mesoporous silica spheres retaining the sizes and monodispersity of the original silica spheres[49] have been directly synthesized through pseudomorphic transformation involving hydrothermal treatment of dense amorphous silica sphere. Solvents used in the synthesis of dense silica spheres are EtOH, iPrOH, or BuOH. Use of solvent influences the degree of condensation in the parent silica sphere which during the hydrothermal transformation determines the extent of silica dissolution and the pore architecture of product sphere both at their surface and in there bulk. Different solvents result in different degree of condensation of amorphous, nonporous silica spheres. There is small change in surface morphologies following hydrothermal reactions. Silica spheres synthesized in EtOH after hydrothermal transformation employing a long chain alkyl ammonium surfactant possess smooth surfaces while sphere synthesized in iPrOH or BuOH possess highly corrugated surfaces. Silica sphere with corrugated surfaces can be used as templates for nanocasting and mesoporous carbon spheres with surface morphologies which resemble to those of mesoporous silica spheres are formed.

Method of nanocasting is used for the production of inverse replica structures of porous material structure and the template used for nanocasting can be deduced through the structural analysis of the product. For polymerization, aluminum was introduced into the sphere as a catalyst and the sphere was then exposed to the mixture of gas phase phenol-paraformaldehyde which was polymerized and carbonized. The silica sphere was removed by extraction with hydrofluoric acid from HT-SiO$_2$-iPrOH performs, an irregular network of interconnected carbon sphere was recovered (Fig. 12.1).

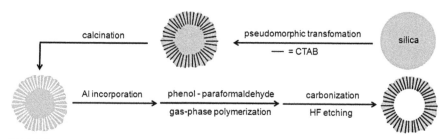

FIGURE 12.1   Schematic illustration of the preparation of hollow carbon spheres by nanocasting. (Reproduced from Reference [50] with permission. © 2011 American Chemical Society.)

Due to radial contraction during the carbonization process, the average diameter of carbon sphere was lower than that of silica perform. Some of the spheres were found to be broken and a hollow core was revealed which was surrounded by 50–120-nm-thick porous shells, a little than mesoporous shells in the perform. SEM images (Fig. 12.2) show that the pore section of the shell structure has average dimensions of ±34 nm for longer axis and ±22 nm for shorter axis. TEM images show that shell has mane-like appearance similar to those observed for silica performs (Fig. 12.2).

The existing porous silica materials provide large specific surface area[50] and any molecule involve in adsorption or reaction have to travel long pathway and the actual surface area may not be fully utilized. To overcome this problem, high surface area fibrous SiO$_2$ spheres have been prepared.[51] This SiO$_2$ sphere is named as KAUST-1. It is open in nature and has large specific surface area and is used as a new catalyst support where diffusion is not the limiting step in heterogeneous catalysis.

Through base etching and dissolution regrowth process[52] solid SiO$_2$ sphere prepared through Stöber protocol undergoes structural transformation into silica hollow spheres or metal silica yolk shells.[53] Monodispersity of porous SiO_2 is very high. The size control of Stöber silica sphere that is precursor solid can be easily achieved without relying on complex polymer/surfactant template and/or space confinement of colloidal vesicles or micelles. A hydrothermal approach is used to chemically modify silica matrix and structurally create interior space and open porosity for conventional Stöber silica spheres.

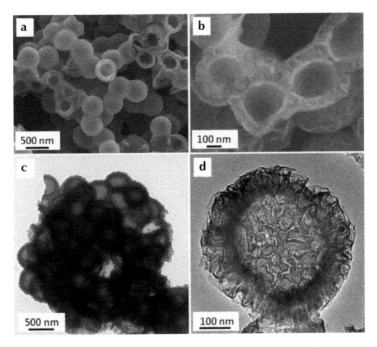

FIGURE 12.2   SEM (a and b) and TEM (c and d) images of nanocast hollow carbon spheres after HF etching (PFC-iPrOH). (Reproduced from Reference [50] with permission. © 2011 American Chemical Society.)

Figure 12.3 shows the surface modification and the synthetic steps for the preparation of new type of porous SiO_2 support, starting from Stöber SiO_2 sphere. Through Stöber protocol,[54] monodispersed SiO_2 were prepared, and the precursor solid sphere was decorated with a layer of cationic preelectrolyte, poly(diallyldimethylammonium chloride) the positively charged PDDA served as the surface protecting agent and the strong base NaOH functioned as an etchant to obtain porosity.[55]

PDDA alters the surface charge of native silica sphere and improves the dispersity of the silica spheres in deionized water.[56] It is known that the surface of the silica spheres obtained by Stöber method is covered with the hydroxyl ions and thus exhibit negative charge change from negative to positive after the adsorption of PDDA. The PDDA-coated SiO_2 sphere disperses better in aqueous solution and these do not aggregate at high temperature.

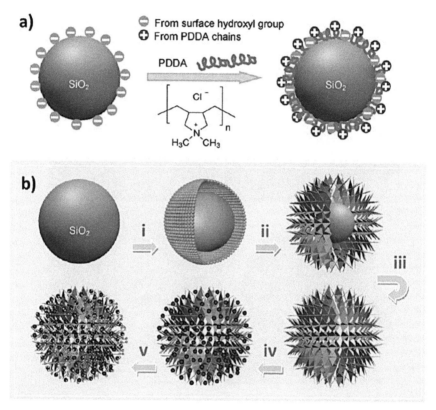

FIGURE 12.3 (a) Modification of SiO_2 surface with PDDA: electrostatic attraction between negative hydroxyl groups and positive charges of PDDA and (b) illustration of this synthetic approach to fabricate integrated nanocatalysts: (i) modify surface of SiO_2 (see (a)) and create yolk (pure SiO_2, dark green)–shell (Zn-doped SiO_2, gray), (ii) reduce yolk size, (iii) form plate-interlaced Zn-doped SiO_2 hollow sphere, (iv) deposit ZnO nanoparticles (blue dots), and (v) add Ru nanoparticles (small yellow dots). Note that (i)–(iii) belong to "top-down" steps while (iv) and (v) are "bottom-up" steps. (Reproduced from Reference [55] with permission. © 2012 American Chemical Society.)

12.1.2   MESOPOROUS SILICA SPHERES

Mesoporous silica material has large surface area, large pore volume, hydrophilic character, chemical and thermal stability, suitable particle morphologies, and toxicological safety but has unsuitable pore structure; therefore, the adsorption rate of protein is very slow and the adsorption capacity is not very high.[57] Adsorption capacity is dependent on the adsorption site on the surface of the material and on the pore site of the adsorbent. The pore size should be large than the hydrodynamic radius of the protein so that the high capacity can be obtained.[58] The bimodal macroporous mesoporous structure is the ideal pore structure where mesopores provide the adsorption location and macropores provide the passage way for the proteins to enter and transfer to the adsorption location. Double template method was used to prepare the macroporous mesoporous structure where the surfactants were used as mesoporous templates and gas, small droplets or particles were used as macroporous templates.[59]

For large-scale applications, morphology of mesoporous silic materials is very important. Silica spheres with uniform sizes and with large diameter are desired. Controllable preparation technologies have been required for spheres with large spherical diameters, large pore size, large volume, and bimodal macroporous mesoporous structure. Narrow size distribution particles could be synthesized by cross-flow membrane, vibrating nozzle, microfluid, etc. Microfluidic devices are important tool for the synthesis of polymer particles. Microfluidic-assisted technologies are used for the production of polymer particles with an improved control over their size, morphologies, size distribution, and compositions.[60]

A gelatin process based on polymerization was used to synthesize micrometer-sized mesoporous silica sphere with uniform and optically adjustable diameter. The synthesized silica spheres exhibited excellent separation performance.[61] In that preparation process, the solidification of droplet was accomplished by the assistance of the acryl amide polymerization reaction. Conditions of polymerization decrease the ability to control the pore structure and size. The pore structure of silica sphere may be destroyed by the alkali trioctylamine, that is, TOA. Therefore, a new method that is temperature induced gelation has been proposed to prepare the silica spheres with bimodal structures in microfludic devise. Methyl cellulose or MC is a chemical compound derived from cellulose, and it possesses a reversible thermosensitive property. MC is water soluble, and in aqueous solution, it forms a thermoreversible hydrogel. This hydrogel is a three-dimensional network of polymer chain cross-linked via either physical or chemical bonds. With respect to the variation of temperature,[62] the sol–gel transition of MC in aqueous solution appears. By controlling the growth rate,[63] MC was used to prepare the size controllable nanocrystals of zeolites. It has been found that in the sol–gel process polyethylene glycol, that is, PEG, adsorbed the silica sols in the acidic aqueous solutions[64] and causes phase separation by the bridging flocculation.[65]Through the temperature induced bridging flocculation, PEG accelerates the coagulation and precipitation of colloidal silica. Silica particles with a diameter of 3–10 μm have been synthesized by Zhang et al.[66]

This possess good dispersibility and smoother outer surface on simultaneous addition of suitable amount of PEG and MC. PEG accelerates the gelation process.[67]

PEG and MC are introduced into the prehydrolysis solution of tetraethylorthosilicate (TEOS) for fast gelation instead of polymerization reaction. A bimodal pore structure in micrometer-sized silica sphere at high temperature about 90°C was formed. The viscosity of the solution increases the sol–gel process (Fig. 12.4).

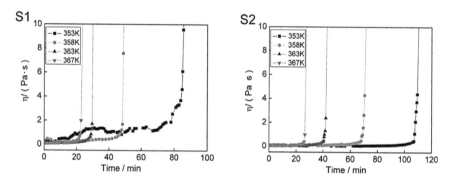

FIGURE 12.4 (See color insert.) Viscosity of the silica sol system with a specific composition (samples S1 and S2) at different temperatures. (Reproduced from Reference [66] with permission. © 2010 American Chemical Society.)

Block copolymer and small surfactant have been used to synthesize novel kind of hierarchically bimodal mesoporous material with tunable pore size and structure in which block copolymer act as a large pore template and a smaller surfactant act as a small pore template.[68] The particles obtained possessed irregular morphology and broadened particle size distribution. A simple method for the synthesis of shell structured and monodispersed dual microporous silica sphere (DMSS) with large pores in the core and smaller pores in the shell by using amphiphilic block copolymers polystyrene-*b*-poly acryl acid (PS-*b*-PAA) and CTAB as dual temperature are used. Anionic block copolymer is used as a template to form large mesoporous structure. The size and the thickness of the shell can be adjusted by changing the length of the PS blocks and the precursor composition, respectively. Dual mesoporous silica spheres prepared by this method have uniform particle size and well-defined core–shell structure. These functional silica spheres were prepared by encapsulating hydrophobic nanometer in the large pores. TEM image of DMMS prepared with block copolymer is shown in Figure 12.5.

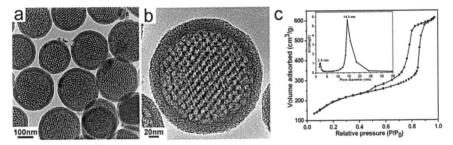

FIGURE 12.5 TEM images (a and b) as well as N_2 adsorption–desorption isotherms and BJH pore diameter distribution curve at adsorption branch (inset) (c) of DMSS-PS100. (Reproduced from Reference [68] with permission. © 2010 American Chemical Society.)

The morphology of the PS-*b*-PAA micelles colloid can be changed from spheres to rod-like aggregates by the addition of inorganic or small molecule surfactants.[69] In the experiment, rod-like aggregates were assumed to be formed in the presence of surfactant CTAB and ammonia. It was confirmed by the TEM observation of the precursor solution, before adding ethanol and TEOS, at the same time, the hydrophilic PAA blocks of rod-like aggregates couple with the CTAB micelles via Coulomb force and electrostatic interaction between CTA^+ and PAA^- to form CTAB-coated rod-like aggregate in solution. After the addition of ethanol and TEOS, the self-assembled hybrid micelles composed of silicate oligomers and CTAB-coated rod-like and PS-*b*-PAA-coated rod-like aggregates were formed first. The rod-like silicate composite material pack together in an ordered fashion to form core part dual mesoporous structures then the electrostatic interaction between the positively charged CTA^+ and negatively charged silica spheres and also the condensation of the rod silicate micelles facilitates the deposition of material which results in the formation of mesoporous silica with spherical morphology. Only limited amount of CTAB can coat the micelle to form the composite micelle, and the remaining amount of CTAB molecule is expelled out of the core portion of the structure, and when the core part becomes large, then the immediate outer region of the core becomes rich in CTAB. Cooperative self-assembly occurs between CTAB molecule and additional

TEOS and outer shell of smaller pores surrounding the cores were formed. CTAB plays an important role in the formation of mesoporous material, and the solvent ethanol plays an important role in the formation of stable CTAB-coated rod-like aggregate (Fig. 12.6).

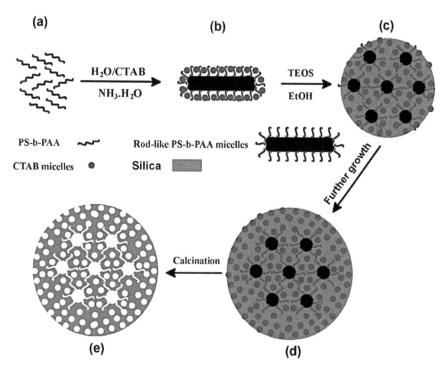

FIGURE 12.6   Schematic illustration for the formation of dual-mesoporous silica spheres. (Reproduced from Reference [68] with permission. © 2010 American Chemical Society.)

The pore size of mesoporous silica spheres lies in the range of 2–50 nm.[70] As this pore size is comparable with the diameter of the enzyme, therefore the mesoporous silica is being used for the enzyme immobilization.[71] On the mesoporous silica sphere, the enzymes were immobilized and multilayer shell was assembled on the surface sphere by the layer-by-layer electrostatic assembly of oppositely charged species, PE, and silica nanoparticles to encapsulate the enzyme. The emulsion biphase chemistry has demonstrated the possibility of controlling shape on micrometer to centimeter length and periodic mesostructure at the molecular mesoscale (15–100 Angstrom). By using shear fluid flow, the morphology of final product can be structured.

A small sampling includes the systematic tuning of shape flat films to mesoporous silica fibers and then curved asymmetric shapes to hollow porous diatoms spheres by the tuning of stirring speed in the synthesis mixture. Mesoscale morphology is not dependent on supporting substrates or pre organized shape molding surfaces, it is limited only by the shapes that can be created using fluid flow and hydrodynamics. Emulsions and microemulsions involve long range force.[72] It involves shape fluctuation and interaction same as the thermal energy kt, structure modulation at micrometer, or longer wavelength scales are readily introduced by the application of small applied external fields. High volume composite synthesis takes place through the hydrodynamic route. It requires low temperature conditions. Using surfactant stabilized emulsions based chemistry,[73] transparent, mesoporous, marble-like sphere can be synthesized high surface area spheres are prepared by this method.

Silica sphere is of uniform size and can be controlled from 0.1 to 2 mm. There is a narrow pore size distribution in the mesoporous silica sphere, and there is a large surface area greater than 100 m^2/g, and it also possesses small pore size in 1–5 nm range. Solvents used for chromatographic separation and solid phase catalyst depend upon mechanical strength of the particles, size, porosity etc. For required space flow velocity and molecular selectivity, high surface area and monodispersed pore size are needed. These properties are obtained by packing and annealing together uniformly sized spheres at high temperature.

Silica is gelled into sphere by two given ways:

1. Through spray drying they form small droplets of sol.
2. By spraying droplets into the immiscible liquid and then gelling through heating 6–10 or by chemical action.

Silica spheres of different sizes are obtained by varying the size of silica sol droplet in an immiscible liquid before it gels. Silicon esters like tetramethyl orthosilicate (TMOS and TEOS) or a concentrated sodium silicate solution act as silica source. The hydrogel bodies of few mm of silica spheres are obtained and are washed and dried at a low temperature, that is, at 120°C, and then calcined at a high temperature of around 450°C. Another method to obtain silica sphere is through the controlled hydrolysis of silicon esters.[74] But the silica sphere prepared by this method have irregular shapes, broad pore size distribution, and relative low specific surface area (30–690 m/g)[75]and can be prepared in a single step as given in Figure 12.7.

FIGURE 12.7 Size and shape of spheres using CTAB as surfactant: (a) as-made sample; (b) calcined at 500°C sample. (Reproduced from Reference [74] with permission. © 1997 American Chemical Society.)

The organic surfactant as the pore structure directing agent is combined with an optional source of base catalyst like NaOH, KOH, or tetramethylammonium hydroxide and water. One or a combination of organic compound of silicon is added to it. The mixture is stirred at the constant speed of about 300 rpm using a magnetic stirrer for 15–30 h at room temperature. The sphere product is recovered from the liquid. By varying the stirring conditions and reaction volumes the size can be controlled. The stirring speed influences the sphere formation; soft gel particles are obtained with low stirring speeds that is <200 rpm, spheres are obtained with medium stirring speed (200–400 rpm), and fine powders are obtained with high stirring speed (Fig. 12.8).

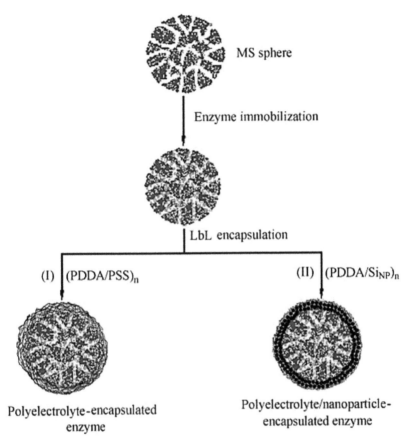

MS sphere

Enzyme immobilization

LbL encapsulation

(I) (PDDA/PSS)$_n$

(II) (PDDA/Si$_{NP}$)$_n$

Polyelectrolyte-encapsulated
enzyme

Polyelectrolyte/nanoparticle-
encapsulated enzyme

FIGURE 12.8 Schematic representation of enzyme encapsulation using MS spheres as supports. The enzyme is first immobilized and subsequently encapsulated by a (I) PE or (II) PE/nanoparticle multilayer shell. (Reproduced from Reference [75] with permission. © 2005 American Chemical Society.)

12.2 HOLLOW SILICA SPHERE

Hollow silica spheres have wide applications in the field of chromatography, shield for enzyme or proteins, delivery vehicles of drugs, photonic crystals, waste removal, and bimolecular release system.[75–82] Hollow silica spheres have high chemical and thermal stability, large surface area, low density, and good compatibility with other materials. Organic or inorganic templates are used in the process of fabrication of hollow silica spheres and after the removal of these templates through calcinations or chemical etching hollow silica spheres are formed. Self-templating method[83] is also used for the synthesis of hollow silica spheres.

A two-step method has been adopted by Hah et al.[84] and Wang et al.[85] to synthesize monodispersed phenyl-functionalized without using templates. The hollow silica spheres thus obtained were made to dissolve in organic solvents including acetone, tetrahydrofuran chloroform, etc. These phenyl functionalized hollow silica spheres can be used in nanoreactors, coating technology, catalysis, etc. These silica spheres exhibited high adsorption capability for noble metal ions such as Pd^{2+}, Pt^{2+}.[86] Molnar et al. first synthesized mesoporous silica host and then grafted the sulfonic groups on the mesoporous silica using chlorosulfonic acid and then materials were obtained which possessed higher acid catalysis activation and selectivity. Through post grafting or cocondensation route,[87,88] organic functional group can be incorporated into silica materials.

One step process is used to synthesize thiol functionalized hollow silica sphere and the spheres thus obtained possess large number of thiol groups. These are used in the removal of heavy metal ions,[89–91] catalysis,[92–98] proton conductivity membrane,[99,100] etc. Mechanism of the thiol functionalized hollow silica sphere is shown in Figure 12.9. MPTS is coated on the positive charged PS particles or formed thiol functionalized hollow silica spheres and have attributed to the combination with the interaction of the positive and negative charge and condensation and hydrolysis of MPTS. Amount of ammonia is important in the formation of these hollow spheres as in the presence of low concentration of ammonia, PS cannot dissolve. When the ammonia concentration is increased to the certain threshold, then the thiol functionalized hollow silica spheres were formed. TEM images of PS spheres (Fig. 12.10) depicts that on increasing the concentration of ammonia core–shell composites, particles with thiol functionalized silica shells are formed.

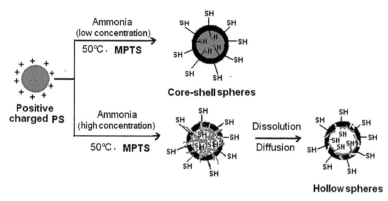

FIGURE 12.9 Schematic representation of the formation of hollow spheres (Reproduced from Reference [99] with permission. © 2008 American Chemical Society.)

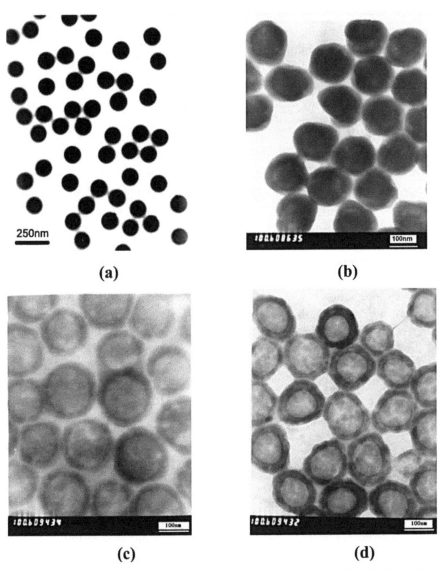

FIGURE 12.10 TEM images of the PS spheres (a); thiol-functionalized hollow silica spheres with different additional amounts of ammonia of (b), 3.0 (c), and 5.0 mL (d). (Reproduced from Reference [99] with permission. © 2008 American Chemical Society.)

Silica-based mesophases are widely used in fabrication of advanced materials and biomineralization. The inorganic moiety of the hybrid material can be silicate, metal oxide, polyoxometalate, metal halide, etc.[101–105] Weak covalent interactions occur between surfactants and inorganic moieties.

The polymerization of silicate species hardens the silicate–surfactant composites;[106] therefore, external driving forces are needed to destabilize the composites. The composites can be dissembled under a strong alkaline condition or by a HF solution, the silica moiety is etched and is usually too fast too control. Etching of silica proceeds slowly under neutral or generic acidic conditions. It has been found that controlled dissolution of silica is performed at high temperature under neutral or acidic conditions.[107–109]

The partial disassembly of $mSiO_2$/CTAB sphere leads to the formation of hollow structure,[110–116] which is used for the synthesis. An anion exchanged mechanism is proposed to explain the disassembly of SiO_2/CTAB sphere. Positive charge of CTAB micelle and negative charge of silica wall are distributed according to the charge-density matching requirements.[117] Disolution of kinetics of silica is slow, and surfactant molecules are also restricted; therefore, dissolution of the silica wall into the bulk water is limited. In present condition, the binding interaction of permanganate ion to the surfactant moieties in the sphere is strong, and therefore, silicate–surfactant interaction is reduced and thus there is acceleration in dissolution of silica into bulk water. The exchange between permanganate of hollow mesoporous silica spheres (HMSs). It is a self-templating method and is used for the synthesis of hollow silica structure[118] and silicate ion is represented by the given equations:

$$mSiO_2/CTAB(s) \rightleftharpoons silicate - CTAB(complex) \qquad (12.1)$$

$$Silicate\ CTAB(complex)\ MnO\ (aq) \rightleftharpoons MnO\ CTAB$$

$$(complex)\ silicate(aq) \qquad (12.2)$$

The proposed mechanism is as follows (Fig. 12.11); for the synthesis of HMS-2, molecules on the $mSiO_2$/CTAB surface have two functions:

1. it protects the silica surface from oxidation and dissolution
2. it serves as the physical barrier, slows down the mass-transfer rate at the solid–liquid interface.

The formation process of HMS-2 is explained by surface diffusion mechanism. Presence of molecules on $mSiO_2$ blocks the normal outward diffusion of the silicate. The silica wall start dissolving between the core and the shell and small voids are left behind. Through surface diffusion modes, etching of silica wall takes place and new openings are observed to release the dissolved silicate. In acidic $KMnO_4$, no appreciable disassembly of mSiO$_2$/CTAB sphere has been observed. It is because the acidic medium suppress the dissolution of the silica wall (Fig. 12.12).

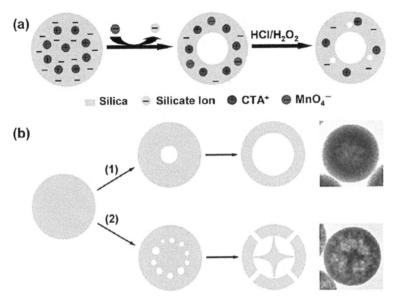

FIGURE 12.11 (a) Schematic illustration of the anion-exchange-driven disassembly of mSiO$_2$/CTAB spheres. (b) Typical structures and proposed formation routes for HMS-1 (1) and HMS-2 (2). (Reproduced from Reference [117] with permission. © 2012 American Chemical Society.)

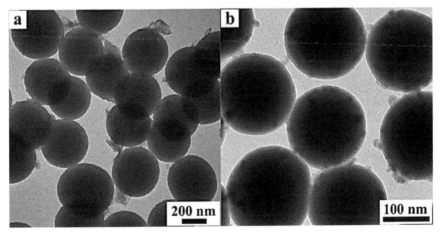

FIGURE 12.12 (a) TEM images of mSiO$_2$/CTAB spheres treated with an acidic KMnO$_4$ solution (in sulfuric acid, pH ~1.8). (b) Stöber silica spheres treated with a KMnO$_4$ solution (Reproduced from Reference [117] with permission. © 2012 American Chemical Society.)

Porous silica hollow spheres can also be synthesized by using surfactant template technique.[119–123] Through surfactant-assisted approach, porous-silica-based magnetic composite particles have been synthesized, and their products are being used in drug delivery and magnetic separation.[124–127] The presence of surfactant is important in the formation of porous hollow silica sphere and silica-based composite particles. A new method for the formation of composite hollow sphere with controllable surface morphology is as shown in Figure 12.13.

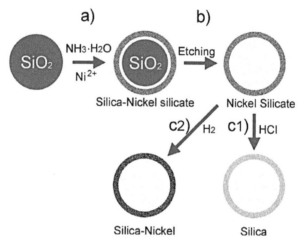

FIGURE 12.13   Schematic depiction of the synthesis procedure of porous nickel silicate, silica, and silica–nickel composite hollow spheres. (Reproduced from Reference [130] with permission. © 2010 American Chemical Society.)

Silica spheres can also be used as physical templates to prepare hollow spheres through coating and subsequently removing themselves in alkaline conditions. Ammonia provides hydroxide ions based on ionization of ammonia which dissolves the surface of the silica sphere by breaking the silicon–oxygen bond and forms the silicate ions. The silicate ions react with the nickel ions to form nickel silicate particles. On removing the silicate ion with etching, the nickel silicate hollow sphere could be obtained.

Formation of silica–nickel composite hollow sphere is depicted as follows:

1.   Through Stöber synthesis, silica spheres were prepared and placed in ammonia solution which contained nickel ions.
2.   The silica core was removed by dissolving in sodium hydroxide solution, the nickel silicate hollow sphere were produced which are

transformed into silica hollow spheres by leaching of nickel ions in acidic solution.

12.2.1 NANOCOATING

Nanocoating relies on the combination of the properties of the two or more materials involved. One of the materials determines the surface properties of the particles, and the other is encapsulated by the shell[131] and does not contribute to surface properties at all but is responsible for the optical, catalytic, and magnetic property of the system. The interaction between the core and shell determines the potential application of the material. Coating of the nanoparticles on flat or spherical surface avoids the spread of the nanoparticles to the environment due to bonding between the substrate and the coated nanoparticle. Thus, nanomaterials can be made to move large distances through nanocoating. Supported catalyst allows the fine dispersion and stabilization of small metallic particles.[132]

Nanoparticles of noble metals are coated on the polystyrene substrate through electrostatic deposition[133] as reported by Dokoutchaev et al. Chen et al. reported the preparation of platinum colloids[134] on the PS nanospheres and the catalytic properties in hydrogenation through surface grafted poly(N-isopropylacrylamide) via the reduction of $PtCl_6^{2-}$ by ethanol. The macroporous polymers[135] with tri-isobutylphosphine sulfide are synthesized and are characterized for the selective adsorption of the gold and palladium. Sonochemical method is used for the deposition task because of their ability to combine the synthesis of the various nanomaterials and their deposition on various substrates in a single operation.

One advantage of the method is the ability to control the particle size of the product by varying the concentration of the precursor in the solution. Sonochemical phenomenon is being used for the deposition of several different nanoparticles on different types of substrates. Metallic, that is, gold, silver, and nickel, metal oxides, that is, magnesium oxide, and highly magnetic, that is, air stable Fe nanoparticles, can be uniformly deposited on a variety of substrates.[136–140]

In previous research, as reported by Breen et al., zinc sulfide[141] films were coated on carboxyl modified PS microsphere, through a sonochemical reaction conducted in an aqueous solution containing zinc acetate and sulfide released through the hydrolysis of thioacetamide.

Through X-ray diffraction (Fig. 12.14) and high resolution transmission electron microscopy (Fig. 12.15), the crystallinity of sonochemically coated silver, gold, palladium, platinum particles on PS surface can be detected.

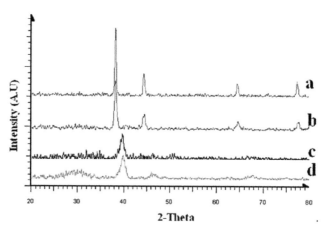

FIGURE 12.14 XRD patterns of (a) Ag, (b) Au, (c) Pt, and (d) Pd nanoparticles deposited on PS spheres sonochemically. (Reproduced from Reference [142] with permission. © 2005 American Chemical Society.)

The deposition is related to the microjet and shock waves created near the solid surface after the collapse of the bubble.[142] These jets push the nanoparticles toward the PS surface at a very high speed. On hitting, the surface morphologies and reactivity changes and finally coating of the particles takes place as shown in Figure 12.15.

12.2.2 DEPOSITION OF SILVER NANOPARTICLES ON SILICA SPHERE

A simple method is used to deposit silver nanoparticles on the silica sphere, and the method is also used to decorate carbon nanotubes.[143,144] The surface of the silica sphere is modified by Sn^{2+} ions, and then, a redox reaction is carried out in which Sn^{2+} ions are oxidized to Sn^{4+}, and at the same time, Ag^+ ions are reduced to Ag, which remains attached on the surface of the silica in the form of nanometer sized particles. This process is equivalent to the "pretreatment steps" in the electroless plating.[145] When AgNP is attached to the silica sphere, then due to agglomeration, size of the particles increases and leads to the decrease in the number of particles; therefore, the catalytic reaction rate of the dye decreases and even stops (Fig. 12.16).

The Sn^{2+} ions are adsorbed on the surface of the spherical silica spheres. The excess of Sn^{2+} ions are removed from the solution through consecutive centrifugation steps. On adding ammonical $AgNO_3$ to the solution, redox reaction occurred which involved the oxidation of surface Sn^{2+} to Sn^{4+} and reduction of Ag^+ to Ag^0.

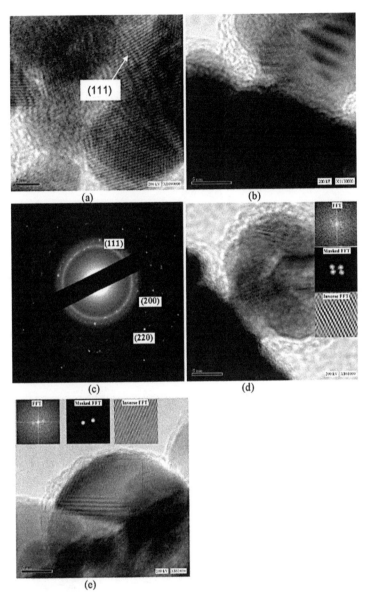

FIGURE 12.15   HR-TEM of (a) the crystalline nature and interlayer spacing between the Au plane (1 1 1), (b) the interface between the coated gold nanoparticles and the PS sphere, (c) the ED obtained from the gold nanoparticles, (d) the crystalline nature of an Au nanoparticle on the surface of a PS (inset: FFT, masked FFT, and inverse FFT is provided for the marked area), surface showing an amorphous layer of oxidation, and (e) the crystalline nature of a Ag nanoparticle on the surface of PS (inset: FFT, masked FFT, and inverse FFT is provided for the marked area); the surface of coated silver shows an amorphous layer of oxidation. (Reproduced from Reference [142] with permission. © 2005 American Chemical Society.)

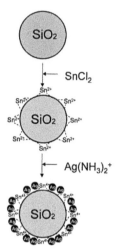

FIGURE 12.16 Sketch of surface reactions produced during the formation of silver nanoparticles on colloidal silica spheres. (Reproduced from Reference [143] with permission. © 2001 American Chemical Society.)

TEM micrographs are shown in Figure 12.17 indicating that after the adsorption of Sn^{2+} ions, an increase in electron density is observed. This increase is due to the presence of metal ions.

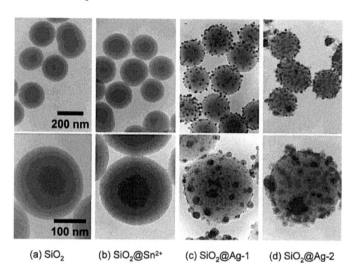

(a) SiO_2 (b) $SiO_2@Sn^{2+}$ (c) $SiO_2@Ag$-1   (d) $SiO_2@Ag$-2

FIGURE 12.17 TEM micrographs of silica spheres during the successive deposition steps. Upper and lower photos were taken at lower and higher magnifications, respectively. Scale bars correspond to all four photos in each row. (Reproduced from Reference [143] with permission. © 2001 American Chemical Society.)

12.2.3 DEPOSITION ZnO NANOPARTICLES ON SILICA SPHERES

Semiconductor materials are being used in the field of solar energy conversion, photocatalysis, and optoelectronic industry which have stemmed from the size dependent optical properties.[146–151] Number of synthetic routes have been developed which control the size and distribution of semiconductor crystal on a solid support. One of the most advance developments is the coating of the semiconductor cluster on the solid support. This coating is being accomplished by either of the two given strategies:

1. Transfer of solution phase synthesized semiconductor nanocluster to a core substrate then made to undergo thermal treatment so as to obtain core–shell material.
2. Direct synthesis of semiconductor nanocluster and solid support.

Liu et al. have electrodeposited zinc oxide nanoparticles on the gold single crystal.[152] Dhas et al. have reported the synthesis of ZnS semiconducting NPs on the submicron-sized silica near at room temperature by the ultrasound irradiation.[153] Currently, self-assembled monolayer approach is used to bind semiconductor nanocluster to metal or in organic surface through layer-by-layer assembly. Ag/TiO$_2$ core–shell nanoparticles have been synthesized as reported by Pastoriza-Santos and Kiktysh.[154]

ZnO semiconducting nanoparticles can be synthesized on submicron-sized SiO$_2$ through simplified controlled doubled jet precipitation method.[155,156] The formation of SiO$_2$/ZnO composite with different surface morphologies is shown in Figure 12.18.

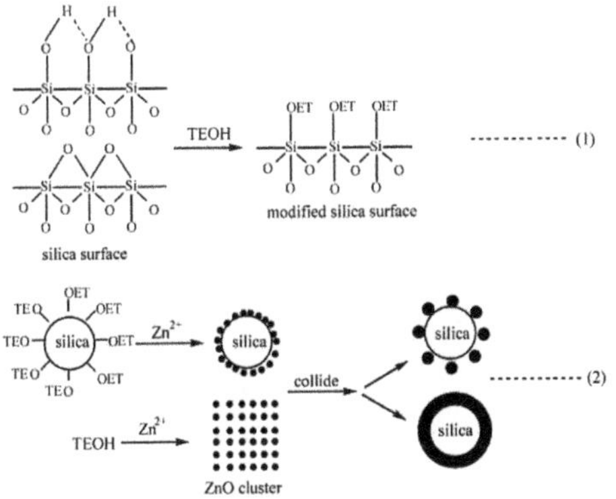

FIGURE 12.18 Scheme for preparing SiO$_2$/ZnO composite nanoparticle (TEOH represents triethanolamine, N(CH$_2$CH$_2$OH)$_3$). The produced particle is heated at 700°C to remove the triethanolamine, resulting in a core–shell material composed of a silica core and ZnO shell. (Reproduced from Reference [153] with permission. © 2003 American Chemical Society.)

12.2.4 DEPOSITION OF AuNPs ON SILICA SPHERES

Silica substrate provides high surface area when nanoparticles are coated on it. These possess chemical and physical properties which were different from those of bulk phase and individual molecules. These are used in the field of catalysis, chemical engineering, pharmaceutics, biology, etc. Efforts are being made to create new class of material through the modification of surface.[157] A range of wet chemistry methods are being employed to modify or coat colloids with nanoparticles of noble metals. A colloid reduction chemical route for the formation of solid-core/gold nanoshell particles[158] have been reported by Oldenburg et al.

At present, an additional method, that is, ultrasound-driven synthesis for coating gold-nanosized particles on silica spheres is being described. Through ultrasound, gold nanoparticles can be coated on silica spheres. Power ultrasound effects the chemical changes which is due to the cavitation phenomenon that involves the formation, growth, and implosive collapse of bubbles in liquid.[159]

Sonication in the presence of silica sphere traps the nanometer scale. This phenomenon is used to prepare metal nanoparticles and nanorods,[160] and the process has been extended to produce core–shell type materials.[161] Previously, this method has been used to disperse the gold nanoparticles within the pores of mesoporous silica.[162] Spherical colloidal particles act as support for the deposition of smaller noble nanoparticles. The substrate for the gold deposition has been obtained by modifying the Stöber method. Gold-deposited silica samples have been characterized by X-ray diffraction, transmission electron microscopy, high resolution scanning electron micros-copy, thermogravimetric analysis, Fourier-transform infrared, photocaustic, and UV–visible spectroscopy (Fig. 12.19).

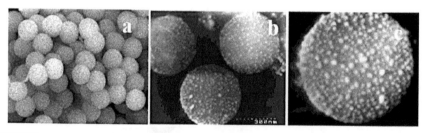

FIGURE 12.19 Scanning electron micrographs of (a) bare silica submicrospheres, (b) silica submicrospheres deposited with gold nanoparticles (sample 2 h), and (c) silica submicrosphere deposited with gold nanoparticles (sample 2 h) shown at higher resolution. (Reproduced from Reference [160] with permission. © 2003 American Chemical Society.)

Noble metals have coupled oscillations of conduction electrons, when they interact with the external electromagnetic wave of certain wavelength known as surface plasmon. The frequency of surface plasmon depends upon the nature of metal, shape, and size of nanoparticles and the nature of the surrounding medium. It has been found that force centered cubic structure of metallic sphere leads to improvement in optical features. Previously, photonic crystal has been made by infiltration of nanoparticles· electroless deposition or electrochemical deposition (Fig. 12.20).

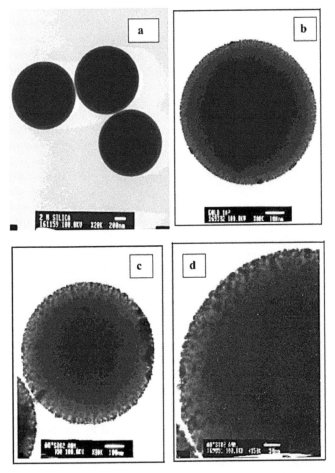

FIGURE 12.20 Transmission electron micrographs of (a) the bare substrate silica submicrospheres (800 nm), (b) as-prepared sample of a gold-deposited silica submicrosphere, (c) crystalline gold nanoparticles deposited on silica submicrospheres, and (d) an Au-coated silica submicrosphere shown at high resolution. (Reproduced from Reference [160] with permission. © 2003 American Chemical Society.)

Gold nanoshells are also used for the formation of opals. Metal nanoparticles can be incorporated into the colloids through the synthesis of composite core–shell colloid sphere comprising the metallic nanoparticle at the core and amorphous silica as the shell (Fig. 12.21).

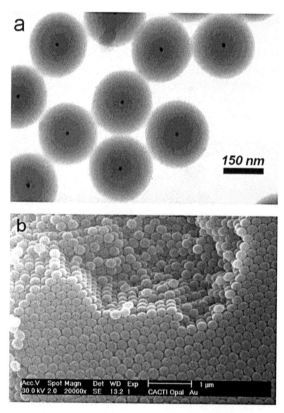

FIGURE 12.21   (a) TEM micrograph showing the core–shell morphology of the composite colloid spheres used for opal formation and (b) representative SEM micrograph showing the high degree of order obtained with the natural sedimentation method. (Reproduced from Reference [162] with permission. © 2002 American Chemical Society.)

The optical properties of opal were determined by the Bragg diffraction and the surface plasmon adsorption of the metal cores. Two systems have been used to form dense coating on the enzyme loaded sphere.

1. PE multilayer shell through the alternate deposition of poly (diallyldimethyl ammonium chloride) PDDA and poly(sodium-4-styrenesulfonate) (PSS)

2. Composite shell through the alternate deposition of PDDA

Noble metals can also be coated on silica spheres. The potential application of the material is determined by the interaction between the core and the shell. The coated nanomaterials provide a very high surface area and possess very different physical and chemical properties distinctive from those of bulk phase and individual molecules.

12.3 CONCLUSION

Silica spheres are being synthesized by Stöber and modified Stöber method, templating and nontemplating technique, and sol–gel process. During synthesis, when the silica precursor reach the certain size, then the precursors and surfactants assemble into small mesoporous silica spheres which then emerge in the solution. On adding the initial silica precursor, particle size expansion takes place. The expansion of the mesoporous size depends upon the concentration of the metal species present into the silica sphere. The silica spheres or its derivatives can find applications in removal of heavy metal ions, catalysis, proton exchange membranes. Nanocoating on silica spheres protects the metal particles from aggregation, thus avoiding poisoning of catalyst during the catalytic reactions, and is also used as therapeutics against the cancer. Through the sedimentation of silica spheres, high quality solid colloid carrying gold nanoparticle in the center has been prepared. Optical properties of the opals have been determined by Braggs diffraction.

KEYWORDS

- **Stöber synthesis**
- **template approach**
- **hollow mesoporous silica spheres**
- **porous silica spheres**
- **applications**
- **drug delivery**

REFERENCES

1. Rao, C. N. R.; Cheetham, A. K. J. Deposition of Gold Nanoparticles on Silica Spheres: A Sonochemical Approach. *Mater. Chem.* **2001**, *11*, 2887.
2. Drexler, K. Eric Molecular Engineering: An Approach to the Development of General Capabilities for Molecular Manipulation. *Proc. Natl. Acad. Sci. U.S.A.* **1981**, *78* (9), 5275–5278.
3. Dubchak, S.; Ogar, A.; Mietelski, J. W.; Turnaw, K. Influence of Silver and Titanium Nanoparticle on Arbuscular My Corhiza Colonization and Accumulation of Radio Caesium in Helianthus Annuus. *J. Agric. Res.* **2010**, *8* (1), 103–108.
4. Rai, M.; Yadav, A.; Grade A. D. D. Evanoff, Jr.; Chumanov, G. Synthesis and Optical Properties of Silver Nanoparticles and Arrays. *Chem. Phys. Chem.* **2001**, *6*, 1221–1231.
5. Sharma, V. K.; Ria, A. Y.; Lin, Y. Silver Nanoparticles: Green Synthesis and Their Antimicrobial Activities. *Adv. Coll. Interface Sci.* **2009**, *145*, 83–96.
6. Balban, D.; Sevmour, L. W. Control of Tumour Vascular Permeability. *Adv. Drug Deliv. Rev.* **1998**, *34*, 109–119.
7. Tomar, A.; Garg, G. Short Review on Application of Gold Nanoparticles. *Global J. Pharmacol.* **2013**, *7* (1), 34–38.
8. Fan, T. X.; Chow, S. K.; Zhang, D. Biomorphic Mineralization: From Biology to Material. *Progr. Mater. Sci.* **2009**, *54*, 542–659.
9. Bhattacharya, J.; Choudhary, U.; Biwach, O.; Sen, P.; Dasgupta, Synthesis, Characterization, Single Crystal Analysis A. *Nanomed. Nanotechnol. Biol. Med.* **2006**, *2*(3), 191–199.
10. Norval, M.; Lucas, R. M.; Cullen, A. P., et al. The Human Health Effects of Ozone Depletion and Interactions with Climate Change. *Photochem. Photobiol. Sci.* **2011**, *10*, 199–225.
11. Rammal, A.; Brisach, F.; Henry, M.; Chimie, C. R. Synthesis, Characterization and Antibacterial Properties of Titanium Dioxide Nanoparticles. *Chem. Rev.* **2002**, *5*, 59.
12. Wu, H.; Yang, R.; Song, B.; Han, Q.; Li, J.; Zhang, Y.; Fang, Y.; Tenne, R.; Wang, C. Biocompatible Inorganic Fullerene-Like Molybdenum Disulfide Nanoparticles Produced by Pulsed Laser Ablation in Water. *ACS Nano* **2011**, *5*, 1276–1281.
13. Huang, D.; Liao, F.; Molesa, S.; Redinger, D.; Subramanian, V. J. Gold Nanoparticles and Nanocomposites in Clinical Diagnostics Using Electrochemical Method. *Electrochem. Soc.* **2003**, *150*, G412–G417.
14. Korbekandi, H.; Iravani, S. Silver Nanoparticles. In *The Delivery of Nanoparticles*; Hashim Abbass, A., Ed., In Tech; 2012, ISBN: 978-953-51-0615-9. Kreibig, U.; Vollmer, M. *Optical Properties of Metal Clusters*; Springer; Berlin, 1995.
15. Liz-Marzan, L. M.; Gedanken, A.; Calderon-Moreno, J.; Mastai, Y. Colloidal Metal Deposition onto Functionalized Polystyrene Microsphere *Chem. Mater.* **2003**, *15*, 1378.
16. Huang, D.; Liao, F.; Molesa, S.; Redinger, D.; Subramanian, V. Gold Nanoparticles; Properties and Applications. *J. Electrochem. Soc.* **2003**, *150*, G412–G417.
17. IIer, K. R. The Chemistry of Silica; Hambleton, F. H.; Mitchell, S. A.; Hockey, J. A. *J. Phys. Chem.* **1969**, *73*, 3947.
18. Caruso, F. Synthesis of Luminescent Nanoporous Silica Spheres Functionalized with Folic Acid for Targeting to Cancer Cells. *Adv. Mater.* **2001**, *13*, 11–22.
19. Hartmann, M. Ordered Mesoporous Material for Bioadsorption and Biocatalysis. *Chem. Mater.* **2005**, *17*, 4577.

20. Stein, A. Aerosol-Assisted Synthesis of Magnetic Mesoporous Silica Spheres for Drug Targeting. *Adv. Mater.* **2003**, *15*, 763.
21. Ruiz-Hernández E.A. Arcos, D. Izquierdo-Barba, I.Terasaki· O. and M. Vallet-Regí, M. M. Aerosol-Assisted Synthesis of Magnetic Mesoporous Silica Spheres for Drug Targeting. *Chem. Mater.* **2007**, *19*, 3455–3463.
22. Vallet-Reigi, M. Preoperative Strategies for Controlling Structure and Morphology of Metal Oxides. In *Perspective in Solid State Chemistry*; Rao, K. J., Ed.; John Wiley and Sons: New York, NY, 1995, Vol. 37.
23. Kersge, C. T.; Leonowicz, M. E.; Roth, W. J.; Vartuli, J. C.; Beck, J. S. Ordered Mesoporous Molecular SIEVES Synthesized by Liquid Crystal Template Mechanism. *Nature* **1992**, *359*, 710–712.
24. Moroz, A. Synthetic Opals based on Silica-Coated Gold. Nanoparticles. *Lett. Europhys.* **2000**, *50*, 466.
25. Busch, K.; John, S. Synthetic Opals Based On Silica-Coated Gold Nano Particles. *Phys. Rev. E* **1998**, *58*, 3896.
26. Yang, H.; Zhao, D. J. Solvent Effect on Morphologies of Mesoporous Silica Spheres Prepared by Pseudomorphic Transformations. *Mater. Chem.* **2005**, *15*, 1217–1231.
27. Thommes, M. Physical Adsorption Characterization of Ordered and Amorphous Mesoporous Materials. In *Nanoporous Materials; Science and Engineering*; Lu, G. Q., Zhao, X. S., Eds.; Imperial College Press; Oxford, 2004; Chapter 11, p 137.
28. Tu, B.; Zhou, W. Z.; Zhao, D. Y. Simultaneous Chemical Modification and Structural Transformation of Stober Silica Spheres for Integration of Nanocatalysts. *Angew. Chem. Int. Ed.* **2003**, *42*, 3146.
29. Kwant, G. J.; Prins, W.; van Swaaij, W. P. M. Preparation of Hard Mesoporous Silica Spheres. *Chem. Eng. Sci.* **1994**, *49*, 4299.
30. Vail. J. G. *Soluble Silicates*; Reinhold: New York, 1952; Vol. 2, p 560.
31. Walash, D.; Mann, S. Synthesis of Uniform Porous Silica Microspheres with Hydrophilic Polymer as Stabilizing Agent. *Nature* **1995**, *377*, 320.
32. Iler, R. K. *The Chemistry of Silica: Solubility, Polymerization, Colloid and Surface Properties, and Biochemistry*; Wiley: New York, 1979.
33. Brinker, C. J.; Scherer, G. W. *Sol–Gel Science: The Physics and Chemistry of Sol–Gel Processing*; Academic Press: Boston, 1990.
34. Bergna, H. E. molecular Aspects of Silica Gel Formation. *Surf. Sci. Ser.* **2006**, *131*, 9.
35. Stein, A.; Schroden, R. C. *Curr. Opin. Solid State Mater. Sci.* **2001**, *5*, 553.
36. Kuai, S.; Badilescu, S.; Bader, G.; Bruning, R.; Hu, X.; Truong, V.-V. Preparation of Large Area 3D Ordered Macroporous Titania Films by Silica Colloidal Crystal Templating. *Adv. Mater.* **2003**, *15*, 73.
37. Grün, M.; Lauer, I.; Unger, K. K. The Synthesis of Micrometer and Submicrometer-Size Spheres of Ordered Mesoporous Oxide MCM-41. *Adv. Mater.* **1991**, *9*, 254.
38. Grün, M.; Unger, K. K.; Matsumoto, A.; Tsutsumi, K. Novel Pathways for the Preparation of Mesoporous MCM-41 Materials: Control of Porosity and Morphology. *Microporous Mesoporous Mater.* **1999**, *27*, 207.
39. Schumacher, K.; Von Hohenesche, C. F.; Unger, K. K.; Ulrich, R.; Chesne, A. D.; Wiesner, U.; Spiess, H. W. The Synthesis of Spherical Mesoporous Molecular Sieves MCM-48 with Heteroatoms Incorporated into the Silica Framework. *Adv. Mater.* **1999**, *11*, 1194.
40. Lebedev, O. I.; Tendeloo, G. V.; Collart, O.; Cool, P.; Vansant, E. F. Structure and Microstructure of Nanoscale Mesoporous Silica Spheres. *Solid State Sci.* **2004**, *5*, 489.

41. Liu, S.; Cool, P.; Collart, O.; Van Der Voort, P.; Vansant, E. F.; Lebedev, O. I.; Van Tendeloo, G.; Jiang, M. The Influence of the Alcohol Concentration on the Structural Ordering of Mesoporous Silica: Cosurfactant versus Cosolvent. *J. Phys. Chem. B* **2003**, *107*, 10405.

42. Chang, S. M.; Lee, M.; Kim, W. S. Preparation of Large Monodispersed Spherical Silica Particles Using Seed Particle Growth. *J. Colloid Interface Sci.* **2005**, *286*, 536.

43. Lee, M. H.; Beyer, F. L.; Furst, E. M. Synthesis of Monodisperse Fluorescent Core–Shell Silica Particles Using a Modified Stober Method for Imaging Individual Particles in Dense Colloidal Suspensions. *J. Colloid Interface Sci.* **2005**, *288*, 114.

44. Büchel, G.; Unger, K. K.; Matsumoto, A.; Tsutsumi, K. A Novel Pathway for Synthesis of Submicrometer-Size Solid Core/Mesoporous Shell Silica Spheres. *Adv. Mater.* **1998**, *10*, 1036.

45. Yoon, S. B.; Kim, J. Y.; Kim, J. H.; Park, Y. J.; Yoon, K. R.; Park, S. K.; Yu, J. S. Synthesis of Monodisperse Spherical Silica Particles with Solid Core and Mesoporous Shell: Mesopore Channels Perpendicular to the Surface. *J. Mater. Chem.* **2007**, *17*, 1758.

46. Unger, K. K.; Kumar, D.; Grün, M.; Büchel, G.; Lüdtke, S.; Adam, T.; Schumacher, K.; Renker, S. Synthesis of Spherical Porous Silicas in the Micron and Submicron Size Range: Challenges and Opportunities for Miniaturized High-Resolution Chromatographic and Electrokinetic Separations. *J. Chromatogr. A* **2000**, *892*, 47.

47. Yano, K.; Fukushima, Y. Synthesis of Mono-Dispersed Mesoporous Silica Spheres with Highly Ordered Hexagonal Regularity Using Conventional Alkyltrimethylammonium Halide as a Surfactant. *J. Mater. Chem.* **2004**, *14*, 1579.

48. Chang, S. M.; Lee, M.; Kim, W. S. Preparation of Large Monodispersed Spherical Silica Particles Using Seed Particle Growth. *J. Colloid Interface Sci.* **2005**, *286*, 536.

49. Sayari, A.; Kruk, M.; Jaroniec, M.; Moudrakovski, I. L. New Approaches to Pore Size Engineering of Mesoporous Silicates. *Adv. Mater.* **1998**, *10*, 1376.

50. Yoo, W.; Stein, A.; Solvent Effects on Morphologies of Mesoporus Silica Spheres Prepared by Pseudomorphic Transformations. *Chem Mater.* 2011, 23, 1761–1767

51. Gellermann, C.; Storch, W.; Wolter, H. Synthesis and Characterization of the Organic Surface Modifications of Monodisperse Colloidal Silica. *J. Sol-Gel Sci. Technol.* **1997**, *8*, 173–176

52. Kuai, S.; Badilescu, S.; Bader, G.; Bruning, R.; Hu, X.; Truong, V.-V. *Adv. Mater.* **2003**, *15*, 73. Han, Y.; Lee, S. S.; Ying, J. Y. Simple Fabrication of a Highly Sensitive Glucose Biosensor Using Enzymes Immobilized in Exfoliated Graphite Nanoplatelets Nafion Membrane. *Chem. Mater.* **2007**, *19*, 2292–2298.

53. Chen, H.; Hu, T.; Zhang, X.; Huo, K.; Chu, P. K.; He, J. Biphase Stratification Approach to Three-Dimensional Dendritic Biodegradable Mesoporous Silica Nanospheres. *Langmuir* **2010**, *26*, 13556–13563.

54. Martin, T.; Galarneau, A.; Renzo, F. D.; Brunel, D.; Fajula, F.; Heinisch, S.; Cretier, G.; Rocca, J.-L. Morphology Control of Porous Materials and Molecular Sieve Membrane Applications. *Chem. Mater.* **2004**, *16*, 1725–1731.

55. Yao, K.; Zeng, H.; Simultaneous Chemical Modification and Structural Transformation of Stober Silica Spheres for Integration of Nanocrystals. *Chem. Mater.* 2012, 24, 140–148.

56. Polshettiwar, V.; Cha, D.; Zhang, X. X.; Basset, J. M. High Surface Area Silica Nanospheres (KCC-1) with a Fibrous Morphology. *Angew. Chem. Int. Ed.* **2010**, *49*, 9652.

57. Deng, Y. H.; Cai, Y.; Sun, Z. K.; Liu, J.; Liu, C.; Wei, J.; Li, W.; Liu, C.; Wang, Y.; Zhao, D. Y. Adsorption and Separation of Reactive Aromatic Isomers and Generation and Stabilization of Their Radicals within Cadmium(II)−Triazole Metal−Organic Confined Space in a Single-Crystal-to-Single-Crystal Fashion. *J. Am. Chem. Soc.* **2010**, *132*, 8466.

58. Kamata, K.; Lu, Y.; Xia, Y. N. Synthesis and Characterization of Monodispersed Core–Shell Spherical Colloids with Movable Cores *J. Am. Chem. Soc.* **2003**, *125*, 2384.

59. Ge, J.; Zhang, Q.; Zhang, T.; Yin, Y. Formation of Hollow Silica Colloids through a Spontaneous Dissolution-Regrowth Process. *Angew. Chem. Int. Ed.* **2008**, *47* (17), 8924.

60. Zhang, Q.; Zhang, T.; Ge, J.; Yin, Y. Permeable Silica Shell through Surface-Protected Etching. *Nano Lett.* **2008**, *8*, 2867.

61. Polshettiwar, V.; Cha, D.; Zhang, X. X.; Basset, J. M. Nano Fibrous Silica Sulphuric Acid as an Efficient Catalyst for the Synthesis of b-Enaminone. *Angew. Chem. Int. Ed.* **2010**, *49*, 9652.

62. Katiyar, A.; Ji, L.; Smirniotis, P. G.; Pinto, N. G. Adsorption of Bovine Serum Albumin and Lysozyme on Siliceous MCM-41. *Microporous Mesoporous Mater.* **2005**, *80*, 311.

63. Deere, J.; Magner, E.; Wall, J. G.; Hodnett, B. K. Adsorption and Activity of Cytochrome Activity of Cytochrome C on Mesoporous Silicates. *Chem. Commun.* **2001**, *5*, 465–465.

64. Lu, Y. F.; Fan, H. Y.; Stump, A. Aerosol-Assisted Self-Assembly of Mesostructured Spherical Nanoparticles. *Nature* **1999**, *398*, 223.

65. Serra, C. A.; Chang, Z. Q. Microfluidic-Assisted Synthesis of Polymer Particles. *Chem. Eng. Technol.* **2008**, *31*, 1099.

66. Zhai,Z.; Wang, Y.; Lu,Y. ; Luo,G. Preperation of Monodispersed Uniform Silica Sphere with large pore size for fast Adsorption of Proteins. *Ind. Eng. chem mater. Res*, 2010,Vol. 49, No.9.

67. Yu, Y.; Zhang, L.; Eisenberg, A. *Macromolecules* **1998**, *31*, 1144–1154.

68. Niu, D.; Ma, Z.; Li, Y.; Shi, J. Synthesis of Core–Shell Structured Dual Mesoporus Silica Spheres with Tunable Pore Size and Controllable Shell Thickness *J. Am. Chem. Soc.* **2010**, 132, 15144–15147.

69. Zhang, L.; Eisenberg, A. Morphogenic Effect of Added Ions on Crew-Cut Aggregates of Polystyrene-*b*-poly(acrylic acid) Block Copolymers in Solutions. *Macromolecules* 1996, *29(27)*, 8805–8815.

70. Rubio, J.; Kitchener, J. A. The Mechanism of Adsorption of Poly(Ethylene Oxide) Flocculant on Silica. *J. Colloid Interface Sci.* **1976**, *57*, 132.

71. Van de Ven, T. G. M. Association-Induced Polymer Bridging by Poly(Ethylene Oxide)-Cofactor Flocculation Systems. *Adv. Colloid Interface Sci.* **2005**, *114–115*, 147.

72. Zhang, Z. T.; Yang, L. M.; Wang, Y. J.; Luo, G. S.; Dai, Y. Y. Morphology Controlling of Micrometer-Sized Mesoporous Silica Spheres Assisted by Polymers of Polyethylene Glycol and Methyl Cellulose. *Microporous Mesoporous Mater.* **2008**, *115*, 447.

73. Takeuchi, M.; Kageyama, S.; Suzuki, H.; Wada, T.; Notsu, Y.; Ishii, F. Rheological Properties of Reversible Thermo-setting In Situ Gelling Solutions with the Methylcellulose-Polyethylene Glycolcitric Acid Ternary System. *Colloid Polym. Sci.* **2003**, *281*, 1178.

74. Huo, Q.; Feng, J.; Sciith, F.; Stucky, G. Preperation of Hard Mesoporus Silica Spheres. *Chem. Mater.* **1997**, 14–17.

75. Wang, Y.; Caruso, F. Mesoporus Silica Spheres as Supports for Enzyme Immobilization and Encapsulation. *Chem. Mater.* 2005, 17, 953–961.

76. Yu, Y.; Eisenberg, A. Control of Morphology through Polymer–Solvent Interactions in Crew-Cut Aggregates of Amphiphilic Block Copolymers. *J. Am. Chem. Soc.* **1997**, *119*, 8383–8384.

77. Yanagisawa, T.; Shimizu, T.; Kuroda, K.; Kato, C. The Preparation of Alkyltrimethylammonium-Kanemite Complexes and their Conversion to Microporous Materials. *Bull. Chem. Soc. Jpn.* **1990**, *63*, 988.

78. Yiu, H. H. P.; Wright, P. A.; Botting, N. P. Non-Destructively Shattered Mesoporous Silica for Protein Drug Delivery. *Microporous Mesoporous Mater.* **2001**, *44–45*, 763.

79. Takahashi, H.; Li, B.; Sasaki, T.; Miyazaki, C.; Kajino, T.; Inagaki. Catalytic Activity in Organic Solvents and Stability of Immobilized Enzymes Depend on the Pore Size and Surface Characteristics of Mesoporous Silica. *Chem. Mater.* **2000**, *12*, 3301.

80. (a) Kisler, J. M.; Dähler, A.; Stevens, G. W.; O'Connor, A. J. *Microporous Mesoporous Mater.* **2001**, *44–45*, 769. (b) Diaz, J. F.; Balkus Jr., K. J. *J. Mol. Catal. B: Enzymol.* **1996**, *2*, 115. (c) Deere, J.; Magner, E.; Wall, J. G.; Hodnett, B. K. *Chem. Commun.* **2001**, *0*, 465.

81. Lei, C.; Shin, Y.; Liu, J.; Ackerman, E. J. *J. Am. Chem. Soc.* **2002**, *124*, 11242.

82. Deere, J.; Magner, E.; Wall, J. G.; Hodnett, B. K. *J. Phys. Chem. B* **2002**, *106*, 7340.

83. Li, Z. Z.; Wen, L. X.; Shao, L.; Chen, J. F. *J. Controlled Release* **2004**, *98*, 245–254.

84. Chen, J. F.; Ding, H. M.; Wang, J. X.; Shao, L. *Biomaterials* **2004**, *25*, 723–727.

85. Sharma, R. K.; Das, S.; Maitra, A. *J. Colloid Interface Sci.* **2005**, *284*, 358–361.

86. Wang, J.; Ding, H.; Tao, X.; Chen, J. *Nanotechnology* **2007**, *18*, 245705.

87. Zhu, Y. F.; Shi, J. L.; Chen, H. R.; Shen, W. H.; Dong, X. P. *Microporous Mesoporous Mater.* **2005**, *84* (1–3), 218–222.

88. Li, Z. Z.; Xu, S. A.; Wen, L. X.; Liu, F.; Liu, A. Q.; Wang, Q.; Sun, H. Y.; Yu, W.; Chen, J. F. *J. Controlled Release* **2006**, *111* (1–2), 81–88.

89. Zhou, J.; Wu, W.; Caruntu, D.; Yu, M. H.; Martin, A.; Chen, J. F.; O'Connor, C. J.; Zhou, W. L. *J. Phys. Chem. C* **2007**, *111* (47), 17473–17477.

90. Hah, J. S.; Kim, B. J.; Jeon, S. M.; Koo, Y. E. *Chem. Commun.* **2003**, *14*, 1712–1713. Wang, Q. B.; Liu, Y.; Yan, H. *Chem. Commun.* **2007**, *23*, 2339–2341.

91. Kang, T.; Park, Y. G.; Yi, J. H. *Ind. Eng. Chem. Res.* **2004**, *43*, 1478–1484.

92. Liao, J. F.; Wu, Q. Z.; Y., Q.; Wang, C. T.; Li, H. Y.; Li, Y. G. *Acta Chim.* **2006**, *64* (24), 2419–2424.

93. Shin, S.; Jang, J. *Chem. Commun.* **2007**, *41*, 4230–4232. Feng, X.; Fryxell, G. E.; Wang, L. Q.; Kim, A. Y.; Liu, J.; Kemner K. M. *Science.* **1997**, *276*, 923.

94. Nam, K. H.; Gomez-Salazar, S.; Talavrides, L. L. *Ind. Eng. Chem. Res.* **2003**, *42*, 1955–1964.

95. Feng, X.; Fryxell, G. E.; Wang, L. Q.; Kim, A. Y.; Liu, J.; Kemner, K. M. *Science* **1997**, *276*, 923.

96. Nam, K. H.; Gomez-Salazar, S.; Talavrides, L. L. *Ind. Eng. Chem. Res.* **2003**, *42*, 1955–1964.

97. Das, D.; Lee, J.; Cheng, S. *J. Catal.* **2004**, *223* (1), 152–160.

98. Díaz, I.; Márquez-Alvarez, C.; Mohino, F.; Pérez-Pariente, J.; Sastre, E. *J. Catal.* **2000**, *193*, 295–302.

99. Yuan, J.; Wan, D.; Yang, Z. A Facile Method for the fabrication of Thiol Functionalized Hollow Silica Spheres . *J. Chem. C* **2008**, *12*, 17156–17160.

100. Sua, Y. H.; Wei, T. Y.; Hsu, C. H.; Liu, Y. L.; Sun, Y. M.; Lai, J. Y. *Desalination* **2006**, *200*, 656–657.

101. Sellinger, A.; Weiss, P. M.; Nguyen, A.; Lu, Y.; Assink, R. A.; Gong, W.; Brinker, C. J. *Nature* **1998**, *394*, 256–260.
102. Huo, Q.; Margolese, D. I.; Ciesla, U.; Feng, P.; Gier, T. E.; Sieger, P.; Leon, R.; Petroff, P. M.; Schüth, F.; Stucky, G. D. *Nature* 1994, 368, 317–321.
103. Huo, Q. S.; Margolese, D. I.; Ciesla, U.; Demuth, D. G.; Feng, P. Y.; Gier, T. E.; Sieger, P.; Firouzi, A.; Chmelka, B. F.; Schuth, F.; Stucky, G. D. *Chem. Mater.* **1994**, *6*, 1176–1191
104. Li, H. L.; Sun, H.; Qi, W.; Xu, M.; Wu, L. X. *Angew. Chem. Int. Ed.* **2007**, *46*, 1300–1303.
105. Trikalitis, P. N.; Rangan, K. K.; Bakas, T.; Kanatzidis, M. G. *Nature* **2001**, *410*, 671–675.
106. Hu, Y. X.; Zhang, Q.; Geobl, J.; Zhang, T. R.; Yin, Y. D. *Phys. Chem. Chem. Phys.* **2010**, *12*, 11836–11842.
107. Yu, Q. Y.; Wang, P. P.; Hu, S.; Hui, J. F.; Zhuang, J.; Wang, X. *Langmuir* **2011**, *27*, 7185–7191.
108. Wong, Y. J.; Zhu, L.; Teo, W. S.; Tan, Y. W.; Yang, Y.; Chen, H. *J. Am. Chem. Soc.* **2011**, *133*, 11422–11425.
109. Hu, Y. X.; Zhang, Q.; Geobl, J.; Zhang, T. R.; Yin, Y. D. *Phys. Chem. Chem. Phys.* **2010**, *12*, 11836–11842.
110. Yu, Q. Y.; Wang, P. P.; Hu, S.; Hui, J. F.; Zhuang, J.; Wang, X. *Langmuir* **2011**, *27*, 7185–7191.
111. Wong, Y. J.; Zhu, L.; Teo, W. S.; Tan, Y. W.; Yang, Y.; Chen, H. J. *Am. Chem. Soc.* **2011**, *133*, 11422–11425.
112. Zhang, Q.; Zhang, T. R.; Ge, J. P.; Yin, Y. D. *Nano Lett.* **2008**, *8*, 2867–2871.
113. Zhang, T. R.; Ge, J. P.; Hu, Y. X.; Zhang, Q.; Aloni, S.; Yin, Y. D. *Angew. Chem. Int. Ed.* **2008**, *47*, 5806–5811.
114. Zhang, Q.; Wang, W. S.; Goebl, J.; Yin, Y. D. *Nano Today* **2009**, *4*, 494–507.
115. Fang, X. L.; Chen, C.; Liu, Z. H.; Liu, P. X.; Zheng, N. F. *Nanoscale* **2011**, *3*, 1632–1639.
116. Wong, Y. J.; Zhu, L.; Teo, W. S.; Tan, Y. W.; Yang, Y.; Chen, H. *J. Am. Chem. Soc.* **2011**, *133*, 11422–11425.
117. Hui, Q.; Wang , P.; Wang, X. Anion Exchange-Driven Disassembly of a SiO₂/ CTAB Composite Mesophase: The Formation of Hollow Mesoporous Silica Spheres. *Inorg. Chem.* 2012, *51*, 9539–9543.
118. Zhang, Q.; Zhang, T. R.; Ge, J. P.; Yin, Y. D. *Nano Lett.* **2008**, *8*, 2867–2871. Wang, J. G.; Xiao, Q.; Zhou, H. J.; Sun, P. C.; Yuan, Z. Y.; Li, B. H.; Ding, D. T.; Shi, A. C.; Chen, T. H. *Adv. Mater.* **2006**, *18* (24), 3284–3288. Fujiwara, M.; Shiokawa, K.; Sakakura, I.; Nakahara, Y. *Nano Lett.* **2006**, *6* (12), 2925–2928.
119. Chang, C. L.; Guo, J. H.; Yin, Y. D. *J. Phys. Chem. C* **2009**, *113*, 3168–3175.
120. Zhang, Q.; Ge, J. P.; Goebl, J.; Hu, Y. X.; Lu, Z. D.; Yin, Y. D. *Nano Res.* **2009**, *2*, 583–591.
121. Suh, W. H.; Jang, A. R.; Suh, Y. H.; Suslick, K. S. *Adv. Mater.* **2006**, *18* (14), 1832.
122. Wang, J. W.; Xia, Y. D.; Wang, W. X.; Poliakoff, M.; Mokaya, R. *J. Mater. Chem.* **2006**, *16* (18), 1751–1756.
123. Yu, M. H.; Wang, H. N.; Zhou, X. F.; Yuan, P.; Yu, C. Z. *J. Am. Chem. Soc.* **2007**, *129*, 14576–14577.
124. Eggeman, A. S.; Petford-Long, A. K.; Dobson, P. J.; Wiggins, J.; Bromwich, T.; Dunin-Borkowski, R.; Kasama, T. *J. Magn. Magn. Mater.* **2006**, *301* (2), 336–342.
125. Suh, W. H.; Suslick, K. S. *J. Am. Chem. Soc.* **2005**, *127* (34), 12007–12010.
126. Wu, P. G.; Zhu, J. H.; Xu, Z. H. *Adv. Funct. Mater.* **2004**, *14* (4), 345–351.

127. Zhou, W.; Gao, P.; Shao, L.; Caruntu, D.; Yu, M.; Chen, J.; O'Connor, C. *J. Nanomed.* **2005**, *1* (3), 233–237.

128. Iler, R. K. *The Colloid Chemistry of Silica and Silicates*; Cornell University Press: Ithaca, NY, 1955; Chapter 2.

129. Ren, N.; Wang, B.; Yang, Y. H.; Zhang, Y. H.; Yang, W. L.; Yue, Y. H.; Gao, Z.; Tang, Y. *Chem. Mater.* **2005**, *17*(10), 2582–2587.

130. Wang, Y.; Tang , C.: Deng, Q.; Liang , C. ; Ng, D.; Kwong, F.: Wang , H.; Cai, W.; Zhang, L.; Wang, G. A Versatile Method for Controlled Synthesis of Porous Hollow Spheres . *Langmuir* **2010**, *2618*,14830–14834.

131. Auer, E.; Freund, A.; Pietsch, J.; Tacke, T. *Appl. Catal. A* **1998**, *173*, 259.

132. Dokoutchaev, A.; James, J. T.; Koene, S. C.; Pathak, S.; Prakash, G. K. S.; Thompson, M. E. *Chem. Mater.* **1999**, *11* (9), 2389.

133. Chen, C. W.; Serizawa, T.; Akashi, M. *Chem. Mater.* **1999**, *11* (5), 1381.

134. Sanchez, J. M.; Hidalgo, M.; Valiente, M.; Salvado, V. *J. Polym. Sci., A: Polym. Chem.* 2000, *38* (2), 269.

135. Pol, V. G.; Srivastava, D. N.; Palchik, O.; Palchik, V.; Slifkin, M. A.; Weiss, A. M.; Gedanken, A. Sonochemical Deposition of Silver Nanoparticles on Silica Spheres. *Langmuir* **2002**, *18*, 3352.

136. Pol, V. G.; Gedanken, A.; Calderon-Moreno, J. Using Sonochemistry for the Fabrication of Nanomaterials. *Chem. Mater.* **2003**, *15*, 1111.

137. Zhong, Z. Y.; Mastai, Y.; Koltypin, Y.; Zhao, Y. M.; Gedanken, A. Spherical Ensembles of Gold Nanoparticles on Silica: Electrostatic and Size Effects. *Chem. Mater.* **1999**, *11* (9), 2350.

138. Pol, V. G.; Motiei, M.; Gedanken, A.; Calderon-Moreno, J.; Mastai, Y. Sonochemical Deposition of Air-Stable Iron Nanoparticles on Monodispersed Carbon Spherules. *Chem. Mater.* **2003**, *15*, 1378.

139. Breen, M. L.; Dinsmore, A. D.; Pink, R. H.; Qadri, S. B.; Ratna, B. R. Mesoscopic Monodisperse Ferromagnetic Colloids Enable Magnetically Controlled Photonic Crystals. *Langmuir* **2001**, *17*, 903.

140. Sivakumar, M.; Gedanken, A. Nanocrystalline Haematite Thin Films by Chemical Solution Spray. *Ultrason. Sonochem.* **2004**, *11*, 373.

141. Ang, L.-M.; Hor, T. S. A.; Xu, G.-Q.; Tung, C.; Zhao, S.; Wang, J. L. S. Fabrication of Nanostructures by Hydroxylamine Seeding of Gold Nanoparticle Templates. *Chem. Mater.* **1999**, *11*, 2115–2118.

142. Pol, V.; Grisaru, H.; Gedanken, A. Coating Noble Metal Nanocrystals (Ag, Au, Pd and Pt) on Polystyrene Spheres via Ultrasound Irradiation. *Langmuir* **2005**, 21, 3635–3640.

143. Kobayashi Y.; Maceira, V.; Marzan, L.; Deposition of Silver Nanoparticles on Silica Spheres by Pretreatment Steps in Electroless Plating. *Chem. Mater.* **2001**, 13, 1630–1633.

144. Golan, Y.; Margulis, L.; Hodes, G.; Rubinstein, I.; Hutchison, J. L. Surface Sol–Gel Synthesis of Ultrathin Semiconductor Films. *Surf. Sci.* **1994**, *311*, L633.

145. Kamat, P. V.; Shanghavi, B. *J. Phys. Chem.* **1997**, *101*(39), 7675–7679.

146. Peng, X. G.; Schlamp, M. C.; Kadavanich, A. V.; Alivisatos, A. P. Synthesis of Mixed Silver Halide Nanocrystals in Reversed Micelles. *J. Am. Chem. Soc.* **1997**, *119*, 7019.

147. Kortan, A. R.; Hull, R.; Opila, R. L.; Bawendi, M. G.; Steigerwald, M. L.; Carroll, P. J.; Brus, L. E. Excitons in InP/InAs Inhomogeneous Quantum Dots. *J. Am. Chem. Soc.* **1990**, *112*, 1327.

148. Liu, R.; Vertegel, A. A.; Bohannan, E. W.; Sorenson, T. A.; Switzer, J. A. Zinc Oxide Piezoelectric Nano-Generators for Low Frequency Applications. *Chem. Mater.* **2001**, *13*, 508.

149. Dhas, N. A.; Zaban, A.; Gedanken, A. Surface Synthesis of Zinc Sulfide Nanoparticles on Silica Microspheres: Sonochemical Preparation, Characterization, and Optical Properties. *Chem. Mater.* **1999**, *11*, 806.

150. Colvin, V. L.; Schlamp, M. C.; Alivisatos, A. P. Shape Control of Colloidal Semiconductor Nanocrystals. *Nature* **1994**, *370*, 354.

151. Zhong, Q. P.; Matijevic, E. Preparation and Characterization of Well Defined Powders and their Applications In Technology. *J. Mater. Chem.* **1996**, *6*, 443.

152. Her, Y. S.; Matijevic, E.; Chon, M. C. Kinetics and Mechanism of Aqueous Chemical Synthesis of $BaTiO_3$ Particles. *J. Mater. Res.* **1995**, *10*, 3106.

153. Xia, H.; Tang, f. Surface Synthesis of Zinc Oxide Nanoparticles on Silica Spheres: Preparation and Characterization. *J. Phys.Chem. B* **2003**,107, 9175–9178.

154. Oldenburg, S. J.; Averitt, R. D.; Wetcott, S. L.; Halas, N. *J. Chem. Phys. Lett.* **1998**, *288*, 243.Suslick, K. S., Ed. In *Ultrasound: Its Chemical, Physical and Biological Effects*. VCH: Germany, 1988.

155. Dokoutchaev, A.; James, J. T.; Koene, S. C.; Pathak, S.; Prakash, G. K. S.; Thompson M. E. Preparation and Optical Properties of Au-Shell Submicron Polystyrene Particles. *Chem. Mater.* **1999**, *11*, 2389.

156. (a) Stöber, W.; Fink, A.; Bohn, E. *J. Colloid Interface Sci.* **1968**, *26* (62), 164. (b) Kreibig, U.; Vollmer, M. *Optical Properties of Metal Clusters*; Springer: Berlin, 1995.

157. Mulvaney, P. *Langmuir* **1996**, *12*, 788. Link, S.; El-Sayed, M. A. Synthesis and Properties of Monometallic and Bimetallic Silver and Gold Nanoparticles. *J. Phys. Chem. B* 1999, *103*, 84

158. Stober, W.; Fink, A.; Bohn, E. Controlled Growth of Monodisperse Silica Spheres in the Micron Size Range. *J. Colloid Interface Sci.* **1968**, *26*, 62–69.

159. Ludtke, S.; Adam, T.; Unger, K. K. Application of 0.5-µm Porous Silanized Silica Beads in Electrochromatography. *J. Chromatogr. A* **1997**, *786*, 229–235.

160. Pol, V.; Gedanken, A.; Moreno, J. Deposition of Gold Nanoparticles on Silica Spheres: A Sonochemical Approach. *Chem Mater.* **2003**, *15*, 1111–1118.

161. Bogush, G. H.; Zukoski, C.F. Preparation of Spherical Silica Nanoparticles: Stober Silica. *J. Colloid Interface Sci.* **1991**, *142*, 1–34.

162. Garcia-Santamaria, F.; Salgueiriño-Maceira, V.; Lopez, C.; Marzan, L. Synthetic Opals Based on Silica–Coated Gold Nanoparticles. *Langmuir* **2002**, *18*, 4519–4522.

163. Pol, V.; Gedanken, A.; Moreno, J. Deposition of Gold Nanoparticles on Silica Spheres: A Sonochemical Approach. *Mater. Chem.* **2003**, *15*, 1111–1118.

164. Santamaria, F.; Maceira, V.; Lopez, C.; Marzan, L. Synthetic Opal Based on Silica-Coated Gold Nanoparticles. *Langmuir* **2002**, *18*, 4519–4522.

CHAPTER 13

INDUSTRIAL FIXED BED CATALYTIC REACTORS: DRAWBACKS, CHALLENGES, AND FUTURE SCOPE

SUKANCHAN PALIT[1,2*]

[1]*Department of Chemical Engineering, University of Petroleum and Energy Studies, Energy Acres, Post-Office: Bidholi via Prem Nagar, Dehradun 248007, Uttarakhand, India*

[2]*43, Judges Bagan, Post-Office: Haridevpur, Kolkata 700082, India*

E-mail: sukanchan68@gmail.com; sukanchan92@gmail.com

ABSTRACT

Science and technology are moving in a rapid pace surpassing visionary scientific frontiers. Energy engineering and energy technology are today moving toward a path of newer scientific rejuvenation and a greater scientific regeneration. Energy security, depletion of fossil fuel resources, and climate change are the scientific parameters of immense concern in present day human civilization. Thus, there is a need of new innovations in energy engineering, petroleum engineering science, and chemical process engineering. Petroleum refining today stands in the midst of immense scientific vision and deep scientific introspection. The challenge and the vision of petroleum engineering science are today evergrowing and crossing visionary frontiers. In this chapter, the author pointedly focuses on the deep scientific success, the scientific potential, and the vast scientific imagination in the operation of a petroleum-refining unit, mainly industrial fixed bed catalytic reactors. Petroleum refinery and petroleum refining units are the heart of chemical process engineering and petroleum engineering paradigm today. Scientific vision, scientific forbearance, and deep scientific fortitude are the hallmarks of research pursuit in petroleum refining, fixed bed catalytic cracking, and fluidized bed catalytic cracking in present day human civilization. This chapter is a veritable eye-opener toward the immense-scientific intricacies, the scientific profundity, and the energy sustainability needs in the furtherance of petroleum engineering science today.

13.1 INTRODUCTION

Science and technology of petroleum engineering science are today moving at a rapid pace surpassing one visionary paradigm over another. The challenges and the vision of petroleum refining are changing the veritable scientific firmament of energy engineering and energy sustainability. Loss of fossil fuel resources and the needs of energy security are urging the scientific domain to gear toward newer innovation and visionary technologies. Mankind's immense scientific prowess, the needs of human society, and the technological vision will all lead a long and visionary way in the true emancipation and true realization of petroleum refining and petroleum engineering science. The utmost need of the human civilization is energy and environmental sustainability. In such a crucial juncture of scientific history and time, the efficiency of petroleum refining units such as fixed bed catalytic reactors and fluidized bed catalytic reactors is the hallmarks and the visionary cornerstones toward a newer emancipation of petroleum engineering science. Human civilization and human scientific endeavor today stand in the critical juncture of deep scientific introspection and unending environmental catastrophes. Sustainable development as regards energy and environment are thus the need of the hour. Sustainability has a visionary definition given by Dr. Gro Harlem Brundtland, the former Prime Minister of Norway. Her vision of energy and environmental sustainability needs to be redefined and readdressed at this crucial juncture of human history and time. Petroleum refining and energy sustainability are two opposite sides of the visionary coin in the pursuit of science and technology today. The author in this chapter pointedly focuses on the deep scientific success, the vast technological vision, and the much-needed scientific profundity in the research pursuit in industrial fixed bed reactors, fluidized bed catalytic reactors, and the vast world of energy sustainability and petroleum engineering science.

13.2   THE AIM AND OBJECTIVE OF THIS STUDY

The vision and the challenge of petroleum refining, petroleum engineering science, and chemical process engineering are surpassing vast and versatile scientific frontiers. Depletion of fossil fuel resources and the immense environmental issue of climate change are challenging the veritable scientific fabric of petroleum engineering and chemical process engineering. Human scientific progress today stands in the midst of deep introspection and unending scientific imagination. Humanity and human civilization need to be revamped and restructured as regards true realization of energy and environmental sustainability. The aim and purpose of this study are to target the vast scientific intricacies of petroleum refining and petroleum engineering science with the sole purpose of furtherance of science and engineering. Technological validation and scientific profundity are the veritable cornerstones of this well-researched chapter. The author with deep and cogent insight delineates the scientific success, the scientific forbearance, and the unending scientific imagination of petroleum refining, fixed bed catalytic cracking, and fluidized bed catalytic cracking. Human scientific vision today is in a state of immense distress and unimaginable crisis. Climate change and loss of ecological biodiversity are the causes of immense concern to mankind today. This challenge and the vast vision of petroleum engineering science are deeply delineated in this chapter.

Technology and engineering science are moving at a rapid pace toward a newer vision and a newer direction. Human scientific endeavor in the field of petroleum engineering science and chemical process engineering today stands in the midst of deep introspection and a catastrophe due to depletion of fossil fuel resources. Scientific ingenuity, scientific sojourn, and deep scientific profundity are the utmost need of the hour. The aim and the objective of this chapter are to target the scientific challenges behind the operation of fixed bed catalytic reactors as well as fluidized bed catalytic reactors. Petroleum refining and petroleum engineering science are the utmost necessities of human civilization and human scientific endeavor today. Mankind's immense scientific grit and determination, the vast scientific prowess, and the futuristic vision of science and engineering will all lead a long and visionary way in the true emancipation of petroleum engineering science in present day human civilization. The vision and farsightedness of petroleum refining and chemical process engineering are today surpassing vast scientific frontiers and will open up new windows of scientific instinct and innovation in decades to come.

13.3   WHAT DO YOU MEAN BY INDUSTRIAL FIXED BED CATALYTIC REACTORS?

The heart of petroleum refining today lies in the operation of Fluidized Catalytic Cracking Unit and the numerous fixed bed industrial reactors. Chemical process engineering and petroleum engineering science are today in the path of newer scientific regeneration and deep visionary rejuvenation. Mankind's immense scientific prowess, the scientific needs of the human society, and the futuristic vision of energy sustainability and energy security will lead a long and visionary way in the true realization and true emancipation of petroleum engineering science. The vision and the challenge of chemical reaction engineering today are surpassing scientific frontiers. Immense scientific and academic rigor in the field of reaction kinetics is the utmost necessities of human scientific endeavor in petroleum refining today.

A fixed bed reactor is a cylindrical tube filled with catalyst pellets with reactants flowing through the bed and being converted into products. The catalyst may have multiple configuration including one large bed, several horizontal beds, several parallel packed tubes, multiple beds in their own shell.

13.3.1   WHAT DO YOU MEAN BY FLUIDIZED BED REACTORS?

A fluidized bed reactor is a type of reactor device that can be used to carry out a variety of multiphase reactions. In this type of reactor, a fluid (gas or liquid) is passed through a solid granular material (usually a catalyst possibly shaped as tiny spheres) at high enough velocities to suspend the solid and cause it to behave as though it were a fluid. The process, known as fluidization, imparts many important advantages to the fluidized bed reactor. As a result, the fluidized bed reactor is now used in diverse applications. Scientific vision, deep scientific doctrine, and the need for scientific challenges are today leading a long and visionary way in the true emancipation of both chemical process technology and petroleum engineering. In this well-researched chapter, the author deeply comprehends and lucidly explains the need of fluidized bed reactors in petroleum refining. The efficiency and the scientific vision of fluidized bed reactor operation need to be re-envisioned and deeply restructured as science and engineering steps into a new era.

13.4 SCIENTIFIC VISION AND DEEP SCIENTIFIC COGNIZANCE IN THE FIELD OF PETROLEUM REFINING

Scientific vision and deep scientific cognizance are the hallmarks of research endeavor in present day human civilization. Loss of fossil fuel resources and the veritable scientific barriers of petroleum engineering science are urging the scientific fraternity to gear toward newer scientific innovation and deep scientific revelation. Energy and environmental sustainability should be the cornerstones of scientific endeavor in energy engineering today. Petroleum refining technology needs to be re-envisioned and restructured as science and engineering moves forward toward a newer visionary paradigm of science and engineering. Petroleum refining, petroleum engineering science, and chemical process engineering today are huge colossus with a definite vision and deep purpose of its own. This vision of energy security needs to be enhanced with more concentration toward application and purpose. Human civilization and scientific research pursuit today stand in the midst of deep vision and scientific forbearance. Technology and engineering science today stand in the midst of deep crisis and unending scientific catastrophe. The devious nature of global climate change, the immense issue of fossil fuel depletion, and the frequent environmental catastrophes have ushered in a new era in scientific research paradigm and the world of challenges in technology applications. A deep scientific understanding of energy sustainability would be incomplete if one does not delineate the strides in petroleum refining today. Scientific vision in the field of petroleum engineering science is today ushering in a new era in energy security and sustainable development.

13.5 ENERGY SUSTAINABILITY AND THE IMPORTANCE OF PETROLEUM ENGINEERING SCIENCE

Energy sustainability, environmental engineering science, and the vast domain of petroleum engineering paradigm are today in the midst of deep scientific hope and scientific revelation. The successful implementation of energy and environmental sustainability throughout countries around the globe are the utmost need of the hour. Human civilization's immense scientific grit and determination, the technological paradigm, and the futuristic vision of sustainable development will lead a long and visionary way in the true emancipation of energy engineering and petroleum engineering today. The challenges and the wide vision of petroleum engineering and chemical process engineering are groundbreaking and surpassing visionary scientific frontiers. Energy sustainability and environmental sustainability are today the utmost need of the hour in the path toward scientific emancipation. The aim and the mission of this chapter are targeted toward greater scientific discernment toward energy security and deep understanding of energy engineering and petroleum engineering. Petroleum refining and renewable energy technology today stand in the midst of deep scientific awakening and unending futuristic vision. Renewable energy such as solar energy, nuclear energy, and biomass energy are the only alternatives to the greater realization of fossil fuel energy today. The challenge and the vision of renewable energy technology and energy sustainability will today lead a long and visionary way in the true realization of energy engineering and petroleum refining. The primordial facet of this chapter is the greater scientific understanding of the operation of fixed bed reactors and fluidized bed reactors. These reactors are the heart of petroleum refining technology and chemical process engineering. A deep scientific vision and a thorough scientific profundity and discernment are the needs of the present day human civilization in the path toward renewable energy technology. Energy security and environmental sustainability are the pivotal points toward the path of human civilization's progress today. Technology has few answers to the immense-scientific vision of petroleum refining and petroleum engineering science. Environmental sustainability stands as a major scientific imperative toward the furtherance of science and engineering and human progress today.

13.6 FIXED BED CATALYTIC CRACKING AND THE SCIENTIFIC DOCTRINE BEHIND IT

Scientific doctrine and deep scientific cognizance in fixed bed catalytic reactor technology today need to be re-envisioned and re-structured as technology and engineering science moves forward. Human scientific paradigm, the veritable needs of human society, and the needs for energy engineering to human civilization are all the pallbearers toward a newer eon in global research and development initiatives. Scientific doctrine, scientific profundity, and vast scientific vision are the necessities of human scientific endeavor today. The vast challenges and the definite vision of petroleum refining and petroleum engineering science need to be envisioned and readdressed with the furtherance of science and engineering. Fixed bed catalytic cracking is a visionary area of scientific endeavor in petroleum refining. Human scientific regeneration and human scientific rejuvenation in petroleum engineering and energy engineering are witnessing immense scientific challenges. Energy security is a major challenge to human civilization today, thus the need for greater introspection into petroleum engineering. Sustainability vision as regards energy and environment is the utmost needs of human society. This area of scientific endeavor needs to be envisioned and readdressed with the passage of scientific history and time.

13.7 SIGNIFICANT SCIENTIFIC ENDEAVOR IN THE FIELD OF INDUSTRIAL FIXED BED CATALYTIC REACTORS

The operation of industrial fixed bed catalytic reactors today stands in the midst of vision and fortitude. Both fixed bed catalytic cracking and fluidized bed catalytic cracking are the scientific needs and deep scientific cognizance of petroleum refining in present day human civilization. Technology and engineering science of petroleum refining are highly challenged today. The author deeply comprehends the vast scientific ingenuity and the scientific provenance in the operation of industrial fixed bed as well as fluidized bed reactor. The scientific and technological acuity in fixed bed catalytic reactors are today ushering in a new era of cross-boundary research. In this section, the author pointedly focuses on the major achievements and the scientific ardor in petroleum engineering applications in present day human civilization. Sustainability today stands in the midst of deep scientific comprehension and ingenuity. The technological challenges need to be redrafted as science and engineering of petroleum refining and technology surges forward.

Endrigo et al.[1] with deep and cogent insight discussed fixed bed reactors. This paper presents the vast phenomena occurring in fixed bed reactors ranging from the small scale of single pellet, where reaction and diffusion are highly competitive, to the macroscale of whole apparatus, where dispersion and heat transfer play a vital role; the most important models used in describing the behavior of fixed bed reactors, and the dependence of most relevant parameter from the geometrical characteristics of the reactor and the vast physical properties of the reacting gas are also deeply investigated.[1] Technological validation and deep scientific objectives are the pillars of this well-researched chapter. Fixed bed reactors are ushering in a newer era in the field of petroleum engineering and chemical process engineering. Catalytic cracking stands today in the midst of deep emancipation and vast vision. Fixed bed reactor operation is of immense importance in the path toward true emancipation of petroleum refining and other chemical engineering operations.[1] Packed bed of catalyst particles is the most widely used reactor type for gas phase reactants in the production of large-scale basic chemicals and intermediates. Technology and engineering science are today ushering in a new era in the field of chemical technology and petroleum engineering science. Fixed bed reactors have also been extremely used in recent years to treat harmful and toxic substances. The challenge and the vision of fixed bed operation in chemical technology, petroleum engineering science, and energy engineering are today surpassing vast and versatile frontiers. The deep analysis of these reactors spans from the microscale, with its pellet and pore structure where the phenomena of reaction and diffusion occur, to the nanoscale, with its intricate geometry and the characteristics of reactor bed where the phenomena of heat and mass transfer convection, dispersion, and transfer take place. The challenge today lies in the hands of scientists and engineers as science and engineering surges forward.[1] Technology of fixed bed reactors has rapidly developed in the last decade, yet many concepts still remain answered. Human scientific vision, the vast technological profundity, and the subtleties of research pursuit will go a long and visionary way in the true realization of chemical reaction engineering and petroleum refining. The performance of a shaped catalyst (tablet, extruded, sphere) depends on various factors such as the chemical composition of the active compounds, promoters, inhibitors, the supported crystallite size and structure distribution, and the synergistic influence of the support.[1] The influence of transport processes on the activity and selectivity of single pellets is one of the vital information in development of new catalysts and is a vast scientific imperative for the rational design of a reactor. The authors in this widely researched chapter depict with deep insight the contribution of diffusion coefficients, the single pellet model, heat transfer parameters, and a deep deliberation on the intrinsic form of kinetics of rate expression. Science and technology of fixed bed catalytic reactor operation and fluidized bed reactor operation are the veritable scientific requisites of petroleum refining and chemical engineering operations today. This challenge is widely depicted with scientific grit, scientific might, and scientific forbearance in this entire chapter.[1]

Gupta et al.[2] discussed with immense lucidity design parameters for fixed bed reactors of activated carbon developed from fertilizer wastes for the removal of some heavy metal ions. Fixed bed reactors are an important and visionary procedure in the furtherance of diverse areas of engineering science today. The vast challenge, the greatness of science, and the technological profundity will lead a long and effective way in the true application of fixed bed reactors in environmental engineering science and chemical process technology. Activated carbon developed from fertilizer waste has been widely used for the removal of Hg^{2+}, Cr^{6+}, Pb^{2+}, and Cu^{2+}.[2] Deep mass transfer kinetic concepts have been successfully applied for the determination of various parameters necessary for designing a fixed bed absorber. Technology and engineering science of fixed bed reactors are surpassing vast and versatile visionary frontiers.[2] Mass transfer kinetic approach has been successfully applied for the determination of various parameters necessary for designing a fixed bed absorber. The challenge and the vision of chemical technology and applied chemistry of fixed bed absorber are deeply delineated with insight and lucidity in this visionary chapter.[2] Parameters selected are the length of the primary adsorption zone, total time involved for the establishment of primary adsorption zone, mass flow rate to the absorber, time for primary adsorption zone to move down its length, amount of adsorbate adsorbed in primary adsorption zone from breakpoint to exhaustion, fractional capacity, time of initial formation of primary adsorption zone, and percent saturation of column at the breakpoint.[2] Deep scientific prowess and intellect, the challenges and the vision of applied chemistry, and the vast futuristic vision of chemical technology will all lead a long and visionary way in the true realization of fixed bed and fluidized bed reactors. Science and engineering are huge colossus with a vast and versatile vision of its own. Activated carbon adsorption systems, widely used in chemical process industries for last several decades, are now playing a pivotal role in the cleaning up of plant effluents and municipal wastewater. Adsorption isotherms have traditionally been used for the preliminary screening before running more complicated and expensive tests. Technology of unit operations of chemical engineering such as adsorption and absorption are highly advanced today. These adsorption isotherms cannot give accurate scale up data for carbon systems because (1) adsorption in carbon columns is not at equilibrium, (2) granular carbon rarely becomes totally exhausted in commercial processes, (3) the effects of vast carbon recycling systems remain unknown, and lastly (4) the isotherms cannot predict chemical and biological changes.[2] Due to these scientific reasons, the practical applicability of the product in column operations has been extensively investigated in this chapter.[2] Regeneration of the reactor and the vast world of cost estimation of this procedure are also delineated in this chapter.

Collot et al.[3] discussed with lucid insight copyrolysis and cogasification of coal and biomass in bench scale fixed bed and fluidized bed reactors. Today's scientific world and deep scientific genre are ushering in a new era in the field of copyrolysis and cogasification. Today's scientific paradigm's immense prowess, the futuristic vision of fossil fuel energy technology, and the holistic world of energy technology will go a long and visionary way in the true realization of energy sustainability globally. In this experiment, the authors used two reactors; a fixed bed reactor providing intimate contact between neighboring fuel particles and a fluidized bed reactor designed to provide near total segregation of sample particles.[3] The scientific sojourn, the technological forbearance, and the necessities of human civilization are the hallmarks toward a newer era in the field of energy engineering and chemical process technology. New power generation technologies involving combined cycle gasification provide a number of opportunities and advantages over coal fired plants. Here comes the scientific necessity of better and improved energy technology. In this chapter, the authors rigorously discuss the scientific success, the scientific ingenuity, and the immense profundity behind gasification and pyrolysis technology.[3]

Elnashaie et al.[4] in an elaborate and detailed chapter discussed modeling, simulation, and optimization of industrial fixed bed reactors. These industrial fixed bed reactors are extremely important in petrochemical and petroleum refining industries. For long decades in the 20th century, chemical reaction kinetics and chemical reaction engineering are witnessing immense scientific upsurge. Fixed bed reactors are applied in diverse areas of chemical technology and petroleum engineering. Modeling, simulation, and optimization of a fixed bed as well as a fluidized bed reactor are an important scientific imperative toward the furtherance of science and technology.[4] Globally, the world of petroleum engineering science is faced with an immense crisis of depletion of fossil fuel resources. Here comes the need of application of mathematical techniques in design of chemical and petroleum engineering systems. Modeling of catalytic processes involves combined knowledge from many disciplines of applied chemistry and chemical engineering.[4] It involves the following pivotal avenues:

- Surface phenomena responsible for the catalytic activity
- Modeling of the intrinsic catalytic reaction rates
- Thermodynamic equilibrium of the reaction mixture for reversible reactions
- External mass and heat transfer resistances
- Intraparticle mass diffusion and heat conduction for porous catalyst pellets[4]
- Pressure drop across the bed
- Heat evolution and heat absorption
- Integration of all the above steps into the overall reactor model[4]

Science and engineering of catalytic cracking are entering into a new phase of scientific vision and deep scientific forbearance. Catalytic cracking is the heart and backbone of petroleum refining. In this chapter, the author pointedly focuses on the scientific success and deep vision behind modeling, simulation, and optimization of fixed bed catalytic reactors. This chapter poignantly depicts the development of reliable mathematical models for industrial fixed bed catalytic reactors from the very beginning of surface phenomena to the final stage of putting the verified model into a user-friendly software package.[4] This entire chapter delineates the system theory and classification of systems, a brief description of chemisorptions and catalysis, and a systematic study of the intrinsic chemical kinetics of gas–solid catalytic reactions. This chapter also delineates the practical relevance of bifurcation behavior and instability problems in chemical reacting systems. The later portion of this chapter deeply comprehends the single particle problem, the main building block of the overall reactor model. The emphasis in this book is lucidly upon the development of rigorous heterogeneous models for petrochemical reactions where all the unit processes are scientifically well defined.[4] The vast concept of modeling and simulation of a chemical engineering and petroleum engineering system are dealt with immense lucidity and cogent insight. In this entire chapter, the entire definition of mathematical modeling and simulation was vastly pragmatic and serves a specific purpose. In the definition, the authors adopted is that mathematical modeling will involve and encompass the process of building the mathematical model itself, while simulation involves simulating the industrial (or experimental) unit using the developed model. Human scientific prowess, the vast technological profundity, and the vast energy scenario are the torchbearers toward a newer era in the field of petroleum engineering and chemical process technology.[4] Mathematical modeling of diffusion and reaction in petrochemical and petroleum refining systems is an excellent and robust approach for design, research, and development. Human scientific genre and the needs of human scientific stature are the parameters toward a newer visionary domain in the field of modeling and simulation of petroleum engineering systems. Science and engineering are immensely enigmatic and visionary in present day human civilization. Mathematical tools and techniques are the backbones of design of petroleum engineering and chemical engineering systems such as fixed bed reactors and fluidized bed reactors. Mathematical modeling leads to a more rational approach for the design of these robust systems in addition to elucidating many important phenomena associated with the coupling of diffusion and reaction. The authors poignantly studies and deeply comprehends rigorous highly sophisticated mathematical models of varying degrees of complexity which are being used in industrial design and the vast world of academic research.[4] An important point to be noticed with regard to state of the art in these fields is that steady-state modeling is more advanced and robust due to additional complexities associated with unsteady state behavior and the additional physicochemical information which are highly necessary.[4] Modeling of industrial fixed beds is a visionary enigma in the path of science and engineering emancipation. Applied mathematics and mathematical modeling are the primordial scientific issues in the path toward scientific rejuvenation and deep scientific vision. Design of a fixed bed or a fluidized bed reactor is just two examples of the vast domain of application of applied mathematics in design of petroleum engineering and chemical engineering systems. Human scientific ingenuity and deep scientific surges are the utmost need of the hour as human civilization faces one of the greatest and most intricate issues of depletion of fossil fuel resources and global climate change. The last part of the 20th century witnessed the tremendous development of applied mathematics, modeling, simulation, and optimization of engineering systems. Human ingenuity is at its level best. Computerized design packages have been developed and computer control of process units and whole plants has been used widely.[4] Technological profundity and vast scientific vision are assuming immense importance in this decade of 21st century as chemical process engineering and petroleum engineering moves with visionary strides. This has been coupled with a considerable increase in the productivity of industrial units and plants with a drastic reduction of manpower and tight control of the product quality. Science in this first half of 21st century is witnessing immense scientific travails and hurdles. There were frequent advances in the areas of modeling and simulation in the present century. Despite these advances, there is a considerable gap between human academic rigor and industrial applications regarding the use of rigorous models for the design, simulation, and optimization of catalytic reactors.[4] In this chapter, the author deeply comprehends the vast scientific success, the technological ingenuity, and the world of challenges in modeling and simulation of chemical and petroleum engineering systems.[4]

Freund et al.[5] delineated with immense insight and lucidity numerical simulations of single phase reacting flows in randomly packed fixed bed reactors and its experimental validation. Multiphase flow is today a burgeoning area of scientific research pursuit. Chemical reaction engineering, mass, and heat transfer are the areas of scientific endeavor which needs to be envisioned and restructured as chemical process engineering and petroleum engineering moves from one scientific paradigm over another. Randomly packed fixed bed reactors are widely used in the chemical process industries. Their integrated design is usually based on pseudohomogeneous model equations with averaged semiempirical parameters. However, their detailed design concept fails for low tube-to-particle diameter ratios where localized phenomenon dominates. The complete three-dimensional structure of the packing has therefore to be considered in order to resolve the local inhomogeneties. Scientific forbearance, scientific vision, and deep scientific comprehension are the utmost necessities of scientific genre and scientific pursuit in present day human civilization.[5] This chapter unfolds and unravels the intricacies and hurdles of science and engineering of the multiphase flow areas in a packed fixed bed reactor. In the chemical industry, packed beds can be found in diverse applications, being used as reaction, separation, and purification units. Despite interesting developments in this area of scientific endeavor which are the applications of structured packings in recent years, the randomly packed bed is still an enigma to the scientists and engineers in recent decades.[5] The vision of science and technology, the vast challenges in multiphase flow, and the futuristic vision of chemical process engineering will all lead a long and visionary way in the true emancipation of engineering science and scientific validation today. Due to their outstanding importance and significance, the mathematical description and modeling of randomly packed catalytic tubular fixed bed reactors are of immense and intricate concern of scientists and engineers throughout the world. In the case of aspect ratios lower than about 10, the plug flow assumption fails due to nonuniformities of the bed structure which have to be considered. Various rigorous scientific models, usually based on homogenization approaches and semiempirical correlations, were derived.[5] By extending the classical design concept with mean radial porosity and corresponding mean radial velocity profiles, the strong influence of the structural properties of the bed on the heat and mass transfer are deeply taken into account.[5] Scientific validation, deep scientific cognizance, and the world of scientific revelation are the torchbearers toward a newer era in the domain of multiphase flow, chemical reaction engineering, and the region of reactor design. New numerical methods and the increase of computational power allow engineers and mathematicians to simulate in detail single phase reacting flows in such reactors, exclusively based on material properties and the three-dimensional description of the geometry, thus without the use of semiempirical data.[5] The successive simulation steps (packing generation, fluid flow, and species calculation) and their validation with experimental data are deeply discussed with cogent insight in this well-researched chapter.[5]

Catalytic fixed bed reactors are the most important type of reactor for the synthesis of large-scale basic chemicals and the vast intermediates. In these reactors, the reaction takes place in the form of a heterogeneously catalyzed gas reaction on the surface of catalysts that are arranged as a so-called fixed bed in the reactor. Technological profundity, scientific regeneration, and deep scientific discernment are the important parameters toward a newer era in the field of chemical reaction engineering and chemical reaction kinetics. In addition to the synthesis of valuable chemicals, fixed bed reactors have been increasingly in operation in recent years to treat harmful and toxic substances. The challenge and the vision of science and engineering today are surpassing vast and versatile scientific frontiers. Human scientific thoughts, the immense scientific prowess of human civilization, and the vast futuristic vision will all lead a long and visionary way in the true emancipation and true realization of energy engineering and chemical process engineering today. With regard to application and construction, it is immensely convenient to differentiate between fixed bed reactors for adiabatic operation and those for nonadiabatic operation. Technology and engineering science are today advancing at a rapid pace as scientific genre and vast scientific paradigm witness newer regeneration. Catalytic fixed-bed reactors are the equipment of immense necessity in the entire process of petroleum refining. The most common arrangement is the multitubular fixed bed reactor, in which the catalyst is arranged in the tubes, and the heat carrier circulates externally around the tubes. In this entire chapter, the author pointedly focuses on the vast scientific success, the scientific potential, and the utmost necessity of fixed bed reactor in the furtherance of science and engineering of petroleum refining.

Zhang et al.[6] discussed with deep and cogent insight catalytic fast pyrolysis of biomass in a fluidized bed with fresh and spent fluidized catalytic cracking catalysts. The conversion of biomass to bio-oil using fast pyrolysis technology is one of the most promising alternatives to convert biomass into liquid products. However, substituting bio-oil for conventional petroleum fuels directly may be difficult because of the high viscosity, high oxygen content, and strong thermal instability of bio-oil.[6] The focus of this current scientific research pursuit is decreasing the oxygen and polymerization precursor content of the obtained bio-oil to improve its thermal stability and heating value.[6] Environmental protection concerns and the security of energy supplies, especially petroleum resources, have surged the interest in development of alternatives for fossil-based energy carriers and chemicals. Technological profundity and deep scientific cognizance are the cornerstones of research pursuit in the field of biofuels and alternate energy technology today.[6] The challenge and vision of chemical process engineering and biotechnology are today surpassing vast and versatile scientific frontiers.[6] Pyrolysis of biomass is a major research thrust area for the future. Plant biomass is the only current sustainable source that could be converted into liquid fuels. In comparison to petroleum resources, the applications of liquid fuels converted from biomass are a carbon-dioxide-neutral process considering the global carbon balance. Biomass fast pyrolysis can produce considerable liquid fuel, named as bio-oil. Biotechnology and bioengineering are the research thrust areas of present day global research and development initiative. Many studies have focus on the production of bio-oils using fast pyrolysis technology in different pyrolysis reactors and a bio-oil yield of up to 75% was scientifically reported.[6]

Geldart[7] discussed in deep and lucid details types of gas fluidization. The behavior of solids fluidized by gases falls into four clearly recognizable groups, characterized by density difference and mean particle sizes. The most easily recognizable features of the groups are powders in group A exhibit dense phase expansion after minimum fluidization and prior to the commencement of bubbling; those in group B bubble at the minimum fluidization velocity; those in C are difficult to fluidize at all; and those in group D can form stable spouted beds.[7] It is impractical for most research engineers and scientists, particularly those working on a reasonably large scale, to test a wide variety of powders, as this is only one of the variables to be dealt upon.[7] Technology of powder science and the deep scientific cognizance behind it are today revolutionizing and revamping the entire world of material science and chemical process engineering today.[7] Human scientific vision, scientific fortitude, and the vast world of technological validation are veritably changing the face of scientific research pursuit in powder technology today. In this chapter, an attempt has been made to group together these powders having broadly similar properties when fluidized by a gas, so that generalizations concerning powders within a group can be envisioned with deep confidence. Although few of the characteristics of fluidized solids are common to all groups, many are not and intergroup predictions should therefore be avoided or made only with considerable caution. The targets and the vision of science and engineering are today ushering in a new era in the field of powder technology, fluidization engineering, and material science. A simple criterion of classification which differentiates between the two largest groups of powders is presented in deep and intricate details.[7] Human scientific forbearance, scientific aura, and the deep scientific cognizance are slowly opening up new knowledge dimensions in the field of fluidization engineering and material science today. This chapter vastly opens up new areas of revolutionary science and new avenues of scientific instinct in decades to come.[7]

Tsuji et al.[8] discussed with deep and lucid details discrete particle simulation of dimensional fluidized bed. Technological and scientific validation, the vast scientific innovation and instinct, and the futuristic vision will all lead a long and visionary way in the true emancipation and the true realization of fluidization engineering and material science today.[8] Numerical simulation, in which the motion of individual particles was calculated, was performed of a two-dimensional gas-fluidized bed. Contact forces between particles are modeled by Cundall's Distinct Element Method which expresses the forces with the use of spring, dash pot and friction slider.[8]

13.8   SIGNIFICANT SCIENTIFIC ENDEAVOR IN THE FIELD OF FLUIDIZED BED CATALYTIC REACTORS

Engineering science and technological profundity in the field of fluidized bed catalytic cracking unit are witnessing immense scientific challenges and deep scientific vision. Fluidized bed catalytic cracking unit is veritably a part and parcel of petroleum refinery. The challenge and the vision of petroleum engineering science need to be re-envisioned and readdressed as science and engineering surges forward. Human scientific research pursuit today stands in the midst of deep introspection and deep scientific discernment. Scientific vision, the vast technological ingenuity, and the futuristic vision of scientific evolution will all lead a long and visionary way in the true realization and true emancipation of mathematical modeling of petroleum engineering and chemical engineering systems.

Modeling, simulation, and optimization of a fluidized bed reactor are the utmost needs and the immediate vision of petroleum engineering science today. Today, science and technology are huge colossus with a vast vision and a pillar of forbearance. In this chapter, the author pointedly focuses on the immense success, the deep vision, and the unending challenges in the field of modeling and simulation of a fluidized bed reactor.

Theologos et al.[9] discussed with deep and cogent insight simulation and design of fluid catalytic-cracking riser-type reactors. Human civilization's immense scientific prowess and the futuristic vision of petroleum engineering and chemical engineering are the forerunners toward a newer era in the field of validation of science and engineering. Two-phase flow, heat transfer, and reaction in fluid catalytic-cracking riser-type reactors are studied using a three-dimensional mathematical model.[9] Scientific discernment and deep scientific understanding are the needs of the science of modeling today. This study was carried out based on the model of Theologos and Markatos, which incorporates a detailed 10-lump reaction kinetics scheme and accounts for gradual feedstock vaporization inside the reactor. Predictions obtained using the new model are effectively compared against the industrial reactor operating data.[9] Technological and scientific validation of modeling and then simulation of fluidized bed reactor needs to be re-envisioned and re-envisaged as petroleum engineering science and chemical process engineering surges forward.[9] In this chapter, the author deeply discussed a three-dimensional mathematical model and the challenges, the scientific intricacies, and the deep scientific profundity behind the application of mathematical model toward the furtherance of science and engineering. A design study was also carried out to illustrate that the model developed is capable of predicting feed-injector geometry effects on overall reactor performance.[9] Fluid catalytic cracking (FCC) process is an industrial process that converts heavy hydrocarbons to lower molecular weight products and is completed in short contact time riser reactors. The vast vision of science and engineering of petroleum engineering science, the technological validation, and the needs of global energy security are the veritable forerunners toward a newer scientific wonder and a newer scientific vision. In a riser reactor, the catalyst is pneumatically conveyed by hydrocarbon vapors from the bottom to the top of a vertical lift line. During this conveying process, the catalytic cracking reactions are completed through the efficient contact of the catalyst with the hydrocarbons. In this chapter, the authors deeply discussed with cogent insight the entire physical and mathematical model. The main backbone of this chapter is the three-dimensional hydrodynamic model which has been developed to predict two-phase flow, heat transfer, and reaction in fluidized catalytic cracking (FCC), riser type reactors.[9] Technology and engineering science of petroleum refining technology are today ushering in a new era in scientific marvels, scientific ingenuity, and scientific achievements. This chapter is a veritable eye opener toward the future success of science and engineering of petroleum refining.[9]

Bai et al.[10] discussed with deep and lucid insight a simulation model of FCC catalyst regeneration in a riser regenerator.[10] A comprehensive one-dimensional model was developed in this work in detail for the simulation of a riser regenerator.[10] The model effectively incorporates the hydrodynamics characteristics of fluidized bed risers and the detailed reaction kinetics of FCC catalyst regeneration and can be vastly used to predict the axial profiles of temperature, oxygen concentration, and coke content of the catalyst in the riser. Science of chemical reaction engineering and chemical kinetics is veritably moving toward scientific ingenuity and regeneration. In this chapter, a model for riser regenerator is effectively discussed with the sole purpose of furtherance of science and technology.[10] This is another example of modeling and simulation of a riser regenerator. Human scientific ingenuity is in the path of newer scientific revelation. The validity of the present model was confirmed by experimental data from a pilot plant demonstration unit. The simulation results have provided the necessary information for the optimal design and operation of a FCC riser regenerator unit, presently in operation in China.[10] Technological validation, the challenge, and the vision of petroleum engineering science will lead a long and visionary way in the true realization of energy sustainability globally. Human scientific forbearance, the deep scientific cognizance, and the futuristic vision of scientific validation are the pallbearers toward a newer scientific understanding in the field of energy security today. Technology of modeling, simulation, and optimization are highly advanced today and surpassing cross-boundary research. This challenge is envisioned in this chapter. In this chapter, the author targeted hydrodynamics and then regeneration kinetics. The model is highly developed and opens up a new chapter in the field of scientific vision in applied mathematics, petroleum engineering science, and chemical process engineering.[10]

Dasila et al.[11] discussed with deep and cogent insight simulation of an industrial FCC riser reactor using a novel 10-lump kinetic model and some parametric sensitivity studies. A FCC unit has been simulated by integrating FCC riser reactor and regenerator models. This simulation uses a new 10-lump riser reactor kinetic model developed in-house.[11] Scientific vision, deep scientific cognizance, and scientific profundity are the challenges of modeling and simulation of FCC riser reactor today. This technology is vastly re-envisioned and revamped in every research and development initiatives in modeling and simulation. The lumping scheme and reactions are based on more detailed description of the feed in terms of paraffins, napthenes, and aromatics in both light and heavy fractions.[11] An artificial neural network model, also developed in-house, relates routinely measured properties such as specific gravity, ASTM (American Society for Testing and Materials) temperatures, etc. to the detailed feed composition needed for the kinetic model development.[11] Simulation results using the present model, when compared with results from a conventional five-lump model, clearly re-envisioned the improvement in prediction because of detailed feed description calculated from artificial neural network models.[11] FCC is an important secondary process, converting low-priced heavy feed stocks like heavy oil from either the refinery crude unit or vacuum unit and heavy fractions from other conversion units (cooker gas oil and hydrocracker fractionator bottoms, etc.) into lighter, more valuable hydrocarbons such as liquefied petroleum gas and gasoline thus effectively increases the profitability of the entire refinery. Human scientific ingenuity, technological validation, and deep scientific doctrine of petroleum engineering science are the hallmarks of research pursuit in fluidized catalytic cracking unit today. This chapter is a watershed text in the field of modeling and simulation of FCC unit with a clear scientific understanding and scientific vision.[11]

Du et al.[12] delineated with immense lucidity an integrated methodology for the modeling of FCC riser reactor.[12] Scientific provenance, scientific revelation, and scientific vision are the necessities of research pursuit in design of FCC unit today. Modeling description of the riser regenerator is a highly interesting problem in the development of a FCC process.[12] One of the most challenging problems in the modeling of FCC riser reactors is that sophisticated flow-reaction models with high accuracy need long computational time, while simple flow-reaction models give rise to results with fast computation but low accuracy. These drawbacks need to be revamped and readdressed as science surges forward toward a newer visionary era. Scientific vision, the deep scientific introspection, and the challenges behind scientific research pursuit will all lead a long and visionary way in the true realization and true emancipation of energy sustainability in present day human civilization. Design of a petroleum refining unit such as FCC unit is an utmost need of human scientific endeavor today.[12] Petroleum engineering science and chemical process engineering are today ushering in a new visionary era in human scientific research pursuit. An integrated methodology for the modeling of FCC reactor will surely open up new dimensions of research pursuit in modeling and simulation domain in decades to come.[12]

13.9 MODELING, SIMULATION, AND OPTIMIZATION OF FIXED BED CATALYTIC REACTORS

Modeling, simulation, and optimization of fixed bed catalytic reactors are the cornerstones of scientific endeavor in petroleum engineering science today. Human scientific forays, the vast scientific vision, and the scientific forbearance are the torchbearers toward a newer understanding and a newer visionary era in mathematical modeling and simulation. Technology and engineering science of modeling, simulation, and optimization are today surpassing vast and versatile scientific frontiers. In this chapter, the author poignantly depicts the operation of industrial fixed bed catalytic reactors with the sole vision of furtherance of science and engineering of petroleum refining and petroleum technology. Modeling and simulation of fixed bed catalytic reactors and fluidized bed catalytic reactors stand today as a major scientific imperative toward greater realization of petroleum refining, energy technology, and chemical process engineering. Optimization of fixed bed catalytic reactors is the other visionary avenues of research pursuit today. This is a latent and dormant area of scientific endeavor in the last two decades. This chapter deeply comprehends the scientific ingenuity and the scientific success behind mathematical applications in design of a fixed bed catalytic reactor and a fluidized bed catalytic reactor. Technology and engineering science needs to be unraveled to the fullest extent as mankind enters into a newer scientific era and technological genre. In this chapter, the author repeatedly stresses upon the drawbacks, challenges, and the scopes behind the operation of industrial fixed bed reactors with a greater emphasis on the design, modeling, and simulation of the reactor. Human scientific sojourn and human scientific stance are the imperatives toward a greater understanding in the field of petroleum refining and the whole scientific domain of chemical reaction engineering. Technology needs to be redrafted and restructured as oil and gas industry today surges forward toward a newer vision and a newer innovative era. The world of scientific challenges today stands in the midst of deep peril and unending scientific barriers. Modeling, simulation, and optimization today stand as major scientific objectives and deep mission toward design of a fluidized and fixed bed reactor. Modeling and simulation of fixed bed reactor today needs to be revamped and read-dressed as petroleum refining and chemical process engineering surges forward. Pinheiro et al.[13] depicted poignantly and with cogent insight FCC process modeling, simulation, and control. Technological and scientific validations are the utmost needs of scientific endeavor today. Scientific articulation, vast scientific acuity and the need of energy security and energy sustainability are urging the scientific domain to take vast strides of vision and forbearance.[13] The authors deeply focused on the FCC process and reviews recent developments in its modeling, monitoring, control, and optimization. This challenging research pursuit exhibits complex behavior, requiring detailed models to express the nonlinear effects and the rigorous interactions between input and control variables that are observed in industrial practice.[13] The authors deeply discussed the different mathematical frameworks that have been applied in the modeling, simulation, control, and optimization of this key downstream unit. Human scientific ingenuity, the vast importance of energy sustainability and the path toward energy security are the forerunners toward a newer visionary era in energy engineering and petroleum engineering.[13]

13.10 THE VISION AND THE CHALLENGE IN DESIGN OF FIXED BED CATALYTIC REACTORS

The deep scientific vision and the intricate challenges in the design of fixed bed catalytic reactors today needs to be readdressed and re-envisaged as human research pursuit enters into a new era. Human civilization and the deep scientific research pursuit in petroleum refining, fixed bed reactors, and fluidized bed reactors are today ushering in a new era in engineering science and technology. Fixed bed catalytic cracking stands as a major parameter toward the worldwide emancipation of petroleum refining and fossil fuel energy science. Depletion of fossil fuel resources, devastating global climate change, and the frequent environmental catastrophes are urging the human scientific domain to move toward newer innovations, newer technologies, and a greater scientific understanding. Today is the world of nuclear science, space science, and the world of immense strides in computer science. Human civilization and human scientific research pursuit today stand in the midst of deep catastrophes and immense scientific introspection. Technology is highly challenged and is moving toward newer visionary boundaries. Petroleum refining stands as a major scientific imperative toward human progress today.

13.11   DRAWBACKS AND CHALLENGES IN INDUSTRIAL FIXED BED REACTOR OPERATION

Technology and engineering science along with applied mathematics and computer science today are ushering in a new era in the field of petroleum refining and chemical engineering. Industrial fixed reactor and fluidized bed reactor are areas of scientific research endeavor which has extensive chemical engineering applications. The drawbacks and challenges of engineering science of industrial fixed bed reactors are immense and pathbreaking. Human scientific regeneration and deep scientific ingenuity are the utmost need of scientific research pursuit today. Catalyst regeneration in fluidized bed reactors and industrial fixed bed reactors are of immense necessity in the path of engineering science and petroleum refining today. The burgeoning world of modeling and simulation are today applied in the design of a fixed bed as well as fluidized bed reactor. This is the major challenge of design of petroleum refining unit and today is replete with vast scientific intricacies. Depletion of fossil fuel resources is a serious scientific issue facing petroleum engineering science today. This challenge and the underlying vision are of immense importance in the furtherance of petroleum engineering and chemical process technology today. Human scientific prowess and the vast scientific cognizance need to be revamped as petroleum engineering catastrophes and the fossil fuel challenges revamps and readdresses the scientific firmament of human civilization today.

13.12   THE VAST SCOPE OF FIXED BED REACTOR OPERATION

The scope and the vision of fixed bed reactor operation are crossing vast and versatile scientific boundaries. Human scientific stature and scientific profundity today stand in the midst of scientific comprehension and scientific vision. Fixed bed reactor is of utmost need in petroleum refining today. Fluidized bed catalytic cracker and its vast scientific vision are similarly of utmost necessity in the road toward scientific emancipation in petroleum refining and petroleum engineering science. The scope of petroleum refining and energy sustainability is vast and versatile. Human scientific ingenuity, scientific forbearance, and deep scientific profundity are the torchbearers toward a newer era in global domain of energy security and energy sustainability. Chemical reaction engineering and fluidized bed reactor operation are an integral part of petroleum refinery and a visionary avenue of research pursuit in petroleum engineering science today. The vast scope of fixed bed reactor operation and petroleum refinery, human civilization's immense scientific prowess, and the needs for energy sustainability will all lead a long and visionary way in the true realization of human scientific progress today. Fluidized bed reactor operation is another vast domain of energy engineering today. In this chapter, the author depicts with cogent insight the scientific success, the vast scientific potential and the immense scientific profundity behind fixed bed reactor, and fluidized bed reactor operation. Technology and engineering science stand challenged today as human civilization surges forward. This scientific chapter repeatedly pronounces the deep human success, the scientific ingenuity, and the scientific cognizance in the catalytic cracker operation in petroleum refinery.

13.13   SIGNIFICANT SCIENTIFIC ENDEAVOR IN SUSTAINABILITY

Sustainability whether it is economic, social, energy, or environmental should be envisioned and restructured today as human civilization surges forward toward a newer era of vision and innovation. Energy security, ecological balance, and the vision toward sound environmental engineering regulations will all today lead a long and visionary way in the true emancipation of sustainable development. Science and technology need to be reframed today as the growing concerns of environmental protection assume immense importance. Scientific endeavor and the human progress are today in the process of newer rejuvenation and deep regeneration. Sustainability whether it is economic, energy, or environmental is today changing the face of human scientific endeavor. Petroleum engineering and energy engineering are the two opposite sides of the visionary coin of sustainability. Nanoscience and nanotechnology are the other challenging areas of deep scientific vision and deep scientific cognizance today.

Markulev et al.[14] discussed with deep details an economic approach on sustainability. Sustainability is invoked as a desirable objective in a wide range of contexts, yet its meaning is not well defined. Sustainability can imply vastly different policy responses depending on interpretation, particularly in relation to the degree to which environmental and other resources should be consumed over time frame.[14] The vision and the challenge of "sustainability" are today evergrowing as human civilization moves from one scientific paradigm toward another. Technological vision, scientific profundity, and deep scientific cognizance are the pillars of scientific research pursuit in diverse areas of engineering and science today. At its most general level, sustainability refers to the capacity to envision and continue an activity or process indefinitely. It can be effectively related to any number of economic, social, or environmental activities and can have diverse definitions of different areas of sustainability and sustainable development. Human scientific progress and scientific emancipation today entirely depend on successful sustainable development.[14] Economic, social, environmental, and energy sustainability are the pillars of human scientific progress and advancement of human civilization. The most widely cited definition is that of the United Nations World Commission on Environment and Development 1987 (the "Brundtland Commission"). In Australia, the National Strategy for Ecological Sustainable Development defines sustainable development as "using, conserving and enhancing the community's resources so that ecological processes, on which life depends, are maintained, and the total quality of life, now and in the future, can be increased (Commonwealth of Australia 1992).[14] Today, human civilization stands in the crossroads of deep catastrophe and widening scientific introspection. Technology and engineering science are the imperatives of economic growth globally. Technological and scientific validation is in the midst of deep scientific rejuvenation and vast scientific vision. The authors delineated sustainability from an economic perspective.[14] The chapter also delineates wellbeing in a sustainability perspective, intergenerational equity, natural capital, and other forms of capital, how sustainability concepts are applied practically, policy prescriptions in waste management, deep scientific understandings of ecological systems, and the overall holistic understanding of environmental and energy sustainability.[14]

Kates et al.[15] discussed and delineated in detail the meaning of sustainable development and its goals, indicators, values, and practices. A brief history of the concept, along with the interpretive differences and the common ground in definitions, goals, indicators, values, and practice follows.[15] Taken totally, these help explain what is meant by holistic sustainable development. The Brundtland Commission's brief definition of sustainable development "ability to make development sustainable—to ensure that it meets the needs of the present without compromising the ability of future generations to meet their own needs" is surely the standard definition when judged by its widespread use and the frequency of citation.[15] Human economic development, the futuristic vision of sustainability, and the utmost needs of human society will all lead a long way in the true realization of science and engineering today. The authors in this chapter rightly proclaim the success of sustainable development of present day human civilization today. The authors analyzed the United Nations Millennium Declaration, 2000 and thus ushered in a new era in the domain of human scientific progress. In the end of the chapter, the authors pronounce the UN Millennium Development Goals, 2000 and a rigorous definition of sustainability.[15]

Kuhlman[16] redefined sustainability. Sustainability as a policy concept has its origin in the Brundtland Report of 1987. That document envisions the tension between the aspirations of mankind toward a better life on the one hand and the vast limitations imposed by nature on the other. In the course of scientific history and visionary time frame, the concept has been reinterpreted as encompassing three dimensions, namely, social, economic, and environmental. The author deeply and poignantly depicted happiness, well-being and welfare in the perspective of sustainable development. The resources and the future of sustainability as well as the demarcation of weak and strong sustainability are deeply delineated in this chapter.[16]

Science and technology in this century needs to be re-envisioned and revamped as human civilization faces immense environmental and energy catastrophe. Thus, sustainable development is the utmost need of the hour as civilization faces the dual crisis of environmental and energy security. The author in this entire chapter pronounces the immense success of sustainable development and environmental engineering in the path toward emancipation of engineering science and technology.

13.14 FUTURE RECOMMENDATIONS OF STUDIES IN FIXED BED CATALYTIC REACTORS

Fixed bed and fluidized bed catalytic reactors are the utmost needs of petroleum refining and chemical process engineering today. Future recommendations in studies in fixed bed catalytic reactors are targeted toward deep scientific vision and vast scientific challenges. Today, technology and engineering science has have few answers to the immense scientific intricacies in operation of fixed bed as well as fluidized bed reactor operation. Energy engineering, petroleum engineering science, and chemical process engineering are the areas of scientific research pursuit which are replete with immense scientific vision and scientific forbearance. Energy security and energy sustainability are the hallmarks toward a newer visionary era in the field of science and engineering today. Human civilization, human scientific endeavor, and vast human ingenuity today stand in the midst of deep crisis and scientific comprehension. Future recommendations of studies in fixed bed catalytic reactors are today aimed at more effectivity, greater vision and the greater emancipation toward energy security and energy sustainability. Human civilization and human scientific research pursuit today stand in the midst of immense scientific intricacies and unending scientific introspection. Energy crisis and energy sustainability are of immense concern in the path of human civilization today. Economic and scientific development of human civilization today depend on success of sustainability whether it is social, energy, or environmental. Human infrastructural development as well as sustainable development are the parameters of human scientific and economic growth. Catalysis or catalytic cracking are parameters toward a greater emancipation and true realization of petroleum refining today. Modeling, simulation, and optimization of fixed bed reactors and fluidized bed reactors are the necessities of petroleum engineering emancipation in modern society. Human scientific regeneration in the field of petroleum refining and energy technology need to be veritably revisited and thoroughly readdressed with the passage of scientific history and time. Scientific and technological validations are the opposite sides of the visionary coin of scientific research pursuit today. Energy security globally is the utmost need of the hour. Chemical process engineering and petroleum engineering are today ushering in a new era in the field of applied mathematics, modeling, and simulation concepts. Industrial fixed bed catalytic reactors and fluidized bed reactors are today opening up new vistas of innovation and scientific instinct in decades to come. This is an age of immense scientific motivation, scientific profundity, and vision in research pursuit in petroleum engineering and chemical process engineering. The vision and challenge of petroleum engineering applications is today evergrowing and steadfastly ushering in a new era in science and engineering.

13.15   FUTURISTIC VISION AND FUTURE OF SCIENCE AND TECHNOLOGY

Science and technology of petroleum engineering science and chemical process engineering are today groundbreaking and veritably crossing vast and versatile scientific frontiers. Technology has few answers to the scientific intricacies and scientific barriers to the operation of industrial fixed bed reactors and fluidized bed reactors. In this chapter, the author deeply delineates the science of petroleum refining with a deep emancipation toward furtherance of science and engineering. The world today stands between deep scientific peril and visionary scientific discernment. The challenge and the vision of petroleum engineering science need to be deeply readdressed and restructured as regards application of energy sustainability to human society. Today is the world of nuclear science and space technology. Also sustainable development, energy, and environmental sustainability are the important parameters toward human progress and economic development of a nation. Industrial fixed bed catalytic reactors and fluidized bed catalytic reactors are the utmost and visionary needs of petroleum refining in present day human civilization. The challenge stands tremendously bright and evergrowing as human civilization surges forward. Futuristic vision of petroleum engineering, petroleum refining, and energy technology should be targeted toward energy security and sustainable development as regards energy. Science today needs to be re-envisioned and re-envisaged as global water crisis and global energy crisis shatters the deep scientific firmament. Provision of drinking water and electricity are the challenges of human civilization today. Sustainable development and provision of basic human needs such as water and energy are the two opposite sides of the visionary coin. In such a crucial juncture of human history and time, energy sustainability assumes immense importance with the human progress. Environmental sustainability is the major avenue of scientific endeavor in present day human civilization. Thus, this chapter opens up new windows of challenge, scope, and opportunities in the field of sustainability and sustainable development in decades to come.

13.16   CONCLUSION AND FUTURE SCIENTIFIC PERSPECTIVES

Future scientific perspectives in petroleum engineering and chemical process engineering are of immense importance and should be targeted toward holistic sustainable development. Scientific imagination, scientific inspiration, and deep scientific forbearance will all lead a long and visionary way in the true emancipation of sustainability in present day human civilization. Fluidized bed catalytic reactor and fixed bed catalytic reactors are the immense necessities of petroleum refining and chemical process engineering in today's scientific landscape. Human scientific regeneration in the field of petroleum engineering science needs to be restructured with the passage of scientific research endeavor. This chapter widely opens a new window in the field catalytic cracking and operation of fixed bed catalytic reactor. Over the previous decades, fluidized catalytic cracking and fixed bed catalytic cracking created new visionary avenues in both petroleum refining and chemical process engineering. Technology and engineering science of petroleum refining are highly challenged with the ever-growing concerns of fossil fuel depletion. Here comes the utmost need of renewable energy technology. The future of human civilization today lies in the hands of effective renewable energy technologies and a deep scientific understanding of the field of energy engineering. In this chapter, the author repeatedly proclaims the immense scientific potential behind renewable energy technology and the future of fossil fuel technology. Technology has few answers to the scientific travails of energy engineering and the holistic domain of energy security. Mankind's immense scientific prowess, the needs of human society, and the futuristic vision of energy sustainability will surely lead a long and visionary way in the true realization of sustainable development today. Human scientific challenges and the veritable intricacies thus will lead mankind to a newer scientific paradigm in years to come.

KEYWORDS

- fluid
- fixed
- catalytic
- cracking
- vision

REFERENCES

1. Endrigo, P.; Bagatin, R.; Pagani, G. Fixed Bed Reactors. *Catal. Today* **1999**, *52*, 197–221.

2. Gupta, V. K.; Srivastava, S. K.; Mohan, D.; Sharma, S. Design Parameters for Fixed Bed Reactors of Activated Carbon Developed from Fertilizer Waste for the Removal of Some Heavy Metals Ions. *Waste Manage.* **1997**, *17* (8), 517–522.

3. Collot, A.-G.; Zhuo, Y.; Dugwell, D. R.; Kandiyoti, R. Co-Pyrolysis and Co-Gasification of Coal and Biomass in Bench-Scale Fixed-Bed and Fluidized Bed Reactors. *Fuel* **1999**, *78*, 667–679.

4. Elnashaie, S. S. E. H.; Elshishini, S. S. *Modelling, Simulation and Optimization of Industrial Fixed Bed Catalytic Reactors*; Gordon and Breach Science Publishers: USA, 1993.

5. Freund, H.; Zeiser, T.; Huber, F.; Klemm, E.; Brenner, G.; Durst, F.; Emig, G. Numerical Simulations of Single Phase Reacting Flows in Randomly Packed Fixed-Bed Reactors and Experimental Validation. *Chem. Eng. Sci.* **2003**, *58*, 903–910.

6. Zhang, S.; Xiao, R.; Wang, D.; Zhong, Z.; Song, M.; Pan, Q.; He, G. Catalytic Fast Pyrolysis of Biomass in Fluidized Bed with Fresh and Spent Fluidized Catalytic Cracking (FCC) Catalysts. *Energy Fuels* **2009**, *23*, 6199–6206.

7. Geldart, D. Types of Gas Fluidization. *Powder Technol.* **1973**, *7*, 285–292.

8. Tsuji, Y.; Kawaguchi, T.; Tanaka, T. Discrete Particle Simulation of Two-Dimensional Fluidized Bed. *Powder Technol.* **1993**, *77*, 79–87.

9. Theologos, K. N.; Nikou, I. D.; Lygeros, A. I.; Markatos, N. C. Simulation and Design of Fluid Catalytic Cracking Riser-Type Reactors. *AIChE J.* **1997**, *43* (2), 486–494.

10. Bai, D.; Zhu, J.-X.; Jin, Y.; Yu, Z. Simulation of FCC Catalyst Regeneration in a Riser Regenerator. *Chem. Eng. J.* **1998**, *71*, 97–109.

11. Dasila, P. K.; Choudhury, I. R.; Singh, S.; Rajagopal, S.; Chopra, S. J.; Saraf, D. N. Simulation of an Industrial FCC Riser Reactor Using a Novel 10-Lump Kinetic Model and Some Parametric Sensitivity Studies. *Ind. Eng. Chem. Res.* **2014**, *53* (51), 19660–19670.

12. Du, Y. P.; Yang, Q.; Zhao, H.; Yang, C. H. An Integrated Methodology for the Modeling of Fluid Catalytic Cracking (FCC) Riser Reactor. *Appl. Petrochem. Res.* **2014**, *4*, 423–433.

13. Pinheiro, C. I. C.; Fernandes, J. L.; Domingues, L.; Chambel, A. J. S.; Graca, I.; Oliveira, N. M. C.; Cerqueira, H. S.; Ribeiro, F. R. Fluid Catalytic Cracking (FCC) Process Modeling, Simulation, and Control. *Ind. Eng. Chem. Res.* **2011**, *51*, 1–29.

14. Markulev, A.; Long, A. *On Sustainability: An Economic Approach, Productivity Commission Staff Research Note*; Commonwealth of Australia Report, 2013.

15. Kates, R. W.; Parris, T. M.; Leiserowitz, A. A. What is Sustainable Development? Goals, Indicators, Values and Practice. *Environ.: Sci. Policy Sustainable Dev.* **2005**, *47* (3), 8–21.

16. Kuhlman, T.; Farrington, J. What is Sustainability? *Sustainability* **2010**, *2*, 3436–3448.

CHAPTER 14

DIFFUSION AND THERMODIFFUSION IN HYDROCARBON MIXTURES

CECÍLIA I. A. V. SANTOS[1,2*], VALENTINA SHEVTSOVA[2], and
ANA C. F. RIBEIRO[1]

[1]*Department of Chemistry, University of Coimbra, 3004-535 Coimbra, Portugal*

[2]*Microgravity Research Centre, CP-165/62, Université Libre de Bruxelles, 50, Av. F.D. Roosevelt, B-1050 Brussels, Belgium*

Corresponding author. E-mail: cecilia.iav.santos@gmail.com

ABSTRACT

Mass transport by diffusion plays an essential role in chemical engineering processes (solid and liquid extractions, distillation, etc.) but also in chemical reactions and biological systems. Thermodiffusion, also called thermal diffusion or the Ludwig–Soret effect, describes the coupling between a temperature gradient and a resulting mass flux in a multicomponent system. For the determination of diffusion coefficients in liquids, various techniques are usually employed: diaphragm cell, conductometric and optical cells, dynamic light scattering and Taylor dispersion technique. The latter has become a fast and reliable method for the measurement of mutual diffusion coefficients (D_{ik}), with the advantage of relative simplicity of the equipment used. In the case of Soret effect there has been great improvement on experimental techniques, especially modern optical methods, which will be here discussed and the new developments on the optical beam deflection technique are described.

The diffusion and thermodiffusion coefficients attained for some relevant hydrocarbon mixtures with three components are presented and discussed and the results are compared, when available, with literature results obtained using different experimental methods.

14.1 INTRODUCTION

Molecular and thermal diffusions correspond to mixing and separation on the molecular level that are essential for a wide range of physical and chemical processes. Important examples include chemical and electrochemical reactions, sedimentation equilibrium (e.g., in crude oil reservoirs), crystal growth, transport across membranes, cell metabolism, and solidification, among others.[1-4] Molecular diffusion itself is also essential for the interpretation of thermal diffusion (also thermodiffusion, Ludwig–Soret effect or, simply, Soret effect). The effect was first discovered by Ludwig in 1856[5] and later studied by Soret in the late 1900s.[6,7] It describes mass diffusion flows in fluids that are driven by temperature gradients. In response to this temperature gradient, concentration gradients appear in an originally uniform mixture. They then produce isothermal molecular diffusion, which aims at eliminating concentration variations. A steady state is reached when the separating effect of thermal diffusion is balanced by the remixing effect of isothermal diffusion. In binary mixtures, one can distinguish between the positive Soret effect, when the lighter heavier component is driven toward the higher lower temperature region, and the negative Soret effect, when the situation is the opposite.

Molecular and thermodiffusion processes are quantified by mass and thermodiffusion coefficients, D and D_T, respectively. Theoretical models developed in the past[8-12] are able to predict diffusion and thermal diffusion coefficients for certain classes of liquid binary mixtures, including some hydrocarbons with good accuracy (e.g., nonassociating mixtures),[13-15] but they lack to provide information for more complex (multicomponent) systems due to absence of experimental data. Still, the mixtures appearing in nature and industrial applications are essentially multicomponent, presenting a more intricate behavior than pure fluids, due to a complex interplay between heat and mass transfer processes that need to be understood. In a multicomponent system, the diffusive mass transport of a given component is induced not only by its compositional gradient (main or principal diffusion) but also by the compositional gradients of the other components (crossdiffusion) and the temperature gradient. For example, a ternary system has four diffusion coefficients, instead of one as found in a binary mixture. Furthermore, the values of diffusion coefficients in ternary mixtures depend on the order of the components as well as on the frame of reference for which the diffusive fluxes are written, while the binary diffusion coefficient does not depend on the frame of reference. Due to both experimental and mathematical difficulties, diffusion coefficients are available only for a very limited number of multicomponent fluids, mainly in ternary systems, and the results are not always in agreement when measured by different experimental methods.[16–18] Furthermore, although measurements of thermal diffusion binary mixtures are well established,[14,19] this is not the case for higher order mixtures.

Reliable experimental techniques have been developed for the measurement of diffusion and thermodiffusion in binary mixtures, giving rise to consistent experimental data for many of these systems. For the determination of diffusion coefficients in liquids, various techniques are usually employed: diaphragm cell,[20] conductimetric and optical cells,[21,22] dynamic light scattering,[23,24] and Taylor dispersion technique.[25,26] The latter has become a fast and reliable method for the measurement of mutual diffusion coefficients (D_{ik}), with the advantage of relative simplicity of the equipment used. Taylor dispersion technique has been extensively tested for electrolytes and diluted solutions[27,28] and, more recently, for the case of binary and ternary mixtures of organic compounds.[23,29,30]

In the thermodiffusion field, it's possible to find several techniques applied in the experimental front. These embrace from the classical Soret cell,[31] the thermogravitational column,[32] the two-chamber thermodiffusion cell,[33,34] the thermal field-flow fractionation,[35] to optical methods like the microfluidic fluorescence method,[36,37] laser-beam deflection technique,[38,39] thermal diffusion forced Rayleigh scattering,[40–42] as well as the thermal lens technique.[43] Accurate measurements of both molecular and thermal diffusion coefficients for binary mixtures are obtained from optical methods with very good accuracy and reliability since no perturbation is introduced into the diffusive process. Still, many efforts are being directed to study thermodiffusion in multicomponent mixtures.[44,45] The situation with ternary mixtures is more complicated as the sign of the Soret coefficients of the various components could be different, and it destabilizes the system.[46–48] In this respect, orbital laboratories provide an ideal environment for the measurements due to the absence of buoyancy driven convection.

One of the fields where the transport properties play a decisive role is in petroleum reservoirs, since one of the major challenges in optimizing their exploitation is a perfect knowledge of the fluid physics in crude oil reservoirs. Most usual modeling methods are based on pressure–temperature equilibrium diagrams and on gravity segregation of the different components of crude oil. However, improved models which more accurately predict the concentrations of the different components are necessary. The concentration distribution of the different components in hydrocarbon mixtures is mainly driven by phase separation and diffusion. One of the objectives in the oil industry is the prediction, as precisely as possible, of the gas–oil contact in an oil reservoir. In order to achieve this goal, the local composition must be known. One of the reasons of a local compositional variation is molecular segregation in the gravitational field. Aside this important "force," the geothermal gradient (a few degrees per 100 m) may also induce local variations in composition due to the thermodiffusion effect.[24] Moreover, these are complex multicomponent systems that change their shape and composition very slowly over geological time scales and where gravity, diffusion, and chemical reactions also have big impact. So, even if the approach to these systems is normally to overcome through modeling of the transport processes using both macroscopic models[49] and molecular dynamics (MD) simulations,[50,51] experimental data are required to build a realistic model for the fluids. In that sense, the ternary model compounds used by oil industry for numerical modeling of petroleum reservoirs and object of great interest is composed by 1,2,3,4-tetrahydronaphthalene (THN), isobutylbenzene (IBB), and dodecane ($C_{12}H_{26}$), and it includes molecules of different hydrocarbon families (polycyclic, alkane, and aromatic). The different pairs of the components of the mixture are known as the Fontainebleau benchmark systems and were used by the scientific community as binary benchmark and ternary liquids.[13,39,52,53] Later on, this mixture has also been chosen as a model system for the investigation of ternary systems on ground and under microgravity condition. Another mixture that has attracted considerable attention from different areas of research, including oil industry, is the ternary liquid mixture cyclohexane (Chex) + toluene (Tol) + methanol (MeOH), mainly due to the fact that it exhibits a very complex behavior combined with the existence of a miscibility gap.[48,54] This chapter will focus on a comprehensive review of the Fickian diffusion coefficients for binary and ternary mixtures of THN–IBB–$C_{12}H_{26}$ and Chex–Tol–MeOH using Taylor dispersion technique together with results thermodiffusion coefficients, obtained using the optical digital interferometry (ODI) technique among others but also a reflexion on the agreement between transport coefficients obtained with other experimental methods.

14.2 THEORETICAL ASPECTS

14.2.1 CONCEPTS OF DIFFUSION

The gradient of concentration inside a mixture (without convection or migration) produces a flow of matter in opposite direction, which arises from random fluctuations in the positions of molecules in space. This phenomenon, denominated by diffusion, is an irreversible process. In fact, the gradient of chemical potential in the real solution is treated as the true virtual force producing diffusion. However, in ideal solutions, that force can be quantified by the gradient of the concentration at constant temperature. Thus, we may consider the following approaches to describe the isothermal diffusion: the thermodynamics of irreversible processes and Fick's laws.[55]

Mutual diffusion coefficient, D, in a binary system, may be defined in terms of the concentration gradient (without convection or migration) by a phenomenological relationship

$$J = D\frac{\partial c}{\partial x} \tag{14.1}$$

where J represents the flow of matter across a suitable chosen reference plane per area unit and per time unit, in a one-dimensional system, and c is the concentration of solute, in moles per volume unit at the point considered; and eq 14.1 may be used to measure D. The diffusion coefficient may also be measured considering Fick's second law

$$\frac{\partial c}{\partial t} = \frac{\partial}{\partial x}\left(D\frac{\partial c}{\partial x}\right) \tag{14.2}$$

In general, the available methods are categorized into two groups: steady- and unsteady-state methods, according to eqs 14.1 and 14.2. In most of the processes, diffusion is a three-dimensional phenomenon. However, the majority of the experimental methods used analyze the diffusion phenomenon by restricting it to a one-dimensional process, because it is much easier to manage the necessary mathematical treatment in one dimension (being afterward generalized to a three-dimensional space).

The resolution of eq 14.2 for a unidimensional process is simpler if we consider D as a constant. This approximation is applicable only when there are small differences of concentration, which is the case in our experimental method (the Taylor dispersion technique).[55,56] In these conditions, it is legitimate to consider that our measurements of differential diffusion coefficients obtained by the above techniques are parameters with a well-defined thermodynamic meaning.[55,56]

Mutual diffusion refers to the fluxes of solution components caused by composition differences and the resulting chemical potential gradient driving forces, in contrast to self-diffusion, which occurs without mass transport in systems of uniform chemical composition. It is very common in the scientific literature to find misunderstandings concerning the meaning of the parameter frequently just denoted by D and referred to as "diffusion coefficient."[56] It is necessary to distinguish between two distinct processes: self-diffusion D^* (also named as intradiffusion, tracer diffusion, single ion diffusion, or ionic diffusion) and mutual diffusion D (also known as interdiffusion, concentration diffusion, or salt diffusion). Methods such as those based in nuclear magnetic resonance (NMR), polarography, and capillary-tube techniques with radioactive isotopes measure self-diffusion coefficients, not mutual diffusion. However, for bulk substance transport, the appropriate parameter is the mutual diffusion coefficient, D. Theoretical relationships derived between self-diffusion and mutual diffusion coefficients, D^* and D, respectively, have had limited success for estimations of D (as well as theoretical formulae for the calculation of D), and consequently, experimental mutual diffusion coefficients are absolutely necessary. In the two infinite dilution limits of a binary mixture, the Fickian diffusion coefficients coincide with the intradiffusion coefficient of the diluted component.

Diffusion in a ternary solution is described by the extension of the Fick diffusion equations (eqs 14.3 and 14.4):

$$(J_1) = (D_{11})_v \frac{\partial c_1}{\partial x} + (D_{12})_v \frac{\partial c_2}{\partial x} \tag{14.3}$$

$$(J_2) = (D_{21})_v \frac{\partial c_1}{\partial x} + (D_{22})_v \frac{\partial c_2}{\partial x} \tag{14.4}$$

where J_1, J_2, $\partial c_1/\partial x$, and $\partial c_2/\partial x$ are the molar fluxes and the gradients in concentration of solute 1 and solute 2, respectively. Main diffusion coefficients, D_{11} and D_{22}, give the flux of each solute produced by its own concentration gradient. Crossdiffusion coefficients, D_{12} and D_{21}, give the coupled flux of each solute driven by a concentration gradient in the other solute. A positive D_{ik} crosscoefficient $(i \neq k)$ indicates a co-current coupled transport of solute i from regions of higher concentration of solute k to regions of lower concentration of solute k. However, a negative D_{ik} coefficient indicates a counter-current coupled transport of solute i from regions of lower to higher concentration of solute k.

14.2.2 CONCEPTS OF THERMODIFFUSION

The formal theoretical description of diffusive heat and mass transport within the framework of nonequilibrium thermodynamics is readily available in the literature and will be briefly described below.[57–62] Starting points are the phenomenological equations in a barycentric reference system, which constitute linear relations between the flows and the generalized thermodynamic forces. In the binary isobaric case without chemical reaction and $n = 2$ components, the reduced heat flow $J_q' = J_q (h_1 h_2) J_1$ and the mass flow J_1 of the independent component 1 are given by

$$J_q' = L_{qq} \frac{\nabla T}{T^2} L_{q1} \frac{(\partial \mu_1 / \partial c_1)_{p,T}}{c_2 T} \nabla c_1 \tag{14.5}$$

$$J_1 = L_{1q} \frac{\nabla T}{T^2} L_{11} \frac{(\partial \mu_1 / \partial c_1)_{p,T}}{c_2 T} \nabla c_1 \tag{14.6}$$

where J_q is the total heat flow and h_k the partial specific enthalpy of species k, c_1 and $c_2 = 1 - c_1$ are the concentrations of the two components (in mass fractions), and μ_1 is the chemical potential per unit mass of component 1. L_{qq}, L_{11}, and $L_{1q} = L_{q1}$ are Onsager's phenomenological coefficients.

Experiments are preferably described in terms of the directly accessible concentration variables. The Dufour effect, corresponding to the L_{q1}-term, can usually be neglected in liquids. The diffusion coefficient D, the thermodiffusion coefficient D_T, and the thermal conductivity κ are defined as

$$D = \frac{L_{11}}{\rho c_2 T} \left(\frac{\partial \mu_1}{\partial c_1} \right)_{p,T} , \quad D_T = \frac{L_{1q}}{c_{21} c_2 \rho T^2} , \quad = \frac{1}{T^2} \left(L_{qq} \frac{L_{1q}^2}{L_{11}} \right) \tag{14.7}$$

The Soret coefficient $S_T = D_T/D$ is a measure of the concentration gradient that can be sustained by a given temperature gradient in the steady state. In the dilute limit, the thermophoretic drift velocity of the minority component is given by

$$v_T = D_T \nabla T \tag{14.8}$$

After combining the flow eqs 14.5 and 14.6 with conservation laws for energy and mass, the heat and the extended diffusion equation are obtained:

$$\rho c_p \frac{\partial T}{\partial t} = \nabla \cdot (\nabla T) + Q \tag{14.9}$$

$$\frac{\partial c}{\partial t} = \nabla \cdot \left[D \nabla c_1 + c_1 (1 c_1) D_T \nabla T \right] \tag{14.10}$$

where ρ is the density, c_p the specific heat of the fluid, and Q is the heat production rate per unit volume. Absolute and reduced heats of transport $Q^{\prime *}_{k,abs}$ s and $Q^{\prime *}_k = Q^{\prime *}_{k,abs} Q^{\prime *}_{n,abs}$, respectively, are defined as the heats transported by the mass flows in an n-component isothermal system with n being the dependent component and $1...(n-1)$ the independent ones:

$$J'_q = \sum_{k=1}^{n-1} Q^{\prime *}_k J_k = \sum_{k=1}^{n} Q^{\prime *}_{k,abs} J^{abs}_k \qquad (14.11)$$

with $J^{abs}_k = \rho_{kk} = J_k + \rho_k$ is the absolute flow, v_k the velocity, and ρ_k the density of component k in the laboratory system; $v = \sum_{k=1}^{n} (\rho_{kk})/\rho$ is the barycentric velocity.

After switching to molar instead of specific quantities, the Soret coefficient S_T can eventually be written in the form[59]

$$S_T = \frac{\tilde{Q}^{\prime *}_{1,abs} \tilde{Q}^{\prime *}_{2,abs}}{RT^2 \left[1 + \left(\partial \ln \gamma_1 / \partial \ln x_1 \right)_{p,T}\right]} \qquad (14.12)$$

where x_k is the mole fraction of the kth component, M_k its molar mass, γ_k its activity coefficient, and $\tilde{Q}^{\prime *}_{k,abs} = Q^{\prime *}_{k,abs} M_k$ its molar absolute reduced heat of transport. R is the gas constant. Equation 14.12 signifies that the Soret coefficient of a binary mixture is defined only by the difference $\tilde{Q}^{\prime *}_{1,abs} \tilde{Q}^{\prime *}_{2,abs}$ of the heats of transport. At the same time, these quantities satisfy an additional condition following from the Gibbs–Duhem equation $x_1 \tilde{Q}^{\prime *}_{1,abs} + x_2 \tilde{Q}^{\prime *}_{2,abs} = 0$ (62). Under an appropriate change of the reference value for the heats of transport, the latter condition can be eliminated.[60] Indeed, in terms of new variables

$$\tilde{Q}^{\prime *}_{1,abs} = Q_1 + q, \quad Q^{\prime *}_{2,abs} = Q_2 + q \qquad (14.13)$$

where q is some reference value characteristic for the peculiar mixture; the equation for the Soret coefficient takes the final form:

$$S_T = \frac{Q_1 Q_2}{RT^2 \left[1 + \left(\partial \ln \gamma_1 / \partial \ln x_1 \right)_{p,T}\right]} \qquad (14.14)$$

The new coefficients Q_k satisfy the more general relation $x_1 Q_1 + x_2 Q_2 = -q$. If Q_1 and Q_2 were known, this relation would provide a way to determine q. Similarly to $\tilde{Q}^{\prime *}_{k,abs}$, we call these variables Q_k also heats of transport. In contrast, $\tilde{Q}^{\prime *}_{1,abs}$ and $\tilde{Q}^{\prime *}_{k,abs}$ the quantities Q_1 and Q_2 are independent variables.

The notation we have introduced is nowadays widely adopted but also alternative descriptions exist. Some authors prefer to include the term $c_1(1-c_1)$ in the definition of the thermodiffusion and the Soret coefficient[63] and some use the thermodiffusion factor $\alpha_T = TS_T$[64] or the thermodiffusion ratio $k_T = T c_1(1-c_1)S_T$.[65]

14.3   EXPERIMENTAL TECHNIQUES

In this work, we will focus on the description of the experimental techniques that are being used by our and associated groups actively working in the field of mutual and thermal diffusion of hydrocarbon systems, namely, the Taylor dispersion technique and the ODI for the Soret cell.

14.3.1   MUTUAL DIFFUSION MEASUREMENTS

Among experimental methods developed for measurements of mass diffusion coefficients in liquids, the Taylor dispersion technique has been used frequently for measuring binary diffusion coefficients of various solutions and lately applied for three-component systems.[25–30,48,66] The concepts and operation of this experimental method for the measurement of diffusion coefficients are well described in the literature so only the most relevant points are highlighted.

In the Taylor dispersion technique, a small amount of a given solution is injected into a laminar carrier streams of solvent, or of solution at a different concentration, to flow throughout a long capillary tube[56,67–68] (Fig. 14.1). The length of the Teflon dispersion tube for this experimental assembly was measured directly by stretching the tube in a large hall and using two high-quality theodolytes and appropriate mirrors to accurately focus on the tube ends. This technique gave a tube length of 3.2799 (±0.0001) ×10⁴ mm, in agreement with less-precise control measurements using a good-quality measuring tape. The radius of the tube, 0.5570 (±0.00003) mm, was calculated from the tube volume obtained by accurately weighing (resolution 0.1 mg) the tube when empty and when filled with distilled water of known density.

In a pattern run, a sample of 0.063 mL of the solution under study ($c_j \pm \Delta c$) is injected into the laminar carrier stream (c_j) through a 6-port Teflon valve (Rheodyne, model 5020). The flow rate is keeping constant (0.17 mL min⁻¹) with the assistance of a metering pump (Gilson model Minipuls 3) which allows retention times of about 1.1×10^4 s. Both the dispersion tube and the injection valve are placed into an air thermostat bath to keep the temperature constant at 298.15 K (±0.01 K).

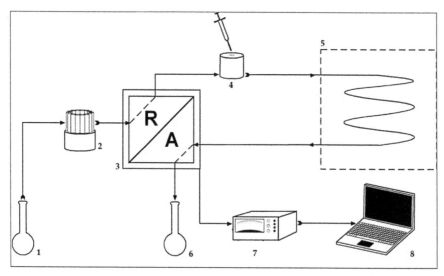

FIGURE 14.1   Schematic of the experimental setup for Taylor dispersion technique. Flow direction is shown by arrows.

Notes: 1—solution reservoir, 2—pump, 3—refractometer, 4—injection valve, 5—thermostatized box with dispersion tube, 6—waste, 7—voltimeter, 8—computer.

The dispersion of the injected samples is monitored by using a differential refractometer (Waters model 2410) at the outlet of the dispersion tube. Voltage values as a function of the elapsed time, $V(t)$, are measured at accurately 5 s intervals by using a digital voltmeter (Agilent 34401 A) provided with an IEEE interface. Binary diffusion coefficients are calculated from the dispersion equation:

$$V(t) = V_0 + V_1 t + V_{max} \left(t_R / t \right)^{1/2} \exp{-12D \left(t - t_R \right)^2 / r^2 t} \qquad (14.15)$$

being the additional fitting parameters: t_R, the mean sample retention time; V_{max}, the peak height; V_0, the baseline voltage; and V_1, the baseline slope.

Extensions of the Taylor technique have been used to determine ternary mutual diffusion coefficients (D_{ik}) for multicomponent solutions. These D_{ik} coefficients are evaluated from the fitting of two or more replicate pairs of peaks for each carrier-stream, to the ternary dispersion equation:

$$V(t) = V_0 + V_1 t + V_{\max} \left(\frac{t_R}{t}\right)^{1/2} \left[W_1 \exp\left(\frac{-12D_1(t-t_R)^2}{r^2 t}\right) + (1W_1)\exp\left(\frac{-12D_2(t-t_R)^2}{r^2 t}\right) \right] \qquad (14.16)$$

The two pairs of refractive-index profiles, D_1 and D_2, are the eigenvalues of the matrix of the ternary D_{ik} coefficients. W_1 and $1-W_1$ are the normalized pre-exponential factors. Details have previously been reported of the method of calculation of the D_{ik} coefficients from the fitted values of D_1, D_2, and W_1 by Deng and Leaist.[69]

In standard experiments, small volumes, ΔV, of a solution of composition, $(c_1 + \Delta c_1)$ and/or $(c_2 + \Delta c_2)$, are injected into carrier solutions of composition (c_1, c_2) at time $t = 0$.[25,27,66] The mixtures in the results reported by our experimental method were prepared using mass fractions and then converted to molar concentration by means of the relation $w_i = c_i(M/\rho)$ where w_i stands for the concentration in mass fraction, c_i is molar concentration, M_i is the molar mass of the constituent i, and ρ is the density of the mixture.

14.3.2 THERMAL DIFFUSION MEASUREMENTS

The optical beam deflection technique is one of the traditional methods used by different teams for the measurement of the Soret effect.[70–73] It has major advantages as the short duration of a typical experiment, the relative simplicity of the setup, and is supported by a well-established theory. The method employs a classical Soret cell with relatively small in height (typically 1 mm) and has been tested against reference data for water–ethanol and the Fontainebleau benchmark mixtures.[71] Nevertheless, this method measures only the concentration difference between hot and cold plates/zones or between central line and some point. Based on this method, a new experimental setup was designed to study heat and mass transfer in liquids with Soret effect. The suggested instrument consists of two principal parts: the optical cell[72] and an interferometric system in combination with equipment for digital recording and processing the phase information. Figure 14.2 shows a cross-section of the diffusion cell (Fig. 14.2a) and a sketch of the ODI setup (Fig. 14.2b). This technique allows performing and reproducing experiments without disturbing the media. It also allows measuring not only Δc between the hot and cold plates but also the concentration and temperature distributions along the diffusion path, giving the possibility to adopt this technique for some other purposes like the study hydrodynamic fluctuations. The reliability of the experiments can be improved by using statistical means.[74]

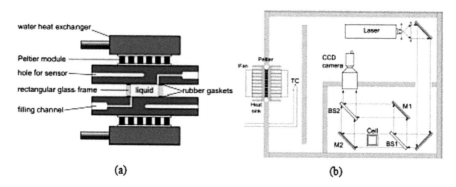

FIGURE 14.2 Optical digital interferometry setup; cross section of the cell (a) and top view of the entire setup (b).

The experiments are performed in a transparent cubic cell filled initially by a homogenous mixture. The thermal gradient is imposed by heating and cooling the top and bottom walls of the cell, respectively. The spatial temperature variation induces mass transfer through the Soret effect. An enlarged collimated laser beam traverses the entire cell perpendicular to the temperature gradient. Both temperature and composition variations contribute to the spatial distribution of the refractive index that modulates the wave-front of the emerging optical beam. The use of an interferometric technique allows the final wave-front shape to be accurately determined by comparison with undisturbed shape. The optical interferometer (Mach–Zehnder interferometer) coupled with a digital recording and processing is used for very accurate determination of a phase shift.[70] Each interferogram is reconstructed by performing a 2D fast Fourier transform (FFT) of the fringe image, filtering a selected band of the spectrum and performing an inverse 2D FFT of the filtered result and phase unwrapping. Knowledge of phase shifts gives information about the local gradients of refractive index. Then, the gradients of composition inside of the fluid are calculated from refractive index gradients and local temperature both for steady state and dynamic regimes. Thus, ODI enabled measurements of Δc between two arbitrary points.

The light source is a He–Ne laser with the wavelength of $\lambda = 632.8$ nm. An expanded and collimated laser beam is split into a reference and an objective arm, with the cell assembly placed within the latter. Then, both beams are deflected by mirrors and merged at the second beam splitter. One of the beams is inclined with respect to the other in order to create a narrow fringe pattern. The resulting interferogram is recorded by a CCD camera with a sensor of 1280×1024 pixels. The resolution of the imaging system is ≈ 50 pixels/mm. In order to reconstruct a wave front (spatial distribution of optical phase), the Fourier transform method is applied to obtain the phase distribution from a single interferogram. To be precise, two interferograms are used in processing, as the interference pattern of interest is always processed against a reference interferogram taken before the refractive index change.

To ensure the thermal stability of the interferometer during the course of an experiment (up to 2–3 days), the whole setup, including a bearing bench plate, is placed inside a box of thermal insulation material and equipped with air-to-air cooling/heating assembly based on a Peltier element and driven by a dedicated PID controller. A set of shields is inserted in the box to prevent air motion over optical paths. The temperature inside the box is kept equal to the mean temperature of the liquid with residual fluctuations of ± 0.02 K.

The unique feature of this method is that it traces the *transient* path of the system over the *entire* two-dimensional cross section of the cell. The quantity obtained after processing an interferogram is a spatial phase distribution $\phi(x, z, t)$, which is transferred into a concentration distribution by the equation

$$\frac{\lambda}{2\pi H}\Delta\varphi(x,z,t) = \left(\frac{\partial n}{\partial c}\right)_{p,T}\Delta c(x,z,t) + \left(\frac{\partial n}{\partial T}\right)_{p,c}\Delta T(x,z,t) \quad (14.17)$$

where the second term is defined during the initial step of the experiment.[39] Consequently, Soret and diffusion coefficients are simultaneously extracted by fitting of $c(x, z, t)$ from eq 14.4, averaged in x-direction, with the one-dimensional analytical solution[72]

$$c(z,t) = c_0 + c_0(1-c_0)S_T\Delta T\left[\frac{1}{2} - \frac{z}{H} - \frac{4}{\pi^2}\sum_{n,Odd}^{\infty}\frac{1}{n^2}\cos\times\left(\frac{n\pi z}{H}\right)\exp\left(\frac{n^2\pi^2}{H^2}Dt\right)\right] \quad (14.18)$$

14.4 OVERVIEW ON HYDROCARBON SYSTEMS

Most of the extensive results existing over transport coefficients in hydro-carbon liquid mixtures refer to binary systems. Only during the last decade, the attention has focused in ternaries mixtures, which are seen as the simplest strictly multicomponent mixtures. This transition has partly been motivated by the multicomponent nature of many relevant natural and technological systems. There are two hydrocarbon systems that have been attracting the interest of research due to the fact that they are quasi-ideal hydrocarbon mixtures and can mimic representative compounds of crude oil that can be modeled with different degrees of success by MD simulations.[75] We refer to 1,2,3,4-THN, IBB, and dodecane mixture (also known as the Fontainebleau benchmark mixture) and toluene, cyclohexane, and methanol mixture. We will do an overview on the experimental results on the transport coefficients (Soret, diffusion, and thermodiffusion coefficients) obtained for both these two hydrocarbon systems.

14.4.1   1,2,3,4-TETRAHYDRONAPHTHALENE, IBB, AND DODECANE MIXTURE

Several international teams and laboratories have dedicated their work to the difficult task of establishing the isothermal diffusion coefficients for binary and ternary mixtures of the 1,2,3,4-THN, IBB, and dodecane ($C_{12}H_{26}$) mixtures. There is fair agreement between all the results reported in the literature, which have been obtained through various experimental methods, with deviations on the order of 8.5% between them.[52]

The determination for the mutual diffusion coefficients for the three possible binary mixtures (THN-IBB, IBB-$C_{12}H_{26}$, and THN-$C_{12}H_{26}$) at the 0.5:0.5 mass fraction symmetric points, and a fourth mixture THN-$C_{12}H_{26}$ with 0.9:0.1 mass fraction, was done using Taylor dispersion technique and compared it against the literature results.[30] The deviations between the results obtained and literature diffusion coefficients, acquired with other experimental methods (e.g., optical methods (see below), thermo-gravitational columns, etc.),[19,52] are, in general, less than 7% and at the most 8.5%.

Table 14.1 gives the mean D values for four binary systems at 298.15 K determined from four to six replicate dispersion profiles, and their relative deviations from the tabulated literature D values.[19,39,52] The results reported on the transport coefficients (Soret, diffusion, and thermodiffusion coefficients) for the same systems,[39] using ODI technique are described in Table 14.2. Comparative plots of our results with those previously published for the same systems, and several experimental methods, are shown on Figure 14.3.

TABLE 14.1 Binary Diffusion Coefficients, D^a, for Mixtures of 1,2,3,4-Tetrahydronaphthalene (THN), Isobutylbenzene (IBB), and Dodecane ($C_{12}H_{26}$) at 298.15 K.

Mass fraction	$C_{12}H_{26}$—IBB	$C_{12}H_{26}$—THN	IBB—THN
0.5:0.5	0.898 ± 0.005	0.592 ± 0.010	0.786 ± 0.005
0.9:0.1	0.980 ± 0.050		

THN, tetrahydronaphthalene; IBB, isobutylbenzene.

$^a D_{ij} \pm SD_{ij}\ 10^{-9}\ \mathrm{m^2\ s^{-1}}$.

TABLE 14.2 Diffusion and Soret Coefficients for 1,2,3,4-Tetrahydronaphthalene (THN), Isobutylbenzene (IBB), and Dodecane ($C_{12}H_{26}$) for Binary 0.5–0.5 Mass Fraction Mixtures.

	Quantity	THN—$C_{12}H_{26}$	THN—IBB	IBB—$C_{12}H_{26}$
ODI measurement[39]	$D_S^a/(10^{-9}\ \mathrm{m^2\ s^{-1}})$	0.616 ± 0.05	0.860 ± 0.28	0.940 ± 0.22
	$D^b/(10^{-9}\ \mathrm{m^2\ s^{-1}})$	0.637 ± 0.05	0.843 ± 0.28	0.923 ± 0.22
	$D^c/(10^{-9}\ \mathrm{m^2\ s^{-1}})$	0.627 ± 0.29	0.852 ± 0.12	0.932 ± 0.12
	$S_T/(10^{-3}\ 1\ \mathrm{K^{-1}})$	9.24 ± 0.01	3.29 ± 0.11	3.98 ± 0.08
Benchmark values[52]	$D_S^a/(10^{-9}\ \mathrm{m^2\ s^{-1}})$	0.621 ± 0.06	0.850 ± 0.6	0.950 ± 0.4
	$S_T/(10^{-3}\ \mathrm{K^{-1}})$	9.5 ± 0.3	3.3 ± 0.3	3.9 ± 0.1

ODI, optical digital interferometry; THN, tetrahydronaphthalene; IBB, Isobutylbenzene.

^{a}Diffusion coefficient measured during Soret separation.

^{b}Diffusion coefficient measured at isothermal conditions.

^{c}Mean value of diffusion coefficient between isothermal and nonisothermal conditions.

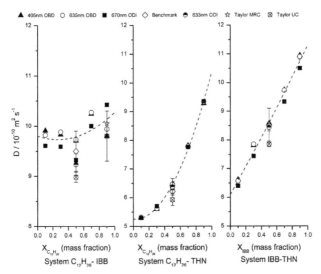

FIGURE 14.3 Binary diffusion coefficients for the systems IBB/THN, $C_{12}H_{26}$/IBB, and $C_{12}H_{26}$/THN.

Direct comparison of the results for the Taylor D values reported here, with D values obtained by other experimental methods, reveals that they are encompassed within the previously found 7% uncertainty interval. If we consider that 1–3% uncertainty is typical for Taylor dispersion measurements, as shown in previous reports for binary systems by Santos et al.,[26,27,76] and Ribeiro et al.,[28,77,78] this suggests an acceptable accuracy in these determinations.

The ternary diffusion coefficients measured for THN (1) + IBB (2) + $C_{12}H_{26}$ (0) system are summarized in Table 14.3. The component zero is for the solvent. The main diffusion coefficients D_{11} and D_{22} were reproducible within $\pm 0.08 \times 10^{-9}$ m^2 s^1. The crosscoefficients were, in general, reproducible within about $\pm 0.10 \times 10^{-9}$ m^2 s^1.

The results obtained for the mixtures under study are in reasonable agreement with available results in the literature, measured using Taylor dispersion, counter flow cell, and sliding symmetric tubes techniques by Mialdun et al.,[29] Sechenyh et al.,[79] and Larrañaga et al.[80] The ternary results here described differ only slightly from literature, and although there is still no agreement on the actual values, the eigenvalues obtained from all the experiments are very close. The eigenvalues differ between 3.5 and 8.8% from those measured with the same experimental method. For the ternary symmetric point, there is only one published result,[45] which was obtained by the Optical Beam Deflection technique, for this system but with a different order of the components ($C_{12}H_{26}$ (1) + IBB (2) + THN (0)). Although this implies different values for the main and crossdiffusion coefficients, the eigenvalues of the diffusion matrix do not depend on the order of the components, and the differences between the calculated eigenvalues calculated from Table 14.3 data and the published OBD data are 17.8% in $\widehat{D}_1$ and 3.4% in $\widehat{D}_2$.

It is observable from the nonzero crossdiffusion coefficient values, D_{ij}, at this finite concentration for the THN (1) + IBB (2) + $C_{12}H_{26}$ (0) system, and by the fact that the main coefficients D_{ii} are not the same as the binary diffusion ones, that there are solute interactions affecting the diffusion of the components, and under these conditions, possible interactions between the solutes leads to a counter flow of IBB (D_{21} <0). The gradient in the concentration of IBB produces cocurrent coupled flows of THN. If we consider that D_{12}/D_{22} values give the number of moles of THN cotransported per mole of IBB driven by its own concentration gradient, we can suggest that a mole of diffusing IBB cotransports 0.14 mol of THN at the concentration studied. From the D_{21}/D_{11} values, at the same compositions, we can anticipate that a mole of diffusing THN counter-transports 0.44 mol of IBB.

TABLE 14.3 Ternary Diffusion Coefficients, D_{ij}^a, for the Mixture of THN (1), IBB (2), and $C_{12}H_{26}$ (0) at Same Mass Fraction (i.e., 0.33:0.33:0.33) at 298.15 K.

Mass fraction of each component in mixture, THN–IBB–$C_{12}H_{26}$	$D_{12}/(10^{-9}\ m^2\ s^{-1})$	$D_{11}/(10^{-9}\ m^2\ s^{-1})$	$D_{21}/(10^{-9}\ m^2\ s^{-1})$	$D_{22}/(10^{-9}\ m^2\ s^{-1})$	$\hat{D}_1/(10^{-9}\ m^2\ s^{-1})$	$\hat{D}_2/(10^{-9}\ m^2\ s^{-1})$
(0.33):(0.33):(0.33)	1.004 ± 0.089	0.078 ± 0.046	-0.448 ± 0.168	0.560 ± 0.060	0.903	0.658
	(-2.6%)[b]	(-76.3%)[b]	(2.6%)[b]	(-13.1%)[b]	(-8.8%)[b]	(-3.5%)[b]
	(+4.0%)[c]	(-68.0%)[c]	(+54.0%)[c]	(-8.5%)[c]		
	(-1.0%)[d]	(45.0%)[d]	(+12.3%)[d]	(-16.3%)[d]		

THN, tetrahydronaphthalene; IBB, isobutylbenzene.

$^a D_{ij} \pm SD_{ij}/(10^{-9}\ m^2\ s^{-1})$.

$^b (D_{ij} - D_{ij,Lit})/D_{ij}\%$, where $D_{ij,Lit}$ represents the values obtained by TDT.[21]

$^c (D_{ij} - D_{ij,Lit})/D_{ij}\%$ where $D_{ij,Lit}$ are the average values calculated from data obtained by TDT.[29,81]

$^d (D_{ij} - D_{ij,Lit})/D_{ij}\%$ where $D_{ij,Lit}$ are the average values calculated from published results by References [29,79,80].

14.4.2 TOLUENE, CYCLOHEXANE, AND METHANOL MIXTURE

The miscibility gap exhibited by the ternary mixture of the hydrocarbons toluene, methanol, and cyclohexane[54] has been appealing to scientists for a long time. A system of this kind has a critical point (consolute point), and it is expected that the diffusion coefficient D should drastically diminish when approaching to the critical point, but thermodiffusion coefficient D_T remains almost constant. Nevertheless, for binary mixtures, the Soret coefficient S_T grows as $\sim 1/D$, and it was observed increase of four orders of in some systems. A similar behavior is expected in the ternary systems.

Before the analysis over the ternary data, a review of the experimental transport properties of the three binary mixtures for the hydrocarbons toluene, methanol, and cyclohexane,[81,82] measured with several different experimental methods, including ODI, is compiled in Tables 14.4–14.6.

Toluene–methanol mixture presents a large region with a positive Soret coefficient, and the results obtained by means of different experimental methods are in good agreement. However, the mixture toluene–cyclohexane displays only negative Soret coefficients, including for point $c_{tol} = 0.4$, measured in the absence of gravity in the International Space Station, while mass diffusion coefficient reveals less favorable agreement between the data from different sources. Finally, the methanol–cyclohexane mixture, that presents a large demixing zone, has, according to literature,[83] negative Soret coefficients on both sides of the demixing zone.

Moving to ternary liquid mixture cyclohexane–toluene–methanol the behavior of the Soret coefficient was analyzed. It was observed that one feature of this mixture is to hold a region of compositions, most of the ternary mixture concentrations, including the demixing zone, where the Soret effect is negative, that is, denser components segregate to the hot regions due to thermal diffusion. It means that in ground conditions of gravity, the negative Soret sign has a disrupting effect on the system not allowing the results on the ternary mixtures to be calculated. However, the region rich in methanol and poor in cyclohexane is gravitationally stable and measurements of Soret can be done in ground laboratory experiments.[48]

For the analysis of the Fickian diffusion, the mutual diffusion coefficients for the five ternary mixtures for the system cyclohexane (1)–toluene (2)–methanol (0) were measured using the Taylor dispersion technique.[84] The composition of the experimental liquid mixtures in mass fraction is given in Table 14.7. The component zero is for the solvent. Table 14.8 gives the mean D values for each mixture at 298.15 K determined from four to six replicate dispersion profiles.

TABLE 14.4 Diffusion, Thermal Diffusion, and Soret Coefficients of Toluene–Methanol Binary Mixtures Measured at $T = 298.15$ K.

Toluene mass fraction	$D/(10^{-9}\ m^2\ s^{-1})$					$D_T/(10^{-9}\ m^2\ s^{-1})$			$S_T/(10^{-3}\ K^{-1})$			
	SST	OBD	ODI	Other (85)	TGC	OBD	ODI	Other (85)	SST	OBD	ODI	Other (85)
0.135	2.87			2.25	9.80			7.31	3.41			3.25
0.189	2.49			2.1	9.00			6.97	3.61			3.32
0.336	1.93	1.82		1.74	7.91	8.06		7.06	4.10	4.44		4.06
0.500			1.14	1.35			5.50	5.52			4.82	4.09
0.552	1.17	1.023		1.20	5.22	4.14		5.03	4.48	4.05		4.19
0.650	0.82	0.89	0.89	0.97	3.39	2.60	3.20	3.51	4.16	2.94	3.60	3.62
0.842				0.90				3.38				-3.75
0.908		1.14		1.26		-8.20		-7.83		-7.19		-6.21

SST, sliding symmetric tubes; OBD, optical beam deflection; ODI, optical digital interferometry; TGC, thermogravitational column.

TABLE 14.5 Diffusion, Thermal Diffusion, and Soret Coefficients of Toluene–Cyclohexane Binary Mixtures Measured at $T = 298.15$ K.

Toluene mass fraction	$D/(10^{-9} \text{ m}^2 \text{ s}^{-1})$		$D_T/(10^{-9} \text{ m}^2 \text{ s}^{-1})$	$S_T/(10^{-3} \text{ K}^{-1})$
	SST	OBD	OBD	OBD
0.150	1.77			
0.200		1.84	−6.20	−3.36
0.400	1.74	1.99	−4.99	−2.50
0.500		1.99	−4.39	−2.20
0.600	1.88	1.97	−3.47	−1.76
0.800		2.33	−2.36	−1.02
0.850	2.09			

SST, sliding symmetric tubes; OBD, optical beam deflection.

TABLE 14.6 Diffusion, Thermal Diffusion, and Soret Coefficients of Methanol–Cyclohexane Binary Mixtures Measured at $T = 298.15$ K.

Methanol mass fraction	$D/(10^{-9} \text{ m}^2 \text{ s}^{-1})$		$S_T/(10^{-3} \text{ K}^{-1})$
	SST	CFC	
0.005		1.71	
0.010		1.23	
0.035		0.63	
0.650		0.37	
0.655		0.49	
0.665		0.64	
0.700		0.75	
0.750			−9.94
0.800	0.98		
0.885		1.75	
0.900	1.98		
0.990		2.32	

SST, sliding symmetric tubes; CFC, counter flow cell.

TABLE 14.7 Composition of the Experimental Liquid Mixtures in Mass Fraction.

	Cyclohexane	Toluene	Methanol
1	0.07	0.63	0.30
2	0.40	0.45	0.15
3	0.20	0.65	0.15
4	0.40	0.20	0.40
5	0.40	0.30	0.30

TABLE 14.8 Diffusion Coefficients for the CHex–Tol–MeoH Mixtures Measured by the Taylor Dispersion Technique at 298.15 K.

Mixture	$D_{11} \pm S_D/$ $(10^{-9}\,m^2\,s^{-1})$	$D_{12} \pm S_D/$ $(10^{-9}\,m^2\,s^{-1})$	$D_{21} \pm S_D/$ $(10^{-9}\,m^2\,s^{-1})$	$D_{22} \pm S_D/$ $(10^{-9}\,m^2\,s^{-1})$
1	2.092 ± 0.012	-0.215 ± 0.042	0.420 ± 0.081	0.954 ± 0.017
2	0.859 ± 0.009	-0.961 ± 0.012	-0.343 ± 0.102	1.635 ± 0.020
3	1.402 ± 0.022	-0.593 ± 0.071	-0.785 ± 0.028	1.419 ± 0.013
4	0.521 ± 0.017	-0.136 ± 0.053	-0.147 ± 0.034	1.911 ± 0.009
5	0.555 ± 0.009	-0.021 ± 0.010	-0.211 ± 0.033	1.876 ± 0.009

The main diffusion coefficients D_{11} and D_{22} were normally reproducible within $\pm 0.02 \times 10^{-9}\,m^2\,s^{-1}$. The crosscoefficients were, in general, reproducible within about $\pm 0.10 \times 10^{-9}\,m^2\,s^{-1}$. For comparison of these results with the literature data, the mass fraction units of the solutions were converted to mole fraction units. The results obtained for the mutual diffusion coefficients of the mixtures are in reasonable agreement with the literature trends, measured using Taylor dispersion, by Grossmann and Winkelmann.[85,86] Another important feature is that crossdiffusion coefficients are very large and negative for all mixtures, meaning that solutes associate and diffuse over counter flow gradients. Additionally, the determinant $|D|$ and eigenvalues D_i of the matrix of mutual diffusion coefficients were calculated and are presented in Table 14.9.

TABLE 14.9 Determinant $|D|$ and Eigenvalues $\hat{D}_1$ and $\hat{D}_2$ for the CHex–Tol–MeoH Mixtures.

Mixture	$\hat{D}_1/(10^{-9}\,m^2\,s^{-1})$	$\hat{D}_2/(10^{-9}\,m^2\,s^{-1})$	$|D|/10^{-9}$
1	2.006	1.039	2.086
2	1.940	0.554	1.075
3	1.572	0.544	1.524
4	2.022	0.410	0.976
5	1.513	0.310	1.037

From the obtained values, it is perceived that eigenvalues are in close agreement to the ones available in the literature (they generally differ from 1 to 13% from those measured with the same experimental method).

For the Taylor dispersion coefficients measured here, two sets of mixtures were selected along concentrations of constant methanol (reference) mole fraction 0.3, which is comprising mixtures 2 and 3 and for concentrations of constant cyclohexane mole fraction 0.3, respectively (mixtures 2, 4, and 5). In all cases, the third component methanol is considered to be the reference. Comparative plots of the results with those previously published for the same system[85,86] are shown in Figures 14.4 and 14.5.

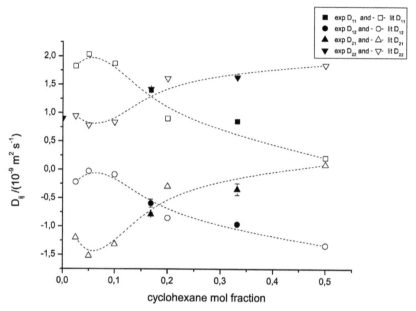

FIGURE 14.4 Mutual diffusion coefficients in the system CHex (1)–Tol (2)–MeOH (0) vs. cyclohexane mole fraction at 298.15 K and constant methanol mole fraction.

Caption: Black symbols correspond to the measured D_{ij} in this work, to mixtures 2 and 3, respectively, withe symbols correspond to literature D_{ij}.

It is noticeable that the increase in the mole fraction of cyclohexane leads to D_{11} decrease and D_{22} increase. Also, from the nonzero crossdiffusion coefficient values, D_{ij}, in this range of composition, there are solute interactions affecting the diffusion of the components, and, under these conditions, possible interactions between the solutes leading to a counter flow of cyclohexane (D_{12} <0) due to the gradient of toluene. Nevertheless, the coupled counter current toluene flow generated by the gradient of cyclohexane tends to zero, when the composition of cyclohexane increases.

If we consider that D_{12}/D_{22} values give the number of moles of cyclohexane cotransported per mole of toluene driven by its own concentration gradient, we can suggest that a mole of diffusing toluene counter-transports 0.6 mole of cyclohexane at the range of compositions studied. From the

D_{21}/D_{11} values, at the same compositions, we can anticipate that a mole of diffusing cyclohexane counter-transports at the most 0.5 mole of toluene and decreases when the composition of cyclohexane in the mixture increases.

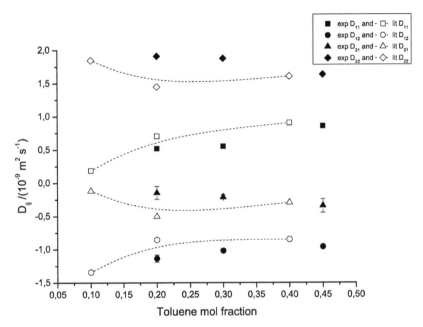

FIGURE 14.5 Mutual diffusion coefficients in the system CHex (1)–Tol (2)–MeOH(0) vs. toluene mole fraction at 298.15 K and constant cyclohexane mole fraction.
Caption: Black symbols correspond to the measured D_{ij} in this work, to mixtures 2, 4, and 5 respectively, withe symbols correspond to literature D_{ij}.

When the cyclohexane composition is kept constant, it is perceptible that the increase in the mol fraction of toluene in the mixtures leads to D_{11} small decrease and D_{22} increase. The crossdiffusion coefficient D_{21}, that represents the coupled diffusion of toluene generated by the gradients of cyclohexane in the mixtures, is smaller and close to zero than the crossdiffusion coefficient D_{12}, which represents the coupled diffusion of cyclohexane generated by the gradients toluene of in the mixtures, higher and more negative.

As before, if we analyze the coupled flows of solutes we can see from D_{12}/D_{22} values that a mole of diffusing toluene counter-transports 0.6 mole of cyclohexane at the range of compositions studied. From the D_{21}/D_{11} values, at the same compositions, we can anticipate that a mole of diffusing cyclo-hexane counter-transports at the most 0.4 mole of toluene, and the mass transport increases when the composition of toluene in the mixture increases.

14.5   FINAL REMARKS

Results on the Soret (S_T), diffusion (D), and thermodiffusion (D_T) coefficients for binary and ternary mixtures of two hydrocarbon systems that have high relevance in research and industry due to their importance as quasi-ideal hydrocarbon mixtures that can mimic representative compounds of crude oil were reviewed under the scope of this work. Soret coefficients for 1,2,3,4-THN, IBB, and dodecane binary mixtures (also known as the Fontainebleau benchmark mixture) are positive and values of the mass diffusion coefficient, measured by several international teams and by means of different experimental methods are in fair agreement.

In what concerns toluene, cyclohexane, and methanol mixture, the Soret coefficients show different signal depending on the binary mixture and are only determinable in a very small zone in the ternary mixture concentrations map. Mutual coefficients for some points of the ternary mixture are discussed and compared with literature data.

There is still a great challenge in the improvement of experimental methodologies and in the development of reliable methods for the determination and prediction of mutual diffusion coefficients of mixtures containing three or more components. Molecular modeling and simulation are evolving and can be a powerful tool to complement experimental efforts.[87]

KEYWORDS

- diffusion
- thermodiffusion
- Soret
- hydrocarbons
- transport properties
- Taylor dispersion technique

REFERENCES

1. Montel, F. Importance de la thermodiffusion en exploration et production pétrolières. *Entropie* **1994**, *86*, 184.

2. Severinghaus, J. P.; Sowers, T.; Brook, E. J.; Alley, R.; Bender, M. Timing of Abrupt Climate Change at the End of the Younger Dryas Interval from Thermally Fractionated Gases in Polar Ice. *Nature* **1998**, *391,* 141.

3. Alexandrov, D. V.; Aseev, D. L. One-Dimensional Solidification of an Alloy With a Mushy Zone: Thermodiffusion and Temperature-Dependent Diffusivity. *J. Fluid Mech.* **2005**, *527,* 57.

4. Mast, C. B.; Braun, D. Thermal Trap for DNA Replication. *Phys. Rev. Lett.* **2010**, *104,* 188102.

5. Ludwig, C. Diffusion zwischen ungleich erwärmten Orten gleich zusammengesetzter Lösung. *Sitzungsberichte der Mathematisch-Naturwissenschaftlichen* **1856**, *20,* 539.

6. Soret, C. Sur l'etat d'equilibre que prend au point de vue de sa concentration une disso-lution saline primitivement homohene dont deux parties sont portees a des temperatures differentes. *Archives des Sciences physiques et naturelles* **1879**, *2,* 46–61.

7. Soret, C. Influence de la temperaturé sur la distribution des sels dans leurs solutions. *Académie des Sciences, Paris* **1880**, *91* (5), 289–291.

8. Bearman, R. J. On the Molecular Basis of Some Current Theories of Diffusion. *J. Phys. Chem.* **1961**, *65,* 1961.

9. Haase, R. *Thermodynamics of Irreversible Processes*; Addison-Wesley Pub. Company: Reading, Massachussetts,1969.

10. Kempers, L. A Thermodynamic Theory of the Soret Effect in a Multicomponent Fluid. *J. Chem. Phys.* **1989**, *90,* 6541.

11. Shapiro, A. A. Fluctuation Theory for Transport Properties in Multicomponent Mixtures: Thermodiffusion and Heat Conductivity. *Phys. A* **2004**, *332,* 151.

12. Ghorayeb, K.; Firoozabadi, A. Molecular, Pressure, and Thermal Diffusion in Non-Ideal Multicomponent Mixtures. *AIChE J.* **2000**, *46,* 883.

13. Gebhardt, M.; Kohler, W.; Mialdun, A.; Yasnou, V. Diffusion, Thermal Diffusion, Soret Coefficients, and Optical Contrast Factors of the Binary Mixtures of Dodecane, Isobutylbenzene, and 1,2,3,4-Tetrahydronaphthalene. *J. Chem. Phys.* **2013**, *138,* 114503.

14. Mialdun, A.; Yasnou, V.; Shevtsova, V.; Königer, A.; Köhler, W.; Alonso de Mezquia, D.; Bou-Ali, M. M. A Comprehensive Study of Diffusion, Thermodiffusion, and Soret Coefficients of Water-Isopropanol Mixtures. *J. Chem. Phys.* **2012**, *136,* 244512.

15. Pan, S.; Yan, Y.; Jabber, R. J.; Kawaji, M.; Saghir, Z. Evaluation of Thermal Diffusion Models for Ternary Hydrocarbon Mixtures. *J. Non-Equil. Thermodyn.* **2007**, *32,* 241.

16. Ivanov, D. A.; Grossmann, T.; Winkelmann, J. Comparison of Ternary Diffusion Coefficients Obtained from Dynamic Light Scattering and Taylor Dispersion. *Fluid Phase Equilib.* **2005**, *228,* 283–291.

17. Bardow, A. On the Interpretation of Ternary Diffusion Measurements in Low-Molecular Weight Fluids by Dynamic Light Scattering. *Fluid Phase Equilib.* **2007**, *251,* 121.

18. Ortiz de Zárate, J. M.; Hita, J. L.; Sengers, J. V. Fluctuating Hydrodynamics and Concentration Fluctuations in Ternary Mixtures. *C. R. Mec.* **2013**, *341,* 399–404.

19. Gebhardt, M.; Köhler, W.; Mialdun, A.; Yasnou, V.; Shevtsova, V. Transport Properties of the Binary Mixtures of the Three Organic Liquids Toluene, Methanol, and Cyclohexane. *J. Chem. Phys.* **2013**, *138,* 114503.

20. Derlacki, Z. J.; Easteal, A. J.; Edge, A. V. J.; Woolf, L. A.; Roksandic, Z. Diffusion Coefficients of Methanol and Water and the Mutual Diffusion Coefficient in Methanol-Water Solutions at 278 and 298 K. *J. Phys. Chem.* **1985**, *89,* 5318.

21. Ribeiro, A. C. F.; Lobo, V. M. M.; Natividade, J. J. S. Diffusion Coefficients in Aqueous Solutions of Cobalt Chloride at 298.15 K. *J. Chem. Eng. Data* **2002**, *47,* 541.

22. Mialdun, A.; Legros, J. C.; Yasnou, V.; Sechenyh, V.; Shevtsova, V. Contribution to the Benchmark for Ternary Mixtures: Measurement of the Soret, Diffusion and Thermodiffusion Coefficients in the Ternary Mixture THN/IBB/nC12 with 0.8/0.1/0.1 Mass Fractions in Ground and Orbital Laboratories. *Eur. Phys. J. E* **2015**, *38*, 27.

23. Sechenyh, V.; Legros, J. C.; Mialdun, A.; Ortiz de Zárate, J. M.; Shevtsova, V. Fickian Diffusion in Ternary Mixtures Composed by 1,2,3,4-Tetrahydronaphthalene, Isobutylbenzene, and *n*-Dodecane. *J. Phys. Chem. B* **2016**, *120* (3), 535.

24. Galliero, G.; Bataller, H.; Croccolo, F.; Vermorel, R.; Artola, P.-A.; Rousseau, B.; Vesovic, V.; Bou-Ali, M.; Ortiz de Zárate, J. M.; Xu, S.; Zhang, K.; Montel, F. Impact of Thermodiffusion on the Initial Vertical Distribution of Species in Hydrocarbon Reservoirs. *Microgravity. Sci. Technol.* **2016**, *28*, 79.

25. Alizadeh, A.; Nieto de Castro, C. A.; Wakeham, W. A. The Theory of the Taylor Dispersion for Technique for Liquid Diffusivity Measurements. *Int. J. Thermophys.* **1980**, *1*, 243.

26. Santos, C. I. A. V.; Esteso, M. A.; Lobo, V. M. M.; Ribeiro, A. C. F. Multicomponent Diffusion in (Cyclodextrin-Drug + Salt + Water) Systems: {2-Hydroxypropyl-β-cyclodextrin (HP-βCD) + KCl + Theophylline + Water}, and {β-Cyclodextrin (βCD) + KCl + Theophylline + Water}. *J. Chem. Thermodyn.* **2013**, *59*, 139.

27. Santos, C. I. A. V.; Esteso, M. A.; Lobo, V. M. M.; Cabral, A. M. T. D. P. V.; Ribeiro, A. C. F. Taylor Dispersion Technique as a Tool for Measuring Multicomponent Diffusion in Drug Delivery Systems at Physiological Temperature. *J. Chem. Thermodyn.* **2015**, *84*, 76.

28. Ribeiro, A. C. F.; Gomes, J. C. S.; Santos, C. I. A. V.; Lobo, V. M. M.; Esteso, M. A.; Leaist, D. G. Ternary Mutual Diffusion Coefficients of Aqueous $NiCl_2$ + NaCl and $NiCl_2$ + HCl Solutions at 298.15 K. *J. Chem. Eng. Data* **2011**, *56*, 4696.

29. Mialdun, A.; Sechenyh, V.; Legros, J. C.; Ortiz de Zárate, J. M.; Shevtsova, V. Investigation of Fickian Diffusion in the Ternary Mixture of 1,2,3,4-Tetrahydronaphthalene, Isobutylbenzene, and Dodecane. *J. Chem. Phys.* **2013**, *139*, 104903.

30. Santos, C. I. A. V.; Shevtsova, V.; Burrows, H. D.; Ribeiro, A. C. F. Optimization of Taylor Dispersion Technique for Measurement of Mutual Diffusion in Benchmark Mixtures. *Microgravity Sci. Technol.* **2016**, *28*, 459.

31. Butler, B. D.; Turner, J. C. R. Flow-Cell Studies of Thermal Diffusion in Liquids. Part 1—Cell Construction and Calibration. *Trans. Faraday Soc.* **1966**, *62*, 3114–3120.

32. Bou-Ali, M. M.; Platten, J. K. Metrology of the Thermodiffusion Coefficients in a Ternary System. *J. Non-equilib Thermodyn.* **2005**, *30*, 385–399.

33. Davarzani, H.; Marcoux, M.; Costeseque, P.; Quintard, M. Experimental Measurement of the Effective Diffusion and Thermodiffusion Coefficients for Binary Gas Mixture in Porous Media. *Chem. Eng. Sci.* **2010**, *65* (18), 5092–5104.

34. Trengove, R. D.; Robjohns, H. L.; Martin, M. L.; Dunlop, P. J. Thermal Diffusion Factors at 300 K for Seven Binary Noble Gas Systems Containing Helium or Neon. *Physica* **1981**, *108A*, 488–501.

35. Schimpf, M. E.; Giddings, J. C. Characterization of Thermal Diffusion in Polymer Solutions: Dependence on Polymer and Solvent Parameters. *J. Polym. Sci.* **1989**, *27*, 1317–1332.

36. Duhr, S.; Braun, D. Thermophoretic Depletion Follows Boltzmann Distribution. *Phys. Rev. Lett.* **2006**, *96*, 168301.

37. Duhr, S.; Braun, S. Why Molecules Move Along a Temperature Gradient. *Proc. Natl. Acad. Sci. U.S.A.* **2006**, *103* (52), 19678–19682.

38. Kolodner, P.; Williams, H.; Moe, C. Optical Measurement of the Soret Coefficient of Ethanol/Water Solutions. *J. Chem. Phys.* **1988**, *88*, 6512–6524.

39. Mialdun, A.; Shevtsova, V.; Measurement of the Soret and Diffusion Coefficients for Benchmark Binary Mixtures by Means of Digital Interferometry. *J. Chem. Phys.* **2011**, *134*, 044524.

40. Polyakov, P.; Zhang, M.; Müller-Plathe, F.; Wiegand, S. Thermal Diffusion Measurements and Simulations of Binary Mixtures of Spherical Molecules. *J. Chem. Phys.* **2007**, *127*, 14502.

41. Köhler, W.; Rosenauer, C.; Rossmanith, P. Holographic Grating Study of Mass and Thermal Diffusion of Polystyrene/Toluene Solutions. *Int. J. Thermophys.* **1995**, *16*, 11–21.

42. Polyakov, P.; Wiegand, S. Systematic Study of the Thermal Diffusion in Associated Mixtures. *J. Chem. Phys.* **2008**, *128*, 34505.

43. Alves, S.; Figueiredo Neto, A. M.; Bourdon, A. Generalization of the Thermal Lens Model Formalism to Account for Thermodiffusion in a Single-Beam Z-Scan Experiment: Determination of The Soret Coefficient. *J. Opt. Soc. Am. B* **2003**, *20* (4), 71.

44. Blanco, P.; Bou-Ali, M. M.; Platten, J. K.; de Mezquia, D. A.; Madariaga, J. A.; Santamaria, C. M. Thermodiffusion Coefficients of Binary and Ternary Hydrocarbon Mixtures. *J. Chem. Phys.* **2010**, *132*, 114506.

45. Koniger, A.; Wunderlich, H.; Kohler, W. Measurement of Diffusion and Thermal Diffusion in Ternary Fluid Mixtures using a Two-Color Optical Beam Deflection Technique. *J. Chem. Phys.* **2010**, *132*, 174506.

46. Ryzhkov, I. I.; Shevtsova, V. Long-Wave Instability of a Multicomponent Fluid Layer with The Soret Effect. *Phys. Fluids* **2009**, *21*, 014102.

47. Shevtsova, V.; Melnikov, D. E.; Legros, J. C. Onset of Convection in Soret-Driven Instability. *Phys. Rev. E* **2006**, *73*, 047302.

48. Shevtsova, V.; Santos, C.; Sechenyh, V.; Legros, J. C.; Mialdun, A. Diffusion and Soret in Ternary Mixtures. Preparation of the DCMIX2 Experiment on the ISS. *Microgravity Sci. Technol.* **2014**, *25*, 275.

49. Wilbois, B.; Galliero, G.; Caltagirone, J.-P.; Montel, F. Macroscopic Model of Multicomponent Fluids in Porous Media. *Philos. Mag.* **2003**, *83*, 2209.

50. Galliero, G.; Duguay, B.; Caltagirone, J.-P.; Montel, F. On Thermal Diffusion in Binary and Ternary Lennard-Jones Mixtures by Non-Equilibrium Molecular Dynamics. *Philos. Mag.* **2003**, *83*, 2097.

51. Galliero, G.; Montel, F. Nonisothermal Gravitational Segregation by Molecular Dynamics Simulations. *Phys. Rev. E* **2008**, *78*, 041203.

52. Platten, J. K.; Bou-Ali, M. M.; Costeseque, P.; Dutrieux, J. F.; Kohler, W.; Leppla, C.; Wiegand, S.; Wittko, G. Benchmark Values for the Soret, Thermal Diffusion and Diffusion Coefficients of Three Binary Organic Liquid Mixtures. *Phil. Mag.* **2003**, *83*, 1965.

53. Croccolo, F.; Bataller, H.; Scheffold, F. A Light Scattering Study of Non Equilibrium Fluctuations in Liquid Mixtures to Measure the Soret and Mass Diffusion Coefficient. *J. Chem. Phys.* **2012**, *137*, 234202.

54. Nagata, I. Liquid–Liquid Equilibria for Four Ternary Systems Containing Methanol and Cyclohexane. *Fluid Phase Equilib.* **1984**, *18*, 83.

55. Robinson, R. A.; Stokes, R. H. *Electrolyte Solutions*; 2nd ed., Butterworths: London, 1959.

56. Tyrrell, H. J. V.; Harris, K. R. *Diffusion in Liquids*; 2nd Ed., Butterworths: London, 1984.

57. Köhler, W.; Morozov, K. I. The Soret Effect in Liquid Mixtures. *J. Non-Equilib. Thermodyn.* **2016**, *41* (3), 151–197.

58. de Groot, S. R.; Mazur, P. *Non-Equilibrium Thermodynamics*; Dover: New York, 1984.
59. Hartmann, S.; Wittko, G.; Köhler, W.; Morozov, K. I.; Albers, K.; Sadowski, G. Thermophobicity of Liquids: Heats of Transport in Mixtures as Pure Component Properties. *Phys. Rev. Lett.* **2012**, *109*, 065901.
60. Hartmann, S.; Wittko, G.; Schock, F.; Groß, W.; Lindner, F.; Köhler, W.; Morozov, K. I. Thermophobicity of Liquids: Heats of Transport in Mixtures as Pure Component Properties—The Case of Arbitrary Concentration. *J. Chem. Phys.* **2014**, *141*, 134503.
61. Denbigh, K. G. The Heat of Transport in Regular Solutions. *Trans. Faraday Soc.* **1952**, *48*, 1.
62. Shukla, K.; Firoozabadi, A. A New Model of Thermal Diffusion Coefficients in Binary Hydrocarbon Mixtures. *Ind. Eng. Chem. Res.* **1998**, *37*, 3331.
63. Haugen, K. B.; Firoozabadi, A. Transient Separation of Multicomponent Liquid Mixtures in Thermogravitational Columns. *J. Chem. Phys.* **2007**, *127*, 154507.
64. Abbasi, A.; Saghir, M. Z.; Kawaji, M. A New Approach to Evaluate the Thermodiffusion Factor for Associating Mixtures. *J. Chem. Phys.* **2009**, *130*, 064506.
65. Lucas, P.; Tyler, A. Thermal Diffusion Ratio of a 3He/4He Mixture Near its λ Transition: The Onset of Heat Flush. *J. Low Temp. Phys.* **1977**, *27*, 281.
66. Legros, J. C.; Gaponenko, Y.; Mialdun, A.; Triller, T.; Hammon, A.; Bauer, C.; Kohler, W.; Shevtsova, V. Investigation of Fickian Diffusion in the Ternary Mixtures of Water–Ethanol–Triethylene Glycol and Its Binary Pairs. *Phys. Chem. Chem. Phys.* **2015**, *17*, 27713.
67. Barthel, J.; Gores, H. J.; Lohr, C. M.; Seidl, J. J. Taylor Dispersion Measurements at Low Electrolyte Concentrations. I. Tetraalkylammonium Perchlorate Aqueous Solutions. *J. Sol. Chem.* **1996**, *25*, 921–935.
68. Callendar, R.; Leaist, D. G. Diffusion Coefficients for Binary, Ternary and Polydisperse Solutions from Peak-Width Analysis of Taylor Dispersion Profiles. *J. Sol. Chem.* **2006**, *35*, 353–379.
69. Deng, Z.; Leaist, D. G. Ternary Mutual Diffusion Coefficients of $MgCl_2 + MgSO_4 + H_2O$ and $Na_2SO_4 + MgSO_4 + H_2O$ from Taylor Dispersion Profiles. *Can. J. Chem.* **1991**, *69*, 1548.
70. Mialdun, A.; Yasnou, V.; Shevtsova, V.; Königer, A.; Köhler, W.; Alonso de Mezquia, D.; Bou-Ali, M. M. A Comprehensive Study of Diffusion, Thermodiffusion, and Soret Coefficients of Water-Isopropanol. *J. Chem. Phys.* **2012**, *136*, 244512.
71. Königer, A.; Meier, B.; Köhler, W. Measurement of the Soret, Diffusion, and Thermal Diffusion Coefficients of Three Binary Organic Benchmark Mixtures and of Ethanol–Water Mixtures Using a Beam Deflection Technique. *Philos. Mag.* **2009**, *89*, 907.
72. Mialdun, A.; Shevtsova, V. Development of Optical Digital Interferometry Technique for Measurement of Thermodiffusion Coefficients. *Int. J. Heat Mass Transfer* **2008**, *51*, 3164.
73. Vigolo, D.; Brambilla, G.; Piazza, R. Thermophoresis of Microemulsion Droplets: Size Dependence of the Soret Effect. *Phys. Rev. E* **2007**, *75*, 040401.
74. Platten, J. K.; Legros, J. C. *Convection in Liquids*; Springer, Berlin, 1984.
75. Harstad, K.; Bellan, J. An All-Pressure Fluid Drop Model Applied to a Binary Mixture: Heptane in Nitrogen. *Int. J. Multiphase Flow* **2000**, *26* (10), 1675–1706.
76. Santos, C. I. A. V.; Esteso, M. A.; Sartorio, R.; Ortona, O.; Sobral, A. J. N.; Arranja, C. T.; Lobo, V. M. M.; Ribeiro, A. C. F. A Comparison Between the Diffusion Properties of Theophylline/β-Cyclodextrin and Theophylline/2-Hydroxypropyl-β-Cyclodextrin in Aqueous Systems. *J. Chem. Eng. Data* **2012**, *57*, 1881–1886.

77. Ribeiro, A. C. F.; Leaist, D. G.; Esteso, M. A.; Lobo, V. M. M.; Valente, A. J. M.; Santos, C. I. A. V.; Cabral, A. M. T. D. P. V.; Veiga, F. J. B. Binary Mutual Diffusion Coefficients of Aqueous Solutions of β-Cyclodextrin at Temperatures from 298.15 to 312.15 K. *J. Chem. Eng. Data* **2006**, *51*, 1368–1371.

78. Ribeiro, A. C. F.; Santos, C. I. A. V.; Lobo, V. M. M.; Esteso, M. A.; Quaternary Diffusion Coefficients of β-Cyclodextrin+ KCl+ Caffeine+ Water at 298.15 K using a Taylor Dispersion Method. *J. Chem. Eng. Data*, **2010**, *55*, 2610–2612.

79. Sechenyh, V.; Legros, J. C.; Shevtsova, V. Development and Validation of a New Setup for Measurements of Diffusion Coefficients in Ternary Mixtures Using the Taylor Dispersion Technique. *C. R. Mecanique* **2013**, *341*, 490–496.

80. Larrañaga, M.; Bou-Ali, M. M.; Soler, D.; Martinez-Agirre, M.; Mialdun, A.; Shevtsova, V. Remarks on the Analysis Method for Determining Diffusion Coefficient in Ternary Mixtures. *C. R. Mecanique* **2013**, *341*, 356–364.

81. Lapeira, E.; Gebhardt, M.; Triller, T.; Mialdun, A.; Köhler, W.; Shevtsova, V.; Bou-Ali, M. M. Transport Properties of the Binary Mixtures of the Three Organic Liquids Toluene. *J. Chem. Phys.* **2017**, *146*, 094507.

82. Bou-Ali, M. M.; Ecenarro, O.; Madariaga, J. A.; Santamaría, C. M.; Valencia, J. J. Measurement of Negative Soret Coefficients in a Vertical Fluid Layer with an Adverse Density Gradient. *Phys. Rev. E* **2000**, *62*, 1420.

83. Story, M. J.; Turner, J. C. R. Flow-Cell Studies of Thermal Diffusion in Liquids. Part 5—Binary Mixtures of CH_3OH with CCl_4, Benzene and Cyclohexane at 25°C. *Trans. Faraday Soc.* **1969**, *65*, 1523.

84. Santos, C. I. A. V.; Shevtsova, V.; Ribeiro, A. C. F. Isothermal Molecular Diffusion in Mixtures Containing Toluene, Cyclohexane and Methanol. *Eur. Phys. J. E* **2017**, *40*, 40.

85. Grossmann, T.; Winkelmann, J. Ternary Diffusion Coefficients of Cyclohexane + Toluene + Methanol by Taylor Dispersion Measurements at 298.15 K. Part 1. Toluene-Rich Area. *J. Chem. Eng. Data* **2009**, *54*, 405.

86. Grossmann, T.; Winkelmann, J. Ternary Diffusion Coefficients of Cyclohexane + Toluene + Methanol by Taylor Dispersion Measurements at 298.15 K. Part 2. Low Toluene Area Near the Binodal Curve. *J. Chem. Eng. Data*, **2009**, *54*, 485.

87. Guevara-Carrion, G.; Gaponenko, Y.; Janzen, T.; Vrabec, J.; Shevtsova, V. Diffusion in Multicomponent Liquids: From Microscopic to Macroscopic Scales. *J. Phys. Chem. B* **2016**, *120* (47), 12193.

PART IV
New Insights

CHAPTER 15

THE ROLE OF CHEMOINFORMATICS IN MODERN DRUG DISCOVERY AND DEVELOPMENT

ASHIMA BAGARIA*

Department of Physics, Manipal University Jaipur, Jaipur, Rajasthan, India

E-mail: ashima.bagaria@jaipur.manipal.edu

ABSTRACT

The major highlight of this review is the impact of chemoinformatics on Modern Drug Discovery and Development process. The differences in traditional approach and modern approach in drug discovery are provided and how chemoinformatics has expedited the process has been discussed. The application of chemoinformatics discussed herein includes compound identification, virtual library generation, and screening; high throughput data mining, and in silico absorption, distribution, metabolism, excretion, and toxicity are also reviewed here.

15.1 INTRODUCTION

The knowledge acquired in the field of chemistry is mainly from observations and the data associated with these observations. In recent times, theoretical chemistry has gained momentum and is able to make concrete predictions, relevant for certain chemical phenomenon. But research in this field has not yet reached to the pinnacle where we could simply base our conclusions on mere predictions. There are many chemical phenomena, which are too complex for a reliable prediction, and we still thrive on experimental approaches. But knowledge like this would be extremely helpful prior to setting up the experimental bench.

The events of using computers for elucidating chemical phenomena date back to the 1960s, where a series of attempts were made in developing a completely new field. The methods were developed for storing and retrieving chemical information from publically accessible databases,[1,2] quantitative structure–activity relationship (QSAR), structure–property relationship,[3] mathematical methods for modeling such relationships, etc. Spectroscopic data were studied extensively for predicting structure information.[4,5] Artificial intelligence algorithms were applied to certain chemical problems.[6] Groups from Harvard, Stony Brook, and Munich developed systems for designing organic syntheses (Computer Assisted Synthesis Design).[7–10]

Since last few decades, study on these approaches has paved way for an interdisciplinary platform where the confluence of chemistry, mathematics, and computer science has brought in a new dimension to interdisciplinary research. Many terms were coined for this filed, "chemical information," "computational chemistry," to name a few. The latter was inclined more toward processing chemical information via new computational techniques, while the former was not able to differentiate between the librarianship and computer methods. The consensus was made on "chemoinformatics," which was defined in many ways, some of which are "The use of information technology and management has become a critical part of the drug discovery process. Chemoinformatics is the mixing of those information resources to transform data into information and information into knowledge for the intended purpose of making better decisions faster in the area of drug lead identification and optimization."[11] "Chem(o)informatics is a generic term that encompasses the design, creation, organization, management, retrieval, analysis, dissemination, visualization, and use of chemical information."[12] "Chemoinformatics is the application of informatics methods to solve chemical problems."[13] Opinion has now shifted toward acceptance of chemoinformatics as a discipline although not everyone agrees about the definition.

Cheminformatics is thus means the use of computational and informational techniques to understand problems of chemistry. Cheminformatics strategies and tools are useful in drug discovery and other efforts where large numbers of compounds are being evaluated for specific properties.

Three major aspects of chemoinformatics are as follows:

1. Information acquisition—process of generating and collecting data experimental and theoretically (simulations).
2. Information management—storage and retrieval of information.
3. Use of information, data analysis, correlation, and application.

15.2 CHEMOINFORMATICS AND MODERN DRUG DISCOVERY AND DEVELOPMENT

The area that has benefited the maximum form chemoinformatics is drug design and development. Various developments have been made in the area of lead discovery and optimization. It is a vital link between theoretical design and drug design in vitro. A huge investment and strong competition in the area of drug design may lead to curtailing the hit and trial approach. A theoretical and computational approach was targeted wherein the active molecules would be tailored accordingly making the drug design process efficient. The usual drug discovery and development process take around 10–15 years and cost approximate 1 billion USD. Figure 15.1 shows the schematics of a traditional drug discovery and development process.

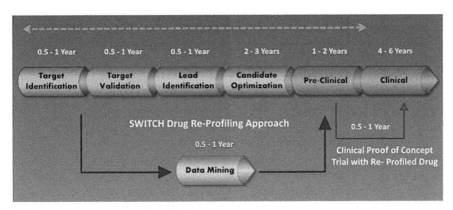

FIGURE 15.1 **(See color insert.)** The traditional drug discovery and development process.

Table 15.1 highlights various disciplines involved in drug designing at various stages. The information explosion created by generation of huge amount of chemical data has raised a demand for organizing and analyzing the chemical information for modern drug discovery and development.

TABLE 15.1 Disciplines Involved in Various Stages of Drug Discovery.

Objective	Discipline
Target identification	Bioinformatics: genomics, proteomics synthesis biology pathway analysis
Lead finding and read optimization	Chemoinformatics: high throughput screening, QSAR, combinatorial chemistry, ADMET

ADMET, absorption, distribution, metabolism, excretion, and toxicity; QSAR, quantitative structure–activity relationship.

According to Figure 15.1, the traditional drug discovery process starts with a particular disease, identification of target, identification of inhibitor molecule, and preclinical testing. The investment of time and money is tremendous and the probability of false positives is huge thereby limiting the research in this field to the traditional way.[14,15]

The modern drug discovery development process is shown in Figure 15.2. The major issues of traditional drug discovery process are replaced with bad identification and optimization. Each phase in modern drug discovery has an interaction component, which transfers data and information to another phase.

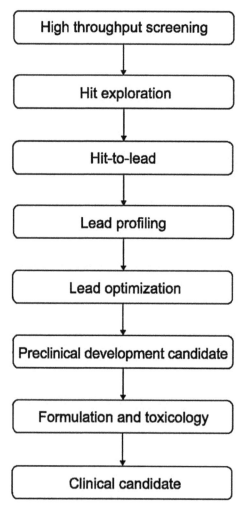

FIGURE 15.2   Flowchart of modern drug discovery design and development.

The major bottlenecks in the traditional drug discovery and development method were the time and cost of testing new entities for the inhibitor toward the target molecules. This leads to the search of new technologies that would curtail both time and cost. The advent of high throughput screening (HTS) and automation has replaced the old handcrafted synthesis and testing approaches.[16,17]

15.2.1 COMBINATORIAL CHEMISTRY

Combinatorial explosion always results in severe problems in ligand-based drug design. With the advantage of HTS, it was thought that the drug discovery process would be accelerated but this could not happen. This size of chemical space is approximately 10^{20}–10^{100}. Combinatorial features of two-dimensional chemical graphs have resulted in this huge chemical space that would contain false positives as well.[18-20] Combining fragments was one approach toward understanding and minimizing this chemical space explosion. But even with the improved computational speed and resources, it is still difficult to overcome this explosion by combining fragments.

With chemical space explosion, chemical diversity of compound libraries needs to be optimized in order to accelerate the drug discovery process. Chemoinformatics approaches are proving to be fruitful in optimizing the libraries. Hence, various technologies catering to structural processing diversity analyses have been developed. These computational approaches were the main component of chemoinformatics and were developed for augmenting chemically diverse computations for structural descriptors, structural similarity algorithms and their classification library, enumerations, and compound selections to name a few.

15.2.2 DRUG-LEAD LIKENESS

Many of the hits found from the chemically diverse libraries may or may not contain drug-like compounds. Thus, it becomes utmost necessary to filter out the nondrug like compounds from these libraries.[21-27] The question to answer here is whether there is any difference between drug-like and lead-like molecules. If this question is answered, it will become easy to further refine the filtering technology of chemically diverse libraries.[28,29]

15.2.3   DRUG DISCOVERY PROCESS AND ABSORPTION, DISTRIBUTION, METABOLISM, EXCRETION, AND TOXICITY PREDICTION

The completion of human genome project has led to the discovery of various new targets for drug design and development, but their structural information is scarce. To overcome this high throughput, protein crystallization has been explored.[30,31] Majority of these targets are membrane proteins, and their three-dimensional structure determination is still very difficult. Hence, the lead optimization is the utmost bottleneck in the drug discovery process.

15.2.4   LEAD SELECTION/IDENTIFICATION/DISCOVERY

For a large member of available targets, combinatorial chemistry leads to a compound explosion that is beyond the capacity of HTS.[32] Thus, there should exist a method that could select the best ones from large pool of compounds thus generated.

This strategy requires

1) Selected compounds should provide complementary diversity to existing libraries.
2) Select from a pool of compound represents diversity.
3) Selection of reagents that maximizes diversity.
4) Selection of compounds that are similar to known liquids but have novel scaffolds.

Many approaches have been carried out to fulfill the above criteria. Neural networks[33] have been used to assess and ascertain the diversity of combinational libraries. Similarly, Sheridan and Kearsley have used genetic algorithm to suggest the libraries.[34] Structural-based clustering methods for compound selections have been used rigorously in compound selections.[35] MDL Unity [Tripos], Daylight two-dimensional descriptors, etc. were used at Abbot Laboratories for the same.

The analysis of the major work of the researchers have revealed that hierarchical clustering[36,37] and two-dimensional descriptors are best methods for filtering out active and inactive biological molecules. Agrafiotis et al. have developed a large number of algorithms for analyzing diversity of compound libraries—stochastic algorithms for maximizing diversity using distance-based measurement is one of the approaches that met the criteria.[38,39]

Studies on library design, involving comparison between random selections, and maximum and minimum dissimilarly based selection show that the later gives more stable QSAR models with high predictive models.[40] Cell-based algorithm approach was also carried out that reduced the dimensionality by identifying the metrics or axes that predicts the information about affinity for a given receptor.[41] Many other chemoinformatics approaches have been successful in finding the optimal solution to compound selection.[42–51]

15.2.5 VIRTUAL LIBRARY GENERATION AND SCREENING

The widespread adoption of automated HTS and ultra-HTS has significantly contributed toward cutting the cost of the drug discovery process. But increased HTS efforts have not significantly increased the success rate of drug discovery and development,[52] rather it has led to various other bottlenecks like lead optimization and target validation.

The molecular space of lead like compounds is still very small and this has paved way for in silico approaches of virtual screening.[53–61]

While generating a virtual library diversity, absorption, distribution, metabolism, excretion, and toxicity (ADMET) properties and accessibility need to be carefully examined. Major challenges that occur while making a virtual library are fulfilling ADMET criteria and synthetic accessibility.

The advent of HTS technology has made virtual screening a major tool for identifying new leads.[67] It is an important computational tool that can filter out lead like compounds from the unwanted compounds in the prepared in silico libraries.[63,64] This in fact is required to maintain the cost effectiveness of the process. To address this, structure coding has to be appropriately chosen so as to it corresponds to the biological activity under consideration.[65,66]

If the target structure is known, then docking strategy is the best known approach for virtual screening (structure based).[67,68] If target structure is unknown but ligands are known, then similarity based approaches often give good leads.[69–71] If neither target nor the ligand is known, then statistical approaches derived from experiment data are the best known method. The structure activity relationship (SAR) measurement is one such method.[72–74] Good et al. have indicated that the integration of HTS and virtual screening are a powerful approach in modern drug discovery.[75]

15.2.6 SEQUENTIAL SCREENING

The purpose of sequential HTS is to make the optimal use of the information obtained from receptor–ligand interaction in order to discover new leads and minimize the cost of drug discovery and development. Figure 15.3 shows the scheme of sequential HTS.

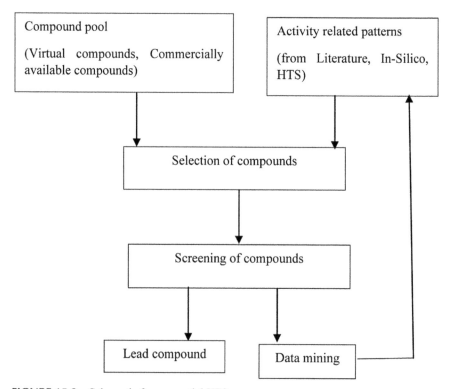

FIGURE 15.3   Schematic for sequential HTS.

15.2.7 IN SILICO ADMET

Over the decades, properties related to ADMET have become the most important issue, assessing the risks of these compounds on humans.[75–78] Thus, it becomes necessary to eliminate false positives, thereby avoiding costs of the drug discovery process. Thrust is applied in the areas of developing computational methods fulfilling the ADMET criteria.

i) *Absorption:* The predominant pathway for absorption of lipophilic compounds is the transcellular passage through the membrane lipid/aqueous environment.[79] Thus, lipophilicity is a key property in drug design. Lipinski rule of five like Log P, molecular weight (<500 Da), H-bonding, free energy, H-bond donor–acceptor are the major descriptors in introducing predictive models for the drug candidates.

ii) *Distribution:* Distribution is the movement of drugs throughout the body. The major issue is of the central nervous system drugs, which should be able to cross the blood–brain barrier (BBB). Thus, the prediction of brain penetration is a necessary requirement for the new lead, and the in vitro techniques for these predictions are very cumbersome. A small data set has been analyzed by the scientists for Log (C_{Brain}/C_{Blood}) or Log BB and Log P (octanol–cyclohexane).[80-82] Recently, many new descriptors have been correlated with Log BB.[83-85] Regression analysis on Log BB and free energy have also generated good result in this aspect.

iii) Human serum albumin (HSA) is the major transporter of the drugs and metabolites in different tissues. Thus, the binding of the drug to serum albumin is the major factor for availability of that drug to target tissues.[86] QSAR studies come handy for drug distribution analysis. Scientists have used genetic algorithms to select the preferred molecular descriptor for distribution analysis. According to their study, hydrophobicity is by way largely responsible in determining the binding strength of metabolites and drugs to HSA.[87] Many groups have built models for BBB penetration prediction using a substructure pattern recognition method.[88] Their results have demonstrated the accuracies of 98.8% and 98.4% for the training set and the data set, respectively.

iv) *Metabolism:* HTS has been used to study metabolism.[89,90] The drugs are broken down to harmless metabolites and are secreted through urine or bile. The five major isoforms of human cytochrome P450 (CYP) that metabolize majority of the approved drugs are CYP1A2, 2C9, 2C19, 2D6, and 3A4.[91] Thus, the elucidation of substrate specificity of different isoforms of CYP will provide and understanding of whether the drug will be metabolized or not. Mishra et al. have developed several models for prediction of CYP1A2, 2C9, 2C19, 2D6, and 3A4 substrates based on approved drug data. They have used support vector machine algorithm[92] for the prediction. The accuracy achieved was 70.6%. Structure metabolism studies have also been carried out studying the drug metabolism.[93] But the high

   degree of structure similarity in combinatorial libraries with a
   common core restricts their application in metabolic analysis.

 v) *Elimination/Excretion:* Nonsteroidal, anti-inflammatory drugs are
   used in long-term treatment, and their accumulation in the body
   may lead to serious side effects. Thus, half-cycle prediction of the
   leads/drug like molecules becomes important. It will determine
   the time for which the drug will persist in the body. Many qualitative predictions of half-life could be possibly made by the use of
   chemoinformatics.[94]

 vi) *Toxicity:* Toxicity is a major concern for the in silico drug design.
   ADME (Absorption, Distribution, Metabolism and Excretion)
   analysis is an essential part of rational drug discovery process.
   Many drugs are withdrawn due to toxicity owing to metabolism,
   excretion. The toxicity prediction approaches are mechanistic,
   which require the involvement of human experts who assesses
   mechanism of interaction with biological systems, and correlative
   that may require statistical analysis of test data sets.[95] The correlative approach has the potential to discover new SARs,[96,97] thereby
   providing a lead to protocol of interaction of the chemicals with
   the biological systems. The early toxicity measuring methods were
   based on QSAR approach, predicting the LD_{50} based on various
   descriptors.[98,99] Artificial neural networks have also been applied in
   toxicity prediction.[100–102]

15.3 CONCLUSION

Chemicals are important in cellular function. This has led to the emergence
of fields like Chemical Biology, Chemoinformatics, and latest being
Systems Chemistry and Systems Chemical Biology. Chemoinformatics has
contributed in various aspects like descriptors and chemical structure database
retrieval, linear notations, canonicalization, multidimensional scaling, etc.
to name a few. One of the major contributions of chemoinformatics is in the
field of modern drug discovery and development. This field unites chemical
information on drug compounds and biological information on drug targets.
The methods in this review are being used in the drug discovery process
and by far have been expatiated the process. This review primarily focuses
on drug discovery issues of traditional methods and modern drug discovery
systems.

KEYWORDS

- **chemoinformatics**
- **drug discovery**
- **in silico**
- **compound identification**
- **virtual library**
- **data mining**
- **toxicity**

REFERENCES

1. Dyson, G. M.; Lynch, M. F.; Morgan, H. L. A Modified IUPAC-Dyson Notation System for Chemical Structures. *Inf. Storage Retr.* **1968**, *4*, 27–83.
2. Tate, F. A. Handling Chemical Compounds in Information Systems. *Ann. Rev. Inf. Sci. Technol.* **1967**, *2*, 285–309.
3. Hansch, C.; Fujita, T. p–σ–π Analysis. A Method for the Correlation of Biological Activity and Chemical Structure. *J. Am. Chem. Soc.* **1964**, *86*, 856–864.
4. Sasaki, S. I.; Abe, T.; Ouki, T.; Sakamoto, M.; Ochiai, S. Automated Structure Elucidation of Several Kinds of Aliphatic and Alicyclic Systems. *Anal. Chem.* **1968**, *40*, 2220–2223.
5. Shelley, C. A.; Hays, T. R.; Munk, M. E.; Roman, H. V. An Approach to Automated Partial Structure Expansion. *Anal. Chim. Acta* **1978**, *103*, 121–132.
6. Lindsay, R. K.; Buchanan, B. G.; Feigenbaum, E. A.; Lederberg, J. *Applications of Artificial Intelligence for Organic Chemistry: The DENDRAL Project*; McGraw-Hill: New York, NY, USA, 1980.
7. Blair, J.; Gasteiger, J.; Gillespie, C.; Gillespie, P. D.; Ugi, I. Representation of the Constitutional and Stereochemical Features of Chemical Systems in the Computer-Assisted Design of Syntheses. *Tetrahedron* **1974**, *30*, 1845–1859.
8. Hendrickson, J. B. Systematic Characterization of Structures and Reactions for Use in Organic Synthesis. *J. Am. Chem. Soc.* **1971**, *93*, 6847–6854.
9. Corey, E. J.; Wipke, W. T. Computer-Assisted Design of Complex Organic Syntheses. *Science* **1969**, *166*, 178–193.
10. Gelernter, H. L.; Sridharan, N. S.; Hart, A. J.; Yen, S.-C. The Discovery of Organic Synthetic Routes by Computer. *Top. Curr. Chem.* **1973**, *41*, 113–150.
11. Brown, F. K. Chemoinformatics: What is it and How does it Impact Drug Discovery. *Annu. Rep. Med. Chem.* **1998**, *33*, 375–384.
12. Hann, M.; Green, R. Chemoinformaticss—A New Name for an Old Problem. *Curr. Opin. Chem. Biol.* **1999**, *3*, 379–383.
13. Gasteiger, J.; Engel, T. *Chemoinformaticss—A Textbook*; Wiley-VCH: Weinheim, 2003; p 600.

14. Arulmozhi, V.; Rajesh, R. *Chemoinformatics—A Quick Review*; 3rd International Conference on Electronics Computer Technology (ICECT 2011), Kanyakumari, India, IEEE, 2011, 978-1-4244-8679-3/11. Vol. 6.

15. Marshall, G. R. *Introduction to Chemoinformatics in Drug Discovery—A Personal View*; Wiley-VCH Verlag GmbH & Co. KGaA: Weinheim, 2004, ISBN: 3-527-30753-2.

16. Gallop, M. A.; Barrett, R. W.; Dower, W. J.; Fodor, S. P. A.; Gordon, E. M. Applications of Combinatorial Technologies to Drug Discovery. 1. Background and Peptide Combinatorial Libraries. *J. Med. Chem.* **1994**, *37*, 1233–1251.

17. Hecht, P. High-Throughput Screening: Beating the Odds with Informatics-Driven Chemistry. *Curr. Drug Discov.* **2002**, 21–24.

18. Bohacek, R. S.; McMartin, C.; Guida, W. C. The Art and Practice of Structure-Based Drug Design: A Molecular Modeling Perspective. *Med. Res. Rev.* **1996**, *16*, 3–50.

19. Kirkpatrick, P.; Ellis, C. Chemical Space. *Nature* **2004**, *432*, 823–823.

20. Ertl, P. Cheminformatics Analysis of Organic Substituents: Identification of the Most Common Substituents, Calculation of Substituent Properties, and Automatic Identification of Drug-Like Bioisosteric Groups. *J. Chem. Inf. Comput. Sci.* **2003**, *43*, 374–380.

21. Bemis, G. W.; Murcko, M. A. The Properties of Known Drugs. 1. Molecular Frameworks. *J. Med. Chem.* **1996**, *39*, 2887–2893.

22. Sadowski, J.; Kubinyi, H. A Scoring Scheme for Discriminating between Drugs and Non-Drugs. *J. Med. Chem.* **1998**, *41*, 3325–3329.

23. Ajay, A.; Walters, W. P.; Murcko, M. A. Can We Learn to Distinguish between "Drug-Like" and "Non-Drug-Like" Molecules? *J. Med. Chem.* **1998**, *41*(18), 3314–3324.

24. Matter, H.; Baringhaus, K.-H.; Naumann, T.; Klabunde, T.; Pirard, B. Computational Approaches Towards the Rational Design of Drug-Like Compound Libraries. *Comb. Chem. High Throughput Screening* **2001**, *4*, 453–475.

25. Xu, J.; Stevenson, J. Drug-Like Index: A New Approach to Measure Drug-Like Compounds and Their Diversity. *J. Chem. Inf. Comput. Sci.* **2000**, *40*, 1177–1187.

26. Lipinski, C. A.; Lombardo, F.; Dominy, B. W.; Feeney, P. J. Experimental and Computational Approaches to Estimate Solubility and Permeability in Drug Discovery and Development Settings. *Adv. Drug Delivery Rev.* **1997**, *23*, 3–25.

27. Clark, D. E.; Pickett, S. D. Computational Methods for the Prediction of 'Drug-Likeness'. *Drug Discov. Today* **2000**, *5*, 49–58.

28. Proudfoot, J. R. Drugs, Leads, and Drug-Likeness: An Analysis of Some Recently Launched Drugs. *Bioorg. Med. Chem. Lett.* **2002**, *12*(12), 1647–1650.

29. Oprea, T. I.; Davis, A. M.; Teague, S. J.; Leeson, P. D. Is There a Difference between Leads and Drugs? A Historical Perspective. *J. Chem. Inf. Comput. Sci.* **2001**, *41*, 1308–1315.

30. Luft, J. R.; Wolfley, J.; Collins, R.; Bianc, M.; Weeks, D.; Jurisica, I.; Rogers, P.; Glasgow, J.; Fortier, S.; DeTitta, G. T. Macromolecular Crystallization in a High Throughput Laboratory—The Search Phase.. *J. Cryst. Growth.* **2001**, *232,* 591–595.

31. Stewart, L.; Clark, R.; Behnke, C. High-Throughput Crystallization and Structure Determination in Drug Discovery. *Drug Discov. Today* **2002**, *7*, 187–196.

32. Chemical Computing Group, Inc., 1010 Sherbrooke Street West, Suite 910, Montreal, QC, Canada, H3A 2R7, Tel: +1 514 393 1055; Fax: +1 514 874 9538.

33. Sadowski, J.; Wagener, M.; Gasteiger, J. Assessing Similarity and Diversity of Combinatorial Libraries by Spatial Autocorrelation Functions and Neural Networks. *Angew. Chem. Int. Ed. Engl.* **1995**, *34*, 2674–2677.

34. Sheridan, R. P.; Kearsley, S. K. Using a Genetic Algorithm to Suggest Combinatorial Libraries. *J. Chem. Inf. Comput. Sci.* **1995**, *35*, 310–320.

35. Brown, R. D.; Martin, Y. C. Use of Structure-Activity Data To Compare Structure-Based Clustering Methods and Descriptors for Use in Compound Selection. *J. Chem. Inf. Comput. Sci.* **1996**, *36*, 572–584.

36. Lior, R.; Maimon, O. *"Clustering Methods". Data Mining and Knowledge Discovery Handbook*; Springer: US, 2005; pp 321–352.

37. Székely, G. J.; Rizzo, M. L. Hierarchical Clustering via Joint Between-Within Distances: Extending Ward's Minimum Variance Method. *J. Classification* **2005**, *22*(2), 151–183.

38. Agrafiotis, D. K.; Lobanov, V. S. An Efficient Implementation of Distance-Based Diversity Measures Based on k-d Trees. *J. Chem. Inf. Comput. Sci.* **1999**, *39*, 51–58.

39. Agrafiotis, D. K. Stochastic Algorithms for Maximizing Molecular Diversity. *J. Chem. Inf. Comput. Sci.* **1997**, *37*, 841–851.

40. Pötter, T.; Matter, H. Random or Rational Design? Evaluation of Diverse Compound Subsets from Chemical Structure Databases. *J. Med. Chem.* **1998**, *41*, 478–488.

41. Pearlman, R. S.; Smith, K. M. Metric Validation and the Receptor-Relevant Subspace Concept. *J. Chem. Inf. Comput. Sci.* **1999**, *39*, 28–35.

42. Munk Jörgensen, A. M.; Pedersen, J. T. Structural Diversity of Small Molecule Libraries. *J. Chem. Inf. Comput. Sci.* **2001**, *41*, 338–445.

43. Xue, L.; Godden, J.; Gao, H.; Bajorath, J. Identification of a Preferred Set of Molecular Descriptors for Compound Classification Based on Principal Component Analysis. *J. Chem. Inf. Comput. Sci.* **1999**, *39*, 699–704.

44. Reynolds, C. H.; Druker, R.; Pfahler, L. B. Lead Discovery Using Stochastic Cluster Analysis (SCA): A New Method for Clustering Structurally Similar Compounds. *J. Chem. Inf. Comput. Sci.* **1998**, *38*, 305–312.

45. Hamprecht, F. A.; Thiel, W.; van Gunsteren, W. F. Chemical Library Subset Selection Algorithms: A Unified Derivation Using Spatial Statistics. *J. Chem. Inf. Comput. Sci.* **2002**, *42*, 414–428.

46. Trepalin, S. V.; Gerasimenko, V. A.; Kozyukov, A. V.; Savchuk, N. Ph.; Ivaschenko, A. A. New Diversity Calculations Algorithms Used for Compound Selection. *J. Chem. Inf. Comput. Sci.* **2002**, *42*, 249–258.

47. Agrafiotis, D. K.; Rassokhin, D. N. A Fractal Approach for Selecting an Appropriate Bin Size for Cell-Based Diversity Estimation. *J. Chem. Inf. Comput. Sci.* **2002**, *42*, 117–122.

48. Mount, J.; Ruppert, J.; Welch, W.; Jain, A. N. IcePick: A Flexible Surface-Based System for Molecular Diversity. *J. Med. Chem.* **1999**, *42*, 60–66.

49. Bayada, D. M.; Hamersma, H.; van Geerestein, V. J. Molecular Diversity and Representativity in Chemical Databases. *J. Chem. Inf. Comput. Sci.* **1999**, *39*, 1–10.

50. Zheng, W.; Cho, S. J.; Waller, C. L.; Tropsha, A. J. *J. Chem. Inf. Comput. Sci.* **1999**, *39*, 738–746.

51. Reynolds, C. H.; Tropsha, A.; Pfahler, L. B.; Druker, R.; Chakravorty, S.; Ethiraj, G.; Zheng, W. Diversity and Coverage of Structural Sublibraries Selected Using the SAGE and SCA Algorithms. *J. Chem. Inf. Comput. Sci.* **2001**, *41*(6), 1470–1477.

52. Valler, M. J.; Green, D. Diversity Screening versus Focused Screening in Drug Discovery. *Drug Discov. Today* **2000**, *5*, 286–293.

53. Walters, W. P.; Stahl, M. T.; Murcko, M. A. Virtual Screening—An Overview. *Drug Discov. Today* **1998**, *3*, 160–178.

54. Bajorath, J. Virtual Screening in Drug Discovery: Methods, Expectations and Reality. *Curr. Drug Discov.* **2002**, 24–27.

55. Joseph-McCarthy, D. An Overview of *In Silico* Design and Screening: Toward Efficient Drug Discovery. *Curr. Drug Discov.* **2002**, 20–23.

56. Cheng, T.; Li, Q.; Zhou, Z.; Wang, Y.; Bryant, S. H. Structure-Based Virtual Screening for Drug Discovery: A Problem-Centric Review. *AAPS J.* **2012**, *14* (1), 133–141.

57. Schneider, G. Virtual Screening: An Endless Staircase? *Nat. Rev. Drug Discov.* **2010**, *9* (4), 273.

58. Grant, J. A.; Gallardo, M. A.; Pickup, B. T. A Fast Method of Molecular Shape Comparison: A Simple Application of a Gaussian Description of Molecular Shape. *J. Comput. Chem.* **1996**, *17* (14), 1653–1666.

59. Drwal, M. N.; Griffith, R. Can Biochemistry Drive Drug Discovery Beyond Simple Potency Measurements? *Drug. Discov. Today: Tech.* **2013**, *10*, 395.

60. Cortés-Cabrera, A.; Gago, F.; Morreale, A. A Reverse Combination of Structure-Based and Ligand-Based Strategies for Virtual Screening. *J. Comput.-Aided Mol. Des.* **2012**, *26* (3), 319–327.

61. Sinko, W.; Lindert, S.; Mccammon, J. A. Accounting for Receptor Flexibility and Enhanced Sampling Methods in Computer-Aided Drug Design. *Chem. Biol. Drug Des.* **2013**, *81* (1), 41–49.

62. Walters, W. P.; Stahl, M. T.; Murcko, M. A. Virtual Screening—An Overview. *Drug Discov. Today* **1998**, *3*, 160–178.

63. Willett, P. Chemoinformatics—Similarity and Diversity in Chemical Libraries. *Curr. Opin. Biotechnol.* **2000**, *11*, 85–88.

64. Leach, A. R.; Hann, M. M. The In Silico World of Virtual Libraries. *Drug Discov. Today* **2000**, *5* (8), 326–336.

65. Kubinyi, H. Similarity and Dissimilarity: A Medicinal Chemists View. *Perspect. Drug Discov. Des.* **1998**, *9–11*, 225–252.

66. Willett, P.; Barnard, J. M.; Downs, G. M. Chemical Similarity Searching. *J. Chem. Inf. Comput. Sci.* **1998**, *38*, 983–996.

67. Diller, D. J.; Merz, K. M., Jr. High Throughput Docking for Library Design and Library Prioritization. *Proteins* **2001**, *43*, 113–124.

68. Abagyan, R.; Totrov, M. High-Throughput Docking for Lead Generation. *Curr. Opin. Chem. Biol.* **2001**, *5*, 375–382.

69. Hopfinger, A. J.; Duca, J. S. Estimation of Molecular Similarity Based on 4D-QSAR Analysis: Formalism and Validation. *J. Chem. Inf. Comput. Sci.* **2001**, *41*, 1367–1387.

70. Willett, P. Chemoinformatics—Similarity and Diversity in Chemical Libraries. *Curr. Opin. Biotechnol.* **2000**, *11*, 85–88.

71. Makara, G. M. Measuring Molecular Similarity and Diversity: Total Pharmacophore Diversity. *J. Med. Chem.* **2001**, *44*, 3563–3571.

72. Willet, P.; Gedeck, P. Visual and Computational Analysis of Structure–Activity Relationships in High-Throughput Screening Data. *Curr. Opin. Chem. Biol.* **2001**, *5*, 389–395.

73. Hopfinger, A. J.; Duca, J. Extraction of Pharmacophore Information from High-Throughput Screens. *Curr. Opin. Biotechnol.* **2000**, *11*, 97–103.

74. Roberts, G.; Myatt, G. J.; Johnson, W. P.; Cross, K. P.; Blower, P. E., Jr. LeadScope: Software for Exploring Large Sets of Screening Data. *J. Chem. Inf. Comput. Sci.* **2000**, *40*, 1302–1314.

75. Good, A. C.; Krystek, S. R.; Mason, J. S. High-Throughput and Virtual Screening: Core Lead Discovery Technologies Move Towards Integration. *Drug Discov. Today* **2001**, *5* (suppl.).

76. Shen, J.; Cheng, F.; Xu, Y.; Li, W.; Tang, Y. Estimation of ADME Properties with Substructure Pattern Recognition. *J. Chem. Inf. Model.* **2010**, *50* (6), 1034–1041.

77. Cheng, F.; Ikenaga, Y.; Zhou, Y.; Yu, Y.; Li, W.; Shen, J.; Du, Z.; Chen, L.; Xu, C.; Liu, G.; Lee, P. W.; Tang, Y. In Silico Assessment of Chemical Biodegradability. *J. Chem. Inf. Model.* **2012**, *52* (3), 655–669.

78. Cheng, F.; Yu, Y.; Shen, J.; Yang, L.; Li, W.; Liu, G.; Lee, P. W.; Tang, Y. Classification of Cytochrome P450 Inhibitors and Noninhibitors Using Combined Classifiers. *J. Chem. Inf. Model.* **2011**, *51* (5), 996–1011.

79. Singer, S. J.; Nicolson, G. L. The Fluid Mosaic Model of the Structure of Cell Membranes. *Science* **1972**, *175*, 720–731.

80. van de Waterbeemd, H.; Kansy, M. Hydrogen-Bonding Capacity and Brain Penetration. *Chimia* **1992**, *46*, 299–303.

81. Young, R. C.; Mitchell, R. C.; Brown, T. H.; Ganellin, C. R.; Griffith, R.; Jones, M.; Rana, K. K.; Saunders, D.; Smith, I. R.; Sore, N. E.; Wilks, T. J. Development of a New Physicochemical Model for Brain Penetration and Its Application to the Design of Centrally Acting H2 Receptor Histamine Antagonists. *J. Med. Chem.* **1988**, *31*, 656–671.

82. Seiler, P. Interconversion of Lipophilicities from Hydrocarbon/Water Systems into the Octanol/Water System. *Eur. J. Med. Chem.* **1974**, *9*, 473–479.

83. Chadha, H. S.; Abraham, M. H.; Mitchell, R. C. Physicochemical Analysis of the Factors Governing Distribution of Solutes Between Blood and Brain. *Bioorg. Med. Chem. Lett.* **1994**, *4*, 2511–2516.

84. Abraham, M. H.; Chadha, H. S.; Mitchell, R. C. Hydrogen Bonding Factors that Influence the Distribution of Solutes between Blood and Brain. *J. Pharm. Sci.* **1994**, *83*, 1257–1268.

85. Abraham, M. H. Scales of Solutes Hydrogen-Bonding: Their Construction and Application to Physicochemical and Biochemical Processes. *Chem. Soc. Rev.* **1993**, *22*, 73–83.

86. Herve, F.; Urien, S.; Albengres, E.; Duche, J.-C.; Tillement, J. Drug Binding in Plasma. A Summary of Recent Trends in the Study of Drug and Hormone Binding. *Clin. Pharmacokinet.* **1994**, *26*, 44–58.

87. Colmenarejo, G.; Alvarez-Pedraglio, A.; Lavandera, J. L. Cheminformatic Models to Predict Binding Affinities to Human Serum Albumin. *J. Med. Chem.* **2001**, *44*, 4370–4378.

88. Honma, W.; Li, W.; Liu, H.; Scott, E. E.; Halpert, J. R. Functional Role of Residues in the Helix B' Region of Cytochrome P450 2B1. *Arch. Biochem. Biophys.* **2005**, *435* (1), 157–165, ISSN 0003-9861.

89. Roberts, S. A. High-Throughput Screening Approaches for Investigating Drug Metabolism and Pharmacokinetics. *Xenobiotica* **2001**, *31*, 557–589.

90. Watt, A. P.; Morrison, D.; Evans, D. C. Approaches to Higher-Throughput Pharmacokinetics (HTPK) in Drug Discovery. *Drug Discov. Today* **2001**, *5*, 17–24.

91. Williams, J. A.; Hyland, R.; Jones. B. C.; Smith, D. A.; Hurst, S.; Goosen, T. C.; Peterkin, V.; Koup, J. R.; Ball, S. E. Drug–Drug Interactions for UDP-Glucuronosyltransferase Substrates: A Pharmacokinetic Explanation for Typically Observed Low Exposure (AUCi/AUC) Ratios. *Drug Metab. Dispos.* **2004**, *32*, 1201–1208.

92. Mishra, N. K.; Agarwal, S.; Raghava, G. P. S. Prediction of Cytochrome P450 Isoform Responsible for Metabolizing a Drug Molecule. *BMC Pharmacol.* **2010**, *10*, 8.

93. Testa, B.; Cruciani, G. Pharmacokinetic Optimization in Drug Research: Biological, Physicochemical and Computational Strategies. In *Verlag Helvetica Chimica Acta (VHCA)*; Wiley-VCH: Zurich, Weinheim, Germany, 2001; pp 65–84.

94. Potter, T.; Lewis, R.; Luker, T.; Bonnert, R.; Bernstein, M. A.; Birkinshaw, T. N.; Thom, S.; Wenlock, M.; Paine, S. In Silico Prediction of Acyl Glucuronide Reactivity. *J Comput Aided Mol Des.* **2011**, *25*(11), 997–1005.

95. Greene, N. Computer Software for Risk Assessment. *J. Chem. Inf. Comput. Sci.* **1997**, *37*, 148–150.

96. Richard, A. M. Application of SAR Methods to Non-Congeneric Databases Associated with Carcinogenicity and Mutagenicity: Issues and Approaches. *Mutat. Res.* **1994**, *305*, 73–97.

97. Hall, L.; Kier, L.; Phipps, G. Structure–Activity Relationship Studies on the Toxicities of Benzene Derivatives: I. An Additivity Model. *Environ. Toxicol. Chem.* **1984**, *3*, 355–365.

98. Gute, B.; Basak, S. Predicting Acute Toxicity (LC50) of Benzene Derivatives Using Theoretical Molecular Descriptors: A Hierarchical QSAR Approach. *SAR QSAR Environ. Res.* **1997**, *7*, 117–131.

99. Tao, S.; Xi X.; Xu F.; Dawson R. A. QSAR Model for Predicting Toxicity (LC50) to Rainbow Trout. Water Res. 2002, *36*(11), 2926–2930.

100. Benfenati, E.; Grasso, P.; Bruschi, M. Predictive Carcinogenicity: A Model for Aromatic Compounds, with Nitrogen-Containing Substituents, Based on Molecular Descriptors Using an Artificial Neural Network. *J. Chem. Inf. Comput. Sci.* **1999**, *39*, 1076–1080.

101. Burden, F. R.; Winkler, D. A. A Quantitative Structure–Activity Relationships Model for the Acute Toxicity of Substituted Benzenes to Tetrahymena Pyriformis Using Bayesian-Regularized Neural Networks. *Chem. Res. Toxicol.* **2000**, *13*, 436–440.

102. Arenas, G. E. A.; Giralt, F. An Integrated SOM-Fuzzy ARTMAP Neural System for the Evaluation of Toxicity. *J. Chem. Inf. Comput. Sci.* **2002**, *42*, 343–359.

INDEX